T0223421

Lecture Notes in Computer Science **8388**

Commenced Publication in 1973
Founding and Former Series Editors:
Gerhard Goos, Juris Hartmanis, and Jan van Leeuwen

Editorial Board

For further volumes:
http://www.springer.com/series/7409

Wooju Kim · Ying Ding
Hong-Gee Kim (Eds.)

Semantic Technology

Third Joint International Conference, JIST 2013
Seoul, South Korea, November 28–30, 2013
Revised Selected Papers

 Springer

Editors
Wooju Kim
Yonsei University
Seoul
Korea, Republic of (South Korea)

Ying Ding
Indiana University
Bloomington, IN
USA

Hong-Gee Kim
Seoul National University
Seoul
Korea, Republic of (South Korea)

ISSN 0302-9743   ISSN 1611-3349   (electronic)
ISBN 978-3-319-06825-1   ISBN 978-3-319-06826-8   (eBook)
DOI 10.1007/978-3-319-06826-8
Springer Cham Heidelberg New York Dordrecht London

Library of Congress Control Number: 2014939606

LNCS Sublibrary: SL3 – Information Systems and Applications incl. Internet/Web and HCI

Printed on acid-free paper

Springer is part of Springer Science+Business Media (www.springer.com)

Preface

The Joint International Semantic Technology Conference (JIST) is a regional federation of semantic technology related conferences. JIST 2013 was a joint effort between the JIST Conference, Korea Intelligent Information System Society, Korean Association of Data Science, and Korea Institute of Science and Technology Information. The mission of JIST is to bring researchers together from the semantic technology community and other areas related to semantic technologies to present their innovative research results or novel applications. This conference addressed not only the Semantic Web, but also issues related to ontologies, metadata, linked open data, OWL/RDF languages, description logics, big data and even art.

The JIST 2013 main technical tracks comprised regular paper sessions, four tutorials, and five workshops including a local workshop. We received submissions for the main technical track from 12 countries, which included not only Asian countries, but also many European and North American countries. Every paper was reviewed by three reviewers on average and then discussed by the chairs.

The topics were interesting and well distributed, ranging from fundamental studies on semantic reasoning and search/query to various applications of semantic technology. Particularly, we found many of the submitted papers were related to the areas of biomedical and multilingual domains. Another notable fact is that the relative importance of learning and discovery issues from the real-world linked data keeps growing every year; this trend seems to reflect the current popularity of big data.

It is far too obvious that the great success of JIST 2013 would not have been possible without the devotion and close cooperation of the Organizing Committee. We all thank the member of the Program Committee and local Organizing Committee for their great support. Special thanks also go to the sponsors and supporting organizations for their financial support.

December 2013

Wooju Kim
Ying Ding
Hong-Gee Kim

Organization

Organizing Committee

General Chair

Hong-Gee Kim — Seoul National University, Korea

Program Chairs

Wooju Kim — Yonsei University, Korea
Ying Ding — Indiana University, USA

In-Use Track Chairs

Jie Tang — Tsinghua University, China
Younghwan Lee — Konkuk University, Korea

Poster and Demo Chairs

Guilin Qi — Southeastern University, China
Kouji Kozaki — Osaka University, Japan

Workshop Chairs

Thepchai Supnithi — NECTEC, Thailand
Jason Jung — Yeungnam Univeristy, Korea

Tutorial Chairs

Haklae Kim — Samsung Electronics, Korea
Hanmin Jung — Korea Institute of Science and Technology Information, Korea

Industry Chairs

Laurentiu Vasiliu — Peracton Co., Ireland
Takahiro Kawamura — Toshiba, Japan

Publicity Chair

Myungdae Cho — Seoul National University, Korea

Local Organizing Chair

June Seok Hong — Kyonggi University, Korea

Program Committee

Bin Xu
Bin Chen
Dejing Dou
Donghyuk Im
Ekawit Nantajeewarawat
Gang Wu
Giorgos Stoilos
Gong Cheng
Guilin Qi
Hanmin Jung
Haofen Wang
Hideaki Takeda
Hong-Gee Kim
Huajun Chen
Hyun Namgung
Ikki Ohmukai
Ioan Toma
Itaru Hosomi
Jae-Hong Eom
Jai-Eun Jung
Jianfeng Du
Juan Miguel
Kangpyo Lee
Kewen Wang
Key-Sun Choi
Koiti Hasida
Kouji Kozaki
Laurentiu Vasiliu
Marco Ronchetti
Masahiro Hamasaki

Mitsuharu Nagamori
Mun Yi
Myungjin Lee
Pascal Hitzler
Pyung Kim
Sa-Kwang Song
Seiji Koide
Seungwoo Lee
Shenghui Wang
Shinichi Nagano
Sung-Kook Han
Takahira Yamaguchi
Takahiro Kawamura
Takeshi Morita
Thepchai Supnithi
Titi Roman
Tony Lee
Umberto Straccia
Wei Hu
Yeye He
Yi Zhou
Yi-Dong Shen
Yoshinobu Kitamura
Youngsang Cho
Young-Tack Park
Yuting Zhao
Yuzhong Qu
Zhiqiang Gao
Zhixian Yan

Local Organizing Committee

Chang Ouk Kim
Haklae Kim
Hanmin Jung
Hee jun Park
Hong-Gee Kim
Jason Jung
Kwang Sun Choi
Myungdae Cho

Myungjin Lee
Pyung Kim
Sangun Park
Seungwoo Lee
Sung-Kook Han
Wooju Kim
Younghwan Lee

Contents

Ontology Construction

Semantic Reasoning

Semantic Search and Query

Ontology Mapping

Learning and Discovery

Semantic Web Services

Semantic Web Services
for University Course Registration

Şengül Çobanoğlu and Zeki Bayram[✉]

Computer Engineering Department, Eastern Mediterranean University,
Famagusta, North Cyprus
{sengul.cobanoglu,zeki.bayram}@emu.edu.tr
http://cmpe.emu.edu.tr/

Abstract. Semantic web services, with proper security procedures, have the potential to open up the computing infrastructure of an organization for smooth integration with external applications in a very controlled way, and at a very fine level of granularity. Furthermore, the process of using the provided functionality, consisting of discovery, invocation and execution of web services may be automated to a large extent. In this paper, we show how semantic web services and service-oriented computing can facilitate this integration in the education domain. Specifically, we use the Rule variant of Web Services Modeling Language (WSML) to semantically specify the functionality of web services for on-line course registration, a goal for consuming the provided functionality, as well as the ontologies needed for a shared terminology between the service and goal.

Keywords: Semantic web services · Ontology · Course registration · Web Service Modeling Language · Web Service Modeling Ontology · Discovery · Service orientation

1 Introduction

Service-oriented architecture [1], consisting of loosely coupled modular components with standard interfaces, can facilitate greatly the process of developing large and complex software in a distributed manner. In this paradigm, due to the standard interfaces of the components, the provided functionality is platform independent. Furthermore, new applications can easily be "assembled" from existing components by providing a user interface and some logic combining the functionalities of the components used and synthesizing the required functionality. The problem with this seemingly ideal approach to software development, which naturally promotes software re-usability and platform independence, is that as the number of modules get larger and larger, finding a web service with the desired functionality becomes harder and harder. This is where semantic web services enter the picture and come to help. Web service functionality can be described semantically using special purpose web service description languages,

W. Kim et al. (Eds.): JIST 2013, LNCS 8388, pp. 3–16, 2014.
DOI: 10.1007/978-3-319-06826-8_1, © Springer International Publishing Switzerland 2014

and desired functionality can also be specified semantically in the form of goals. A matcher can then be used for matching the desired functionality as specified in the goal, with the functionalities provided by various web services. Once a satisfactory match is found, invocation can be made, either in automatic mode, or manually, or somewhere in between.

In this paper, we show the potential beneficial use of semantic web services [2] in the educational domain, more specifically for making course registrations. In previous work, semantic web service applications were investigated in the banking domain [3,4], as well as the health domain [5].

Before we go into the specifics of our application, we present below terminology and concepts relevant to semantic web services. The notion of an ontology is used widely in the semantic web. Ontologies are used for representing shared domain knowledge [6,7]. They form an important foundation of the semantic web on which other components are built [8]. The main component of an ontology is the "concept," which is similar in essence to classes in object-oriented programming languages [9]. Web services are computational units on the World Wide Web (WWW), and can be called through standard interfaces and protocols, such as HTTP [10] and SOAP [11]. Web services also allow an organization to open up part of its internal computing infrastructure to the outside world in a controlled way. A drawback of "normal" web services is their purely syntactic specification, which makes reliable automatic discovery impossible. *Semantic* Web services attempt to remedy the purely syntactic specification of regular web services by providing rich, machine interpretable semantic descriptions of Web services [12] so that the whole process of web service discovery, orchestration or composition, and execution can be automated through appropriate semantic web service frameworks.

Web Service Modeling Ontology (WSMO) [13] is a framework for semantic description of Semantic Web Services based on the Web Service Modeling Framework (WSMF) [14]. Its four main components are ontologies, web services, goals and mediators. Web Service Modeling Language (WSML) [15] is a language for modelling web services, ontologies, and related aspects of WSMO framework, to provide the description of semantic web services so that automatic discovery and invocation becomes possible. Five language variants of WSML exist, which are WSML-Core, WSML-DL, WSML-Flight, WSML-Rule and WSML-Full.

In this paper, we use WSML-Rule to semantically specify an online university course registration web service. We first define an ontology which contains relevant concepts, axioms, relations and instances. Then we proceed to semantically define the web service itself, and then a goal to consume the web service.

The remainder of the paper is organized as follows. Section 2 describes the semantic specification of a web service for course registration in WSML-Rule. This specification includes an ontology, consisting of concepts, instances, relations and axioms of the course registration domain, the functional specification of the web service itself in terms of preconditions, postconditions, assumptions and effects, as well as a goal for consuming the web service. Section 3 contains a discussion of the problems encountered and shortcomings of the WSML language. Finally, Sect. 4 is the conclusion and future work.

2   Specifying Course Registration Functionality in WSML

The web service whose functionality we want to semantically describe implements a course registration use-case, where a student tries to register for a course in a given semester. Several conditions must be satisfied before the registration can take place. In the following sub-sections, we present the ontology, goal and web service specification for the course registration use-case. Note that we used WSML-Rule throughout the specification, since it supports rule-based logic programming, with which we can write axioms that act as integrity constraints, and also do computations.

2.1   Course Registration Ontology

The course registration ontology consists of *concepts, relations, instances, relationInstances* and *axioms*. We have placed concepts and instances into different files and used the importing facility of WSML to better manage the ontology. Figure 1 depicts the prologue of the WSML file containing the ontological structures used in the specification. Note the choice of WSML variant selected (WSML-Rule), the definition of the namespace property, the beginning of the registration ontology, and the importation of two other ontologies containing instances.

```
wsmlVariant _"http://www.wsmo.org/wsml/wsml-syntax/wsml-rule"
namespace { _"http://cmpe.emu.edu.tr/courseRegistration#",
    discovery _"http://wiki.wsmx.org/index.php?title=DiscoveryOntology#",
    dc       _"http://purl.org/dc/elements/1.1#" }

ontology courseRegistration
    importsOntology {courseRegistrationInstances, courseRegistrationRelationInstances}
```

Fig. 1. Prologue of the course registration ontology file

Concepts in the Registration Ontology. Figure 2 graphically depicts the concepts of the ontology. In Fig. 3 we have the concept definitions for University, Address, Faculty, Department, and AcademicProgram and Curriculum. A department may be running more than one program, and for each program, we have a curriculum, with a list of courses.

Figure 4 shows the sub-concept relationships among concepts. For example, EnglishProgram is a sub-concept of AcademicProgram. Note how multiple inheritance is possible.

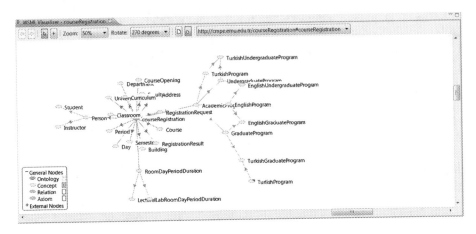

Fig. 2. Concepts of CourseRegistration ontology (graphical representation)

```
concept University
     uname ofType (1 *) _string
     locatedAt ofType  (1 *) Address

concept Address
     street ofType  (0 1) _string
     city ofType  (0 1) _string

concept Faculty
     facultyName ofType  (0 1) _string
     atUniversity ofType  (0 1) University

concept Department
     deptID ofType  (0 1) _string
     deptName ofType  (0 1) _string
     inFaculty ofType  (0 1) Faculty

concept AcademicProgram
     programName ofType  (0 1) _string
     programID ofType  (0 1) _string
     belongsTo ofType  (0 1) Department

concept Curriculum
     academicProgram ofType  (0 1) AcademicProgram
     refCode ofType  (0 1) _string
     courseName ofType  (0 1) Course
```

Fig. 3. Registration ontology concepts (part 1)

Figure 5 depicts concepts needed for course registration specifically. Note the difference between Course and CourseOpening concepts. The former gives information about a course in general, and the latter is used when a course is offered in a given semester. Semester, Day, and Period are *utility* concepts.

```
concept UndergraduateProgram subConceptOf AcademicProgram
concept GraduateProgram subConceptOf AcademicProgram
concept TurkishProgram subConceptOf AcademicProgram
concept EnglishProgram subConceptOf AcademicProgram
concept EnglishGraduateProgram subConceptOf {EnglishProgram, GraduateProgram}
concept EnglishUndergraduateProgram subConceptOf { EnglishProgram, UndergraduateProgram}
concept TurkishGraduateProgram subConceptOf {TurlishProgram, GraduateProgram}
concept TurkishUndergraduateProgram subConceptOf { TurlishProgram, UndergraduateProgram}
```

Fig. 4. Some sub-concept relationships

```
concept Course
      courseCode ofType  (0 1) _string
      courseName ofType _string
      hasPrerequisite ofType (0 *) Course
      lecture_hour ofType  (0 1) _integer
      tutorial_hour ofType  (0 1) _integer
      credits ofType  (0 1) _integer
      ects ofType  (0 1) _integer
      belongsToProgram ofType  (0 1) UndergraduateProgram

concept CourseOpening
      groupNo ofType  (0 1) _integer
      ofCourse ofType  (0 1) Course
      semester ofType  (0 1) Semester
      id_courseOpening ofType  (0 1) _string
      teaching_times ofType  (1 4) RoomDayPeriodDuration
      current_size ofType (0 1) _integer
      year ofType  (0 1) _integer

concept Classroom
      classID ofType  (0 1) _string
      location ofType  (0 1) Building
      capacity ofType (0 1) _integer
      inDepartment ofType  (0 1) Department
      roomNumber ofType  (0 1) _string

concept Semester
concept Day
concept Period
concept Building

concept RoomDayPeriodDuration
      room ofType  (1 1) Classroom
      day ofType  (1 1) Day
      period ofType  (1 1) Period
      duration ofType  (1 1) _integer


concept LectureRoomDayPeriodDuration subConceptOf RoomDayPeriodDuration
concept LabRoomDayPeriodDuration subConceptOf RoomDayPeriodDuration
```

Fig. 5. Registration ontology concepts (part 2)

```
concept Person
      ID ofType   (0 1) _string
      gender ofType   (0 1) _string
      Date_of_Birth ofType   (0 1) _date
      name ofType   (0 1) _string
      lastName ofType   (0 1) _string
      address ofType Address

concept Student subConceptOf Person
      yearEnrolled ofType   (0 1) _integer
      overallCGPA ofType   (0 1) _float
      enrolledIn ofType   (0 2) AcademicProgram
      semesterEnrolled ofType   (0 1) Semester
      tookCourse_ ofType (0 *) Course
            nfp
                  dc#relation hasValue {TookRelation}
            endnfp

concept Instructor subConceptOf Person
      works_in ofType   (1 *) Department
```

Fig. 6. Registration ontology concepts (part 3)

```
concept RegistrationRequest
      student ofType Student
      course ofType Course
      year ofType _integer
semester ofType Semester

concept RegistrationResult
      student ofType Student
      courseOpening ofType CourseOpening
```

Fig. 7. Registration ontology concepts (part 4)

Figure 6 depicts the definition of the Person concept, and its two sub-concepts Student and Instructor. The *yearEnrolled* attribute of Student is the year that the student enrolled university and the *enrolledIn* attribute is the Academic program in which student is enrolled. Of special interest is the *tookCourse* attribute, since it will be indirectly defined through the *TookRelation* axiom, given in Fig. 12.

Finally in Fig. 7, we have two concepts, RegistrationRequest and RegistrationResult that are used to pass information to the web service, and get a confirmation back. The request is for a course in a given year and semester. The confirmation is the registration into a specific course opening.

Relations. Although attributes of a concept act like relationships between the host object and some other object, *n-ary* relationships where $n > 2$ are best represented using explicit relations, which WSML provides as a construct.

Sometimes we use binary relationships for convenience, and it is also possible to establish a connection between attributes of a concept and relations explicitly through axioms.

In Fig. 8, we see the four relations present in the ontology. These are *teaches* for specifying which instructor teaches which course opening, *takes* for students for specifying which student takes which course opening, *tookCourse* for recording which student has already taken which course, and *prerequisite* for recording the prerequisite relationship among courses.

```
relation teaches( ofType Instructor,  ofType CourseOpening)
relation takes( ofType Student,  ofType CourseOpening)
relation tookCourse(ofType Student, ofType Course)
relation prerequisite(ofType Course,ofType Course)
```

Fig. 8. Relations in the course registration ontology

Concept and Relation Instances. Instances are actual data in the ontology. In Fig. 9 we give a representative set of instances for several concepts to enable better understanding of the way in which concepts and relations are used in the ontology. Figure 10 has a representative set of relation instances.

Axioms. An axiom in WSML is a logical statement which can be used to implicitly define concepts and conditions for concept membership (i.e. instances), specify relations on ontology elements as well as assert any kind of constraints regarding the ontology. Axioms that specify forbidden states of the ontology start with "!-", and they function like integrity constraints of databases.

Figure 11 depicts four named axioms. The "registrationRules" axiom defines the "clashes", "prerequisiteNotTaken" and "classSizeExceeded" predicates. The "clashes" predicate is true when two course openings in a given semester are taught in the same physical location at exactly the same time. The "prerequisiteNotTaken" predicate is true when a student is taking a course without having taken its prerequisite course. Note the use of the *naf* operator in WSML, which performs "negation as failure", the operational counterpart of logical negation. The "classSizeExceeded" predicate is true if the number of students registered to a course opening has exceeded the capacity of the room in which the course in taught. The "noClashRoom" axiom is a constraint, forbidding any clashes of rooms, which means the same room cannot be assigned to two courses at exactly the same time slot. Similarly, "prerequisiteTaken" forbids taking a course without first taking is prerequisite course, and "classSizeViolation" forbids exceeding class sizes when registering students.

Figure 12 depicts the remaining axioms of the ontology. The "noClashTeacher" axiom makes sure that the meeting times of two courses taught by a teacher have no overlap, since naturally a person cannot be present in two places at the

```
instance dept_computer_engineering memberOf Department
    deptName hasValue "Computer Engineering"
    inFaculty hasValue faculty_of_engineering
    deptID hasValue "computer_engineering"

instance cmpe_undergrad_eng memberOf EnglishUndergraduateProgram
    programID hasValue "cmpe_undergrad_eng"
    programName hasValue "Computer Engineering Undergraduate English"
    belongsTo hasValue dept_computer_engineering

instance curriculum_cmpe_undergrad_eng memberOf Curriculum
    academicProgram hasValue cmpe_undergrad_eng
    refCode     hasValue  "cmpecurriculum"
    courseName hasValue cmpe318

instance cmpe354 memberOf Course
    belongsToProgram hasValue cmpe_undergrad_eng
    courseName hasValue "Introduction to Databases"
    courseCode hasValue "cmpe354"
    hasPrerequisite hasValue cmpe211
    ....

instance cmpe354_spring_2012_gr1 memberOf CourseOpening
    id_courseOpening hasValue "cmpe354_spring_2012"
    year hasValue 2012
    semester hasValue spring
    ofCourse hasValue cmpe354
    groupNo hasValue 1
    current_size hasValue 4
    teaching_times hasValue {lecture_cmpe128_wednesday_per2_2,
                             lab_cmpelab5_friday_per4_2 }

instance room_cmpe128 memberOf Classroom
    inDepartment hasValue dept_computer_engineering
    classID hasValue "room_cmpe128"
    capacity hasValue 60
    location hasValue cmpe_building
    ....

instance lecture_cmpe128_monday_per2_2 memberOf LectureRoomDayPeriodDuration
    room hasValue room_cmpe128
    day hasValue monday
    period hasValue per2
    duration hasValue 2

instance ayse memberOf Student
    ID hasValue "104059"
    name  hasValue  "Ayse"
    lastName  hasValue  "Akcam"
    ....

instance spring memberOf Semester
instance monday memberOf Day
instance per1 memberOf Period
instance cmpe_building memberOf Building
```

Fig. 9. Various concept instances in the course registration ontology

```
relationInstance  teaches(ali, cmpe354_spring_2012_gr1)
relationInstance  takes(ayse, cmpe354_spring_2012_gr1)
relationInstance prerequisite(cmpe211,cmpe354)
relationInstance tookCourse(zainab,cmpe211)
```

Fig. 10. Various relation instances in the course registration ontology

```
axiom registrationRules
    definedBy
        clashes(?co1,?co2):-
            ?co1 memberOf CourseOpening
            and ?co2 memberOf CourseOpening
            and ?co1 != ?co2
            and ?co1[teaching_times hasValue ?tt1, year hasValue ?y1, semester hasValue ?s1]
            and ?co2[teaching_times hasValue ?tt1, year hasValue ?y1, semester hasValue ?s1].

        prerequisiteNotTaken(?student,?course,?precourse):-
                            takes(?student, ?courseOpening) and
                            ?courseOpening[ofCourse hasValue ?course] memberOf CourseOpening and
                            prerequisite(?pre,?course) and
                            naf tookCourse(?student,?precourse).

        classSizeExceeded:- ?co [teaching_times hasValue ?tt,current_size hasValue ?s] memberOf CourseOpening and
                            ?tt[room hasValue ?room] memberOf RoomDayPeriodDuration and
                            ?room[capacity hasValue ?maxCap] and
                            ?s > ?maxCap.
axiom noClashRoom
    definedBy
            !-clashes(?x,?y).

axiom prerequisiteTaken
    definedBy
            !- prerequisiteNotTaken(?student,?course,?pre).

axiom classSizeViolation
    definedBy
            !- classSizeExceeded.
```

Fig. 11. Some axioms of the registration ontology

same time. Axiom "noClashStudent" imposes a similar constraint for students, although this restriction may be somewhat unrealistic in a university environment. Finally, axiom "TookRelation" establishes a connection between the "tookCourse" relation and the tookCourse_ attribute of Student objects: if a student s and a course c are in the tookCourse relation, then the tookCourse_ attribute value of s is c.

2.2   Goals

A goal is a declarative statement of what the requester can provide, and what it expects as a result of invoking a web service. The "what I can provide" part is specified in the "precondition" of the goal (this is an unfortunate terminology, since it implies that the goal has a precondition), and the "what I expect" part is specified in the "postcondition". Goals are like elaborate queries in a database

```
axiom noClashTeacher
   definedBy
      !- ?t1 memberOf Instructor
         and ?co1 memberOf CourseOpening
         and ?co2 memberOf CourseOpening
         and teaches(?t,?co1)
         and teaches(?t,?co2)
         and ?co1 != ?co2
         and ?co1[teaching_times hasValue ?tt1, year hasValue ?y1, semester hasValue ?s1]
         and ?co2[teaching_times hasValue ?tt2, year hasValue ?y1, semester hasValue ?s1]
         and ?tt1[day hasValue ?d1, period hasValue ?p1]
         and ?tt2[day hasValue ?d1, period hasValue ?p1].

axiom noClashStudent
   definedBy
      !- ?t1 memberOf Student
         and ?co1 memberOf CourseOpening
         and ?co2 memberOf CourseOpening
         and takes(?t1,?co1)
         and takes(?t1,?co2)
         and ?co1[teaching_times hasValue ?tt1, year hasValue ?y1, semester hasValue ?s1]
         and ?co2[teaching_times hasValue ?tt2, year hasValue ?y1, semester hasValue ?s1]
         and ?tt1[day hasValue ?d1, period hasValue ?p1]
         and ?tt2[day hasValue ?d1, period hasValue ?p1]
         and ?co1 != ?co2.

axiom TookRelation
   definedBy
      ?x[tookCourse_ hasValue ?course] memberOf Student:- tookCourse(?x,?course).
```

Fig. 12. More axioms of the course registration ontology

system, except they can be far more complex, since they can involve any logical statement, both in the "precondition" part and the "postcondition" part. A goal is "matched" against web service specifications, and the one that "best" satisfies the requirements of the goal is chosen for execution. The definition of "best" can change, and non-functional requirements can also play a role in which web service is chosen. Execution can be automatic if an appropriate semantic web service framework is used. Ontologies are used to establish a "common understanding" of terminology between goals and web services.

In Fig. 13, we have a goal for registering a specific student, Ali, to the CMPE354 course in the spring 2012 semester. The goal specification starts with stating the variant of WSML that will be used (WSML-rule), namespaces, and imported ontologies which contain concept definitions and instances. In the capability section, non-functional properties state that "heavyweight rule matching" should be used in the discovery process, meaning that full logical inference should be the method utilized. The shared variables section should contain logical variables that are shared in all parts of the goal (e.g. the precondition and postcondition). In this case, since there are no shared variables, this section is empty. The "precondition" of the goal contains the information needed to register the student, i.e. an instance of the "RegistrationRequest" concept. This instance

```
wsmlVariant _"http://www.wsmo.org/wsml/wsml-syntax/wsml-rule"
namespace { _"http://cmpe.emu.edu.tr/courseRegistration#",
            discovery _"http://wiki.wsmx.org/index.php?title=DiscoveryOntology#",
            dc _"http://purl.org/dc/elements/1.1#" }

goal goalCourseRegistration
   importsOntology
      { _"http://cmpe.emu.edu.tr/courseRegistration#courseRegistration",
        _"http://cmpe.emu.edu.tr/courseRegistration#courseRegistrationInstances",
        _"http://cmpe.emu.edu.tr/courseRegistration#courseRegistrationRelationInstances"}

   capability goalCourseRegistrationAliCapability
      nonFunctionalProperties
         discovery#discoveryStrategy hasValue {discovery#HeavyWeightRuleDiscovery,
                                               discovery#NoPreFilter}
      endNonFunctionalProperties

      sharedVariables {}

   precondition
      definedBy
         ?rr[student hasValue ali,
             course hasValue cmpe354,
             year hasValue 2012,
             semester hasValue spring] memberOf RegistrationRequest.

   postcondition
      definedBy
         ?aResult[student hasValue ali,courseOpening hasValue ?co]
                                       memberOf RegistrationResult
               and ?co [ofCourse hasValue cmpe354,
                             year hasValue 2012,
                             semester hasValue spring,
                     groupNo hasValue ?groupno ] memberOf CourseOpening.
```

Fig. 13. Goal for registering a specific student to a specific course in a given year and semester

acts like the parameter to the web service call. The postcondition states that an instance of the "RegistrationResult" concept is required, which acknowledges that the requested registration has been done. The variables *?co* and *?aResult* are logic variables that are meant to be bound to values as a result of a successful invocation of a web service.

2.3   Web Services

The web service specification in WSML describes in logical terms what the web service expects form the requester, the state of the world which must exist before it can be called, as well as results it can return to the requester, and the changes it can cause to the state of the world after it has finished its execution. *Preconditions* specify what information a web service requires from the requester. *Assumptions* describe the state of the world which is assumed before the execution of the Web service. *Postconditions* are statements about the existence of objects in the ontology created by the web service to be returned to he requester.

Effects describe the state of the world that is guaranteed to be reached after the successful execution of the Web service.

In Fig. 14, we have the course registration web service specification. It starts out, similarly to the goal specification, with the WSML variant used and the namespaces. Three ontologies are imported as in the goal. There are five shared variables in the capability. They are implicitly universally quantified. These variables are *?student, ?course, ?year, ?semester, ?oldsize* and *?co*. The precondition instantiates its variables with the data provided by the requester, and through shared variables passes them to the other parts of the specification. For example, the *?course* variable's contents are used in the assumption to find a CourseOpening instance for that course. In the assumption, the current size of the CourseOpening instance is stored in the *?oldsize* variable, which is used in the effect to increase the current size by 1. Note that the effect is not "performing" an action, but merely declaring the state of the world after the web service has finished execution, which is that the current size in the course opening is increased by 1, and that now the student is considered to be registered to the course opening. Lastly, the postcondition states the existence of an instance of the RegistrationResult concept in the ontology that will be "sent" to requester.

3 Discussion

Although WSML-Rule is a relatively comprehensive language, we have still found several constructs that are missing from it, the presence of which would have made the language much more powerful and expressive. Some of these are:

- *Lack of aggregate operators, such as "sum", "count" etc.* Due to this missing feature, we could not check in a declarative way that the capacity of each classroom should not be exceeded through registrations.
- *Lack of an exception mechanism.* Currently there is no way to deal with constraint violations in the ontology. They just get reported as errors.
- *Inability to specify different postconditions and effects depending on the preconditions and assumptions.* Currently, a web service specification is like an "if-then" statement of programming languages: if the preconditions and assumptions are correct, then the web service guarantees the effect and postcondition. What would be much more useful is a "case" structure, where more than one set of precondition-assumption pairs would exist, with their corresponding postcondition-effect pairs. The ability to have more than one capability for a web service would solve this problem.
- *Inability to have more than one outcome for one set of conditions in which the web service is called.* It would be desirable to be able to associate more than one postcondition-effect pair for the *same* precondition-assumption pair. We need to be able to say that the result of calling the web service can be $<postcondition_1, effect_1>$ or $<postcondition_2, effect_2>$ or ... or $<postcondition_n, effect_n>$, but this is not possible. For example, currently we cannot deal with the case that a request to register to a course may fail, in which case the state of the world will not change as a result of the web service call.

```
wsmlVariant _"http://www.wsmo.org/wsml/wsml-syntax/wsml-rule"

namespace { _"http://cmpe.emu.edu.tr/courseRegistration#",
    discovery _"http://wiki.wsmx.org/index.php?title=DiscoveryOntology#",
    dc        _"http://purl.org/dc/elements/1.1#" }

webService web_service_courseRegistration

    importsOntology {courseRegistration, courseRegistrationInstances,
                    courseRegistrationRelationInstances}

    capability web_service_courseRegistrationCapability

        nonFunctionalProperties
            discovery#discoveryStrategy hasValue discovery#HeavyweightDiscovery
        endNonFunctionalProperties

        sharedVariables {?student, ?course,?year, ?semester, ?oldsize, ?co}

        assumption
            definedBy
                ?co[ofCourse hasValue ?course,
                        year hasValue ?year,
                         semester hasValue ?semester,
                         groupNo hasValue ?groupno,
                         current_size hasValue ?oldsize] memberOf CourseOpening.

        precondition
            definedBy
                ?rr[student hasValue ?student,
                    course hasValue ?course,
                    year hasValue ?year,
                    semester hasValue ?semester] memberOf RegistrationRequest.
        effect
            definedBy
                takes(?student,?co) and
                ?co[current_size hasValue (?oldsize+1)].

        postcondition
            definedBy
                ?aResult[student hasValue ?student,
                        courseOpening hasValue ?co] memberOf RegistrationResult.
```

Fig. 14. Specification of the course registration web service

4   Conclusion and Future Research Directions

We have specified semantically, using WSML-Rule, a web service for course registration. This involved the definition of an ontology, with concepts needed to capture domain knowledge and axioms to implement integrity constraints, as well the specification of web service capability and goal to consume the provided service. Our work has revealed several weaknesses of WSML-Rule, which we identified above.

For future work, we are planning to work on remedying the weak points of WSML-Rule. We believe semantic web services, more specifically WSMO and WSML, when properly fixed and updated to answer the discovered deficiencies, can be the backbone of the future semantic web with intelligent agents.

References

1. Erl, T.: Service-Oriented Architecture (SOA): Concepts, Technology, and Design. Prentice Hall, Upper Saddle River (2005)
2. Dieter Fensel, D., Facca, F.D., Simperl, E., Toma, I.: Semantic Web Services. Springer, Heidelberg (2011)
3. Sharifi, O, Bayram, Z.: Database modelling using WSML in the specification of a banking application. In: Proceedings of WASET 2013, Zurich, Switzerland, pp. 263–267 (2001)
4. Sharifi, O, Bayram, Z.: Specifying banking transactions using Web Services Modeling Language (WSML). In: The fourth International Conference on Information and Communication Systems (ICICS 2013), Irbid, Jordan, pp. 138–143 (2001)
5. Sharifi, O, Bayram, Z.: A critical evaluation of WSMO and WSML through an e-health semantic web service specification case study. Submitted for publication (2013)
6. Domingue, J., Cabral, L., Galizia, S., Tanasescu, V., Gugliotta, A., Norton, B., Pedrinaci, C.: IRS-III: a broker-based approach to semantic web services. Web Semantics: Sci. Services Agents World Wide Web 6(2), 109–132 (2008)
7. Gruber, T.R.: A translation approach to portable ontology specifications. Knowl. Acquisition 5(2), 199–220 (1993)
8. Berners-Lee, T., Hendler, J.: Scientific publishing on the semantic web. Nature 410, 1023–1024 (2001)
9. Lara, R., Roman, D., Polleres, A., Fensel, D.: A conceptual comparison of WSMO and OWL-S. In: (LJ) Zhang, L.-J., Jeckle, M. (eds.) ECOWS 2004. LNCS, vol. 3250, pp. 254–269. Springer, Heidelberg (2004)
10. Hypertext Transfer Protocol Overview. http://www.w3.org/Protocols
11. SOAP Specifications. http://www.w3.org/TR/soap
12. Fensel, D., Lausen, H., Polleres, A., Bruijn, J., Stollberg, M., Roman, D., Domingue, J.: Enabling Semantic Web Services: The Web Service Modeling Ontology. Springer, New York (2006)
13. Web Service Modeling Ontology (WSMO). http://www.w3.org/Submission/WSMO
14. Fensel, D., Bussler, C.: The web service modeling framework WSMF. Electron. Commerce Res. Appl. 1(2), 113–137 (2002)
15. D16.1v1.0 WSML Language Reference. http://www.wsmo.org/TR/d16/d16.1

Context-Aware Music Recommendation
with Serendipity Using Semantic Relations

Mian Wang(✉), Takahiro Kawamura, Yuichi Sei, Hiroyuki Nakagawa,
Yasuyuki Tahara, and Akihiko Ohsuga

The University of Electro-Communications,
1-5-1 Chofugaoka, Chofu, Tokyo, Japan
wangmian30@gmail.com, kawamura@ohsuga.is.uec.ac.jp,
{sei,nakagawa,tahara,ohsuga}@uec.ac.jp
http://www.ohsuga.is.uec.ac.jp/

Abstract. A goal for the creation and improvement of music recommendation is to retrieve users' preferences and select the music adapting
to the preferences. Although the existing researches achieved a certain
degree of success and inspired future researches to get more progress,
problem of the cold start recommendation and the limitation to the
similar music have been pointed out. Hence we incorporate concept of
serendipity using 'renso' alignments over Linked Data to satisfy the users'
music playing needs. We first collect music-related data from Last.fm,
Yahoo! Local, Twitter and LyricWiki, and then create the 'renso' relation on the Music Linked Data. Our system proposes a way of finding
suitable but novel music according to the users' contexts. Finally, preliminary experiments confirm balance of accuracy and serendipity of the
music recommendation.

Keywords: Recommendation system · Context awareness · Linked Data

1 Introduction

Since Internet was invented and exploded in the past decades, people have been
used to get multimedia information (i.e. video, book and music) from Internet in
private time. Especially, the music is getting much more important aspect of our
daily lives. Recently some research indicated that listening to the music is more
often than any of the other activities (i.e. watching television, reading books
and watching movies). Also as a powerful communication and self-expression
approach, the music has been appealing a target of researches.

However, the problem now is to organize and manage millions of music
released unstoppable. For solving this problem music recommendation system
comes into our view. The music recommendation system helps users find music
from a large set of music database, and some of which consistently match the
user's preference. Context-aware music recommender systems (CAMRSs) have

W. Kim et al. (Eds.): JIST 2013, LNCS 8388, pp. 17–32, 2014.
DOI. 10.1007/978-3-319-06826-8_2, © Springer International Publishing Switzerland 2014

been exploring many kinds of context information [1], such as weather [2], emotional state [3], running pace [4], location [5], time [6], social media activities [7] and low-level activities [8]. A music system that can detect users' context in real-time and play suitable music automatically thus could save time and effort. So far as we know, however, almost all of the existing systems need to accumulate personal information in advance, which is a time-consuming and inconvenient issue (cold start problem). On the other hand, recent music recommendation systems are practical enough for the recommendation of the similar music. But along with music boom, only the similar music cannot meet consumers' appetite.

Thus, we propose a unique method based on the content-based system to avoid the above cold start problem. Also, we considered that people need more music with the diversity they may like. Serendipity [9] appeared as an important evaluation criterion several years ago. Serendipity means a 'happy accident' or 'pleasant surprise', the accident of finding something good or useful while not specifically searching for it. To this end, our system uses a variety of Linked Open Data (LOD) without analyzing audio content or key, etc.

Linked Data refers to a style of publishing and interlinking structured data on the Web. The significance of Linked Data is in the fact that the value and usefulness of data increases the more it is interlinked with other data. It builds upon standard Web technologies such as HTTP, RDF (Resource Description Framework) and URIs, but rather than using them to serve web pages for human readers, it extends them to share information in a way that can be read automatically by computers. For now, we can see various projects using Linked Data to construct their services and datasets.

Meanwhile, the ongoing research on Twitter has found that [10] six mood states (tension, depression, anger, vigor, fatigue, confusion) from the aggregated Twitter content and compute a six-dimensional mood vector for each day in the timeline. Needless to say, music also has an inseparable connect with emotion. Even contextual features [11] influence expressed emotion at different magnitudes, and their effects are compounded by one another. Lyric is a kind of Emotion-Presentation in music composing. It includes composers' implicit thinking. Therefore, we can try to fathom emotions and associated thinking in each song. Since Social Network Services and Music are all connected with Emotion-Presentation to express people's mental state, links between them are reasonable and feasible.

Motivated by these observations, this paper presents a music recommendation method using Linked Data aiming at the avoidance of the cold start problem and the realization of the serendipity in the recommendation illustrated in Fig. 1.

This paper is organized as follows. Section 2 presents the data collection and their triplication to Linked Data which a basis of our music recommendation. Section 3 describes how to connect the Music Linked Open Data (LOD) and recommend the music based on the data. Section 4 describes preliminary experiment and evaluation of our system. Section 5 discusses related researches. Then, finally Sect. 6 concludes this paper with the future work.

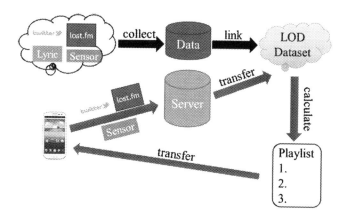

Fig. 1. System approach

2   Data Collection

When talking about music, inevitably involve many new Web Services (i.e. Last.fm, Pandora and so on). In these services, the most valuable data are public music information and personal user information. We choose Last.fm to serve as the base of the music information. Also, we use GPS sensor to collect the user's context information, Twitter to present users' implicit thoughts, and lyrics of English songs and Japanese songs to present composers' implicit thoughts.

2.1   Last.fm

Last.fm is a music-based social networking company using a music recommender system called "Audioscrobbler". It offers events, wiki-created artist profiles and discographies and community forums. By using its RESTful API, developers can read and write access to the full slate of last.fm music data resources - albums, artists, playlists, events, users and more. We use this mainly to gather artist, track and user information.

To keep the coherence of all the dates, we chose six methods listed below.

artist.getTopTracks: Get the top tracks by an artist on Last.fm, ordered by popularity.

geo.getTopArtists: Get the most popular artists on Last.fm by country. It was restricted to Japan for this research.

track.getInfo: Get the metadata for a track on Last.fm using the artist/track name or a musicbrainz id. We acquired artist/track name and track release year through this method.

user.getInfo: Get information about a user profile. Used to catch user's age.

user.getRecentTracks: Get a list of the recent tracks listened to by this user.

user.getRecommendedArtists: Get Last.fm artist recommendations for a user.

2.2   Yahoo! Local

Yahoo! is widely known for its web portal, search engine, and related services. And several years ago, it released Open Local Platform API to developers in Japan region. Yahoo! Open Local Platform (YOLP) is the API & SDK of geographical map and local information that Yahoo! JAPAN provides to the developers. You can take advantage of the rich features of multifold map display, store and facility search, geocoding, route search, and land elevation acquisition in the development of web page and applications for smartphones.

We use Yahoo! Place Info API that one of the most useful APIs for searching region-based store information, events, reviews and other information (POI). This API can translate geographical coordinates into industry code. Part of the industry codes is presented in Fig. 2. We mainly use this API to do translation and its request and response are just like in Fig. 3 showed below.

By using this service, the developers can catch the location information more particular than industry code and name, specific to a particular location.

2.3   Twitter

User-generated content on Twitter (produced at an enormous rate of 340 million tweets per day) provides a rich source for gleaning people's thoughts, which is necessary for deeper understanding of people's behaviors and actions.

Therefore, we collected a quantity of tweets relevant to industry names from Yahoo! Local Search API. While the sum of industry name is 584, input each name as keyword, through the Twitter API to acquire the related data. But a problem we have to solve is the language classification. People all around the world are using Twitter, and they send messages in English and many other languages. So in this paper we focused Japanese tweets alone.

MeCab [12] is a fast and customizable Japanese morphological analyzer. MeCab used in our system has two tasks. One is morphological analysis, and the other is keyword extraction using a TF-IDF algorithm. In morphological analysis, it analyzes all the tweets sentence by sentence, and returns parts of speech. MeCab identify sentence into a noun, verb, adjective, adverb, pronoun, preposition, conjunction, and interjection. After morphological analysis, all the parts of speech are collected, we counted times each occurs and total number of them. Then use this simple but effective algorithm TF-IDF.

2.4   Lyric

Lyric information is very important for identifying the music, since there is no two songs have the same lyric. Hence, we build an environment to catch up English and Japanese music lyrics. For the English music, we receive the lyric data from LyricWiki. It has very high credibility just like Wikipedia. LyricWiki released its API in 2008. The LyricWiki API lets you search for songs, artists, and albums. While in Japanese music, things are more complex. It is impossible to fetch lyric data from the website directly. After a huge amount of effort

業種 コード1 Industry Code 1	業種 コード2 Industry Code 2	業種 コード3 Industry Code 3	業種名1 Industry Name 1	業種名2 Industry Name 2	業種名3 Industry Name 3
1	101	101001	グルメ Gourmet	和食 Japanese Food	懐石料理 Kaiseki Cuisine
1	101	101002	グルメ Gourmet	和食 Japanese Food	会席料理 Banquet
1	101	101003	グルメ Gourmet	和食 Japanese Food	割ぽう Japanese- style Cooking
1	101	101004	グルメ Gourmet	和食 Japanese Food	料亭 High-class Restaurant
1	101	101005	グルメ Gourmet	和食 Japanese Food	小料理 Snack
1	101	101006	グルメ Gourmet	和食 Japanese Food	精進料理 Vegetarian Cooking
1	101	101007	グルメ Gourmet	和食 Japanese Food	京料理 Kyoto Cuisine
1	101	101008	グルメ Gourmet	和食 Japanese Food	豆腐料理 Tofu Dish
1	101	101009	グルメ Gourmet	和食 Japanese Food	ゆば料理 Yuba Cuisine

Fig. 2. Part of industry codes

http://placeinfo.olp.yahooapis.jp/V1/get?lat=35.66521320007564&lon=139.7
300114513391&appid=＜あなたのアプリケーション ID＞&output=json

Your Application ID

```
{
    "ResultSet": {
      "Result": [
          {                        Around Tokyo MidTown
            "Combined": "外苑東通りの東京ミッドタウン付近",
            "Uid": "29586a2bea6bcea0f0934695959d389404b04269",
            "Score": 87.9099807710434,
            "Label": "東京ミッドタウン",   Tokyo MidTown
            "Name": "東京ミッドタウン",   Tokyo MidTown
            "Where": "外苑東通り",   Meijijingu Gaien
            "Category": "ショッピングセンター・モール、複合商業施設"
                     Shopping Center, Mall, Commercial Institution
          },
```

Fig. 3. Example of request(upper) and response(lower) through Yahoo! Local Search API

in intelligence gathering, we found an intermediate tool called Lyrics Master, which is a lyric software made by Kenichi Maehashi. It provides lyric search and download services from over 20 lyric websites. Not only Japanese pop music, but also western and other genre music is supported.

2.5 Linked Data Conversion

Once collected, data is organized in different formats. Before converting into Linked Data in RDF, defining schema is indispensable. We aim to make RDF triples providing descriptions of relevance using the Music Ontology [13] and a large SKOS taxonomy [14].

After collecting data and defining schema, we use an online service called LinkData [15] to convert table data into an RDF file. Each data is converted to

an RDF file and uploaded to the LinkData site. It is set to public and anyone
can access and use them. By using this service, we generated several RDF/Turtle
files. But these files have full of redundancies, hence some post-processing is done
to fix the problem.

The detailed structure of our Music LOD is presented in Fig. 4.

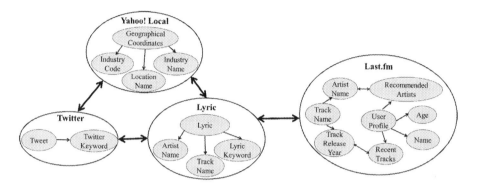

Fig. 4. Data relationship

We collected four datasets altogether: 20,050 triples from Last.fm, 584 triples
from Yahoo! Local, 584,000 triples from Twitter and 400,000 triples from Lyric.
And then we use over 1,000,000 links which can be divided into 15 types to make
relations. Yahoo! Local, Twitter and Lyric connect each other, while Last.fm
connects with Lyric directly. The nodes are connected through properties like
mo:track, *owl:hasValue* and *mo:lyrics*.

3   Proposal of 'renso' Alignment of Music LOD

In the previous section, we created the four kinds of LOD. However, they are
almost isolated and the links among different datasets are few, although we
created the *owl:sameAs* links if the nodes between the datasets are identical.
This is a well-known instance matching issue in the area of ontology alignment.

Ontology alignment means determining corresponding nodes between ontologies. We can define different types of (inter-ontology) relationships among their
nodes. Such relationships will be called alignments. There are several matching
methods like predicating about the similarity of ontology terms, and mappings
of logical axioms, typically expressing the logical equivalence or inclusion among
ontology terms.

Although we use the basic instance matching methods such as the string
similarity using n-gram and the semantic similarity using Japanese Word Net,
the expression of a term may vary especially in Twitter, and the semantics are
richer than the Word Net especially in the lyric. Therefore, we propose new
relations called 'renso', that mean n hops of implicit associations, to create

more connections among the LOD. For example, we can say there is a 'renso' relation between 'Cherry blossoms' and 'Graduation' in Japan. Or, a Japanese proverb says 'if the wind blows the bucket makers prosper', which means any event can bring about an effect in an unexpected way. Thus, by tracing these 'renso' links in the LOD we intend to recommend the music with the serendipity. More precisely, given two data sources A_1 and A_n as input, the goal of 'renso' alignment is to compute the set $M = \{(a_1, a_n) | (a_1, a_n) \in A_1 \times A_n, (a_1, a_n) \in R_a$. Here, $(a_1, a_n) \in R_a$ means a pair of distinct instances a_1 and a_n has a 'renso' relation if they are part of (any kind of) n-ary relation $R \subseteq A_1 \times ... \times A_n$.

3.1   Three Alignments

Based on the experiments, we defined three types of the 'renso' alignments between different LOD for the music recommendation.

The first one is the simplest query method. It mainly connects twitter keywords and lyric keywords, and presented in Fig. 5. In case that the user who has a smartphone in hand is using our system, firstly the GPS sensor in the smartphone locates the user's situation and return a pair of coordinate value. We collect this coordinate and convert it into the industry code through Yahoo! Place Info API, then get a return value like 'Bijutsukan' (Art Museum). We then use 'Bijutsukan' (Art Museum) as a keyword to search tweets through Twitter API, and obtain several twitter keywords with strong relevance to 'Bijutsukan' (Art Museum), then count the number of the keywords appeared in each song lyric. Finally we sort the songs with the number, and get No.1 song which in the example has three same keywords with the Twitter keywords called 'Meganegoshinosora' (Sky over glasses). The same keywords are 'Shashin' (Photograph), 'Yume' (Dream) and 'Megane' (Glass). As a result, the 'renso' relation between the location of the user and the resulted song has been created. These relations can be found and created on the fly and stored in the LOD for the future use. In the recommendation, we can search on the graph of LOD with tracing the 'renso' relations, and find the song using SPARQL querying methods.

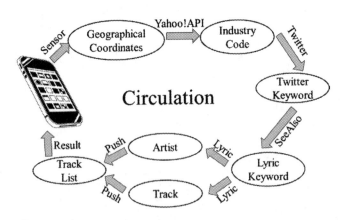

Fig. 5. Alignment 1

Alignment 2 and alignment 3 are on the basis of the above alignment 1, adding user's profile information and the user's listening history with the recommended artist information separately. They go through Yahoo! Local, Twitter, Last.fm and Lyric graphs sequentially.

Definitions of the alignments are as follows.

Considering that there is an infinite set U (RDF URI references), and an infinite set L (literals). A triple $(s, p, o) \in (U \cup L) \times U \times (U \cup L)$ is called an RDF triple (s is called the subject, p the predicate and o the object). An RDF Knowledge Base (KB) K, or equivalently an RDF graph G, is a set of RDF triples.

For an RDF Graph G_1 we shall use U_1, L_1 to denote the URIs, literals that appear in the triples of G_1 respectively. The nodes of G_1 are the values that appear as subjects or objects in the triples of G_1. The main graph G_1 contains four datasets collected separately: F is from Last.fm, Y is from Yahoo Local, T is from Twitter and L is from Lyric.

The algorithm described in chapter 3 can be expressed in the following pseudo-code. First, we define the following additional functions as proposed in

---

Algorithm 1. Define Some Functions

```
1: function SEARCH(x, G₁)                         ▷ Search x in graph G₁
2:     for all g ∈ G₁ do
3:         if g = x then
4:             Push g into array Result
5:         end if
6:     end for
7: return Result
8: end function

9: function MATCH(X, Y)                           ▷ Match between X and Y
10:     for all x ∈ X do
11:         SEARCH(x, Y)
12:     end for
13:     return Result
14:     Calculate times each element appeared
15:     return Result
16: end function

17: function SORT(M)                              ▷ Sort by numeric value
18:     for i ← 1, length(M) do
19:         for j ← length(M), i + 1 do
20:             if thenMⱼ < Mⱼ₋₁
21:                 Exchange Mⱼ and Mⱼ₋₁
22:             end if
23:         end for
24:     end for
25: end function
```

Algorithm 1. Function Search is a function to search a value in a graph. While function Match is to find common elements between two graphs, function Sort is to sort an array by numeric value.

Our starting point is geographical *coordinates(x,y)* received from user's smartphone, while ending point is a music playlist. Our mapping pseudo-code is defined as Algorithm 2 and Algorithm 3. It describes three alignments' implementation procedures. The three alignments are designed to use less profile information in order to avoid cold start problem. Meanwhile, be independent of user profile lead to high serendipity. What we try hard to do is enhance accuracy.

4   Music Recommendation Experiment

In this section, we describe the results of our recommendation system. Adequacy of the 'renso' relation highly depends on the user's feeling, and has difficulty to be measured in an objective way. Instead, we evaluate the accuracy of the music recommendation according to the purpose of this paper. We have conducted an evaluation of accuracy and serendipity, and the results demonstrate significant promise from both perspectives.

4.1   Dataset

All the data were collected for two weeks; their details are presented in Table 1. We cascade all the data and let the prior one to determine the subsequent one.

Based on the industry codes, we defined 584 location keywords. By using these location keywords, we employ Twitter as a platform to search tweets. Every location keyword corresponds to 1,000 tweets. Then we utilize MeCab to extract 50 twitter keywords by analyzing every 1,000 tweets. Also, Last.fm stores many aspects of information. All the data are collected through its API methods. We set Japan as the main place for experiments, request Top 500 artists in Japan and their Top 20 tracks. Afterwards, I use my Last.fm account to acquire any personal information for alignment 2 and alignment 3. Every track corresponds to one lyric data, and 10,000 in total. As well as the tweets, we processed lyric data with MeCab and extracted 40 lyric keywords per song.

Table 1. Data details

Category	Industry code	Tweet	Lyric	Last.fm
Triples	584	584,000	400,000	20,050

4.2   Experiment Setting

We conducted an evaluation to identify the accuracy and the serendipity. We invited 20 test users to participate in this experiment. Six of them are our college students, the others are working adults. Also, three of them are Japanese,

---

Algorithm 2. Three Alignments Part 1

1: **procedure** BASE
2:     $coordinates(x, y) \leftarrow GPSsensor$
3:     $category \leftarrow$ SEARCH$(coordinates(x, y), Yahoo!API)$
4:     $categorykeyword \leftarrow$ SEARCH$(category, Y)$
5:     $TwitterKeyword \leftarrow$ SEARCH$(categorykeyword, T)$
6:     **for all** $twitterkeyword \in TwitterKeyword$ **do**
7:         SEARCH$(twitterkeyword, L)$
8:     **end for**
9:     **return** $Track(TK)$
10: **end procedure**
11:
12: main ***Alignment 1***
13: **Start**
14:     **procedure** BASE
15:     SORT(Track) by LK
16:     **for** $count \leftarrow 0, 2 \in Track$ **do**
17:         SEARCH$(count, F)$
18:     **end for**
19:     **return** $Track(Info)$
20: **End**
21:
22: main ***Alignment 2***
23: **Start**
24:     $User(age) \leftarrow$ SEARCH$(user, Last.fm)$
25:     $Now(year) \leftarrow getYear$
26:     **for** $year \leftarrow Now(year) - $Age(user), Now(year) **do**
27:         SEARCH$(year, F)$
28:     **end for**
29:     **return** $YearTrack$
30:     **for all** $yeartrack \in YearTrack$ **do**
31:         SEARCH$(yeartrack, L)$
32:     **end for**
33:     **return** $Track(Lyric)$
34:     **for all** $tracklyric \in Track(Lyric)$ **do**
35:         SEARCH$(tracklyric, L)$
36:     **end for**
37:     **return** $Track(LK)$
38:     **procedure** BASE
39:     $Count \leftarrow$ MATCH$(Track(TK), Track(LK))$
40:     SORT(Track) by Count
41:     **for** $count \leftarrow 0, 2 \in Track$ **do**
42:         SEARCH$(count, F)$
43:     **end for**
44:     **return** $Track(Info)$
45: **End**

Algorithm 3. Three Alignments Part 2

46: main ***Alignment 3***
47: **Start**
48:     $User(info) \leftarrow$ SEARCH$(user, Last.fm)$
49:     $User(SelectedTrack) \leftarrow User(ListenedTrack) + User(RecommenedTrack)$
50:     **for all** $selectedtrack \in User(SelectedTrack)$ **do**
51:         SEARCH$(selectedtrack, L)$
52:     **end for**
53:     **return** $Track(Lyric)$
54:     **for all** $tracklyric \in Track(Lyric)$ **do**
55:         SEARCH$(tracklyric, L)$
56:     **end for**
57:     **return** $Track(LK)$
58:     **procedure** BASE
59:     $Count \leftarrow$ MATCH$(Track(TK), Track(LK))$
60:     SORT$(Track)$ by Count
61:     **for** $count \leftarrow 0, 2 \in Track$ **do**
62:         SEARCH$(count, F)$
63:     **end for**
64:     **return** $Track(Info)$
65: **End**

---

the others are not. The question are contains ten items which is formed in 3 parts: situation description, question 1 and question 2. Situation description is written in this pattern: "When you are in (location name) right now, our system recommend a track which has no direct connection with (location name) but contain (keyword) and (keyword) in its lyric named (song name)". Question 1 "Do you think this recommendation is unexpected but fun?" is asking for the serendipity, while question 2 "Do you think this recommendation is fit for this place?" is for the accuracy. We wrote the question are with forms in Google Drive and set it public for test users here[1].

The result of the experiment is presented in Fig. 6. We can see that all three alignments achieved about 60 % satisfaction in the accuracy and the serendipity. Among them, alignment 3 presented the highest value; both evaluating indicators are above 65 %. While in Table 2 that shows the user's favorite alignment is alignment 1. Overall, although alignment 3 shows the best performance, the test users preferred the one uses the most simplest but straight recommendation method.

Table 2. The result of favorite alignment

Alignment 1	Alignment 2	Alignment 3
45 %	25 %	30 %

---

[1] https://docs.google.com/forms/d/
1VGHP6OuSJqBxIo3F7pvSJv5LNqZvtZbVHXmI24982fQ/viewform

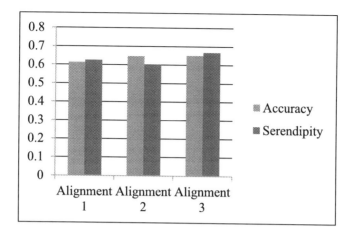

Fig. 6. The result of experiment

4.3   Lesson Learnt

Through this experiment, we found two key points. First, we confirmed there is no cold start problem in our recommendation method, since our base is the large set of Music LOD, where a point is about the data relevance. Twitter is a platform the people use when they want to express their emotions, while the lyric is a transfer mediator for the people from who composed it. They all include implicit thoughts. But the emotion extraction is very difficult, not to mention implicit thoughts. Even so, along with the increase of the relation path's length in the LOD, we could find more implicit connections to reach higher accuracy and serendipity.

Also, unlike the conventional content-based method, we confirmed our system is easier to archive the serendipity in the recommendation. However, meeting both the accuracy and the serendipity at a time is an arduous task. This research is for getting the users' implicit context, but the emotion is more complicated and changeable than any other. For example, in our university there are so many students who love sci-fi and robot animation, and are especially crazy about 'Evangelion' (Japanese TV Animation Program). Thus, 'University of Electro-Communications' comes along with 'Evangelion' frequently on Twitter. Then, we recommend 'Evangelion' theme song for the user who is in our university. It is full of fun, but the accuracy is subtle. In other words, it is difficult enough for satisfying everyone. As a summary of our evaluation, only one of two criteria, accuracy and serendipity can be satisfied in most of the cases.

5   Related Works

Most of the existing music recommendation systems that calculate with the users' preferences present methods to meet long term or short term music needs.

Comparing with constant the users' preferences, users' context (such as location, mood and people around) is changeable and flexible. For example, people who are in the library need quiet and melodious music. The existing commercial music recommendation systems such as Last.fm and Pandora are very popular and excellent, however cannot satisfy these needs very well. Meanwhile, applications for smartphones use rich sensing capabilities to collect context information in real-time meet the needs better. Recently some researches paid more attention on CAMRSs in order to utilize contextual information and better satisfy users' needs.

Twitter Music [16] is the brainchild of 'We Are Hunted'. A smartphone application pulls in music from Rdio, Spotify and iTunes, while using data from your Twitter follower graph to deliver the best possible music for you. It aims to give Twitter a mainstream launch and expand audiences of Spotify and Rdio.

EventMedia [17] is a service targets user who wants to relive past experiences or to attend upcoming events. It is a web-based environment that exploits real-time connections to the event and media sources to deliver rich content describing events that are associated with media, and interlinked with the Linked Data cloud. It is a pure service to satisfy daily users and developers.

SuperMusic [18] is a prototype streaming context-aware mobile music service. It is a context-based recommendation combined with collaborative filtering technologies using location and time information. The users have a chance to vote the recommended music and let the system learn their musical preference and opinion concerning song similarity. But this application cannot show the user's current context categories explicitly, many users get confused about the application's situation concept. For this reason, this inspired us to search for understandable context categories in the music recommendation.

Foxtrot [19] is a mobile augmented reality application with the audio content-based music recommendation. Rather than textual or visual content, audio content is potentially more emotionally engaging. Furthermore, people are allowed to share sound and music which can be tagged with a particular location. Their friends and other people can listen to shared geo-tagged music everywhere. But it is weak in automatic recommendation without geo-tagged music dataset.

6   Conclusions

In this paper, we presented a music recommendation method using the 'renso' alignment that connects various kinds of LOD, aiming at avoidance of the cold start problem and the realization of the serendipity in the recommendation.

We collected four datasets: 20,050 triples from Last.fm, 584 triples from Yahoo! Local, 584,000 triples from Twitter and 400,000 triples from Lyric. They are public in LinkData and developers who want to carry out the related research can reuse them. Based on the datasets and three different alignments, we found that there really is a relation between lyric and tweet in expressing people's implicit thoughts.

As the future work, we plan to include more data sources and design more alignments. Not only from Twitter and Lyric, we will discover the people's implicit thoughts from other online resources, and design more reasonable and effective alignment. Then, we will develop the music recommendation agent for the smartphone in the near future. Furthermore, this recommendation method using the user's implicit thought, that is, the 'renso' alignment could be used for the other media like movie, fashion items and so forth. So we will consider the adaptation of this method to other domains.

Acknowledgments. This work was supported by JSPS KAKENHI Grant Number 24300005C2350003-9C25730038 from Graduate School of Information Systems at University of Electro-Communications. We would like to thank Professor Shinichi Honiden in National Institute of Informatics/University of Tokyo and his group for offering a place for discussing, studying and providing instructions.

References

1. Ricci, F.: Context-aware music recommender systems: workshop keynote abstract. In: Proceedings of the 21st International Conference Companion on World Wide Web, pp. 865–866. ACM, Lyon (2012)
2. Beach, A., Gartrell, M., Xing, X., Han, R., Lv, Q., Mishra, S., Seada, K.: Fusing mobile, sensor, and social data to fully enable context-aware computing. In: Proceedings of the 11th Workshop on Mobile Computing Systems & Applications, pp. 60–65. ACM, New York (2010)
3. Komori, M., Matsumura, N., Miura, A., Nagaoka, C.: Relationships between periodic behaviors in micro-blogging and the users' baseline mood, pp. 405–410. IEEE Computer Society (2012)
4. Elliott, G.T., Tomlinson, B.: PersonalSoundtrack: context-aware playlists that adapt to user pace. In: CHI '06 Extended Abstracts on Human Factors in Computing Systems, pp. 736–741. ACM, New York (2006)
5. Kaminskas, M., Ricci, F.: Location-adapted music recommendation using tags. In: Konstan, J.A., Conejo, R., Marzo, J.L., Oliver, N. (eds.) UMAP 2011. LNCS, vol. 6787, pp. 183–194. Springer, Heidelberg (2011)
6. Cebrián, T., Planagumà, M., Villegas, P., Amatriain, X.: Music recommendations with temporal context awareness. pp. 349–352. ACM, New York (2010)
7. Bu, J., Tan, S., Chen, C., Wang, C., Wu, H., Zhang, L., He, X.: Music recommendation by unified hypergraph: combining social media information and music content. In: Proceedings of the International Conference on Multimedia. pp. 391–400. ACM, New York (2010)
8. North, A.C., Hargreaves, D.J., Hargreaves, J.J.: Uses of music in everyday life. Music Percept. Interdiscip. J. **22**(1), 41–77 (2004)
9. Zhang, Y.C., Séaghdha, D., Quercia, D., Jambor, T.: Auralist: introducing serendipity into music recommendation. In: Proceedings of the 5th ACM International Conference on Web Search and Data Mining, pp. 13–22. ACM, New York (2012)
10. Johan, B., Alberto, P., Huina, M.: Modeling public mood and emotion: twitter sentiment and socio-economic phenomena. CoRR. abs/0911.1583 (2009)
11. Scherer, K.A., Zentner, M.R.: Emotional effects of music: production rules. In: Juslin, P.N., Sloboda, J.A. (eds.) Music and Emotion: Theory and Research, pp. 361–392. Oxford University Press, New York (2001)

12. MeCab: Yet Another Japanese Dependency Structure Analyzer. https://code.google.com/p/mecab/
13. Music Ontology. http://musicontology.com/
14. SKOS Simple Knowledge Organization System. http://www.w3.org/2004/02/skos/
15. LinkData. http://linkdata.org/
16. Twitter Music. https://music.twitter.com/
17. Khrouf, H., Milicic, V., Troncy, R.: EventMedia live: exploring events connections in real-time to enhance content. In: ISWC 2012, Semantic Web Challenge at 11th International Semantic Web Conference (2012)
18. Lehtiniemi, A.: Evaluating superMusic: streaming context-aware mobile music service. In: Proceedings of the 2008 International Conference on Advances in Computer Entertainment Technology, pp. 314–321. ACM, New York (2008)
19. Ankolekar, A., Sandholm, T.: Foxtrot: a soundtrack for where you are. In: Proceedings of Interacting with Sound Workshop: Exploring Context-Aware, Local and Social Audio Applications, pp. 26–31. ACM, New York (2011)

Ontology-Based Information System

Martins Zviedris$^{(\boxtimes)}$, Aiga Romane, Guntis Barzdins,
and Karlis Cerans

Institute of Mathematics and Computer Science,
University of Latvia, Raina blvd. 29, Riga 1459, Latvia
{martins.zviedris,guntis.barzdins,
karlis.cerans}@lumii.lv, aiga.romane@inbox.lv

Abstract. We describe a novel way for creating information systems based on ontologies. The described solution is aimed at domain experts who would benefit from being able to quickly prototype fully-functional, web-based information system for data input, editing and analysis. The systems backbone is SPARQL 1.1 endpoint that enables organization users to view and edit the data, while outside users can get read-only access to the endpoint. The system prototype is implemented and successfully tested with Latvian medical data ontology with 60 classes and imported 5 000 000 data-level triples.

Keywords: OWL ontology · SPARQL · Instance editing · Prototyping

1 Introduction

J.F. Sowa proposed that "Every software system has an ontology, implicit or explicit" [1]. It means that every software system is created to work with a knowledge base that can be described using ontologies and it may be possible to create an information system based on a formal ontology.

In our work we perceive the ontology definition as two separate parts. First is the factual part that includes known facts about a knowledge base, for example, a statement that some individual is Person. In the presented paper we mainly focus on this part.

Second are the reasoning and constraint rules for the factual part, for example, a rule "if an individual is male and has children then he is a father" or a constraint that a person can have only two parents. We see that these rules can be implemented using SPARQL queries, but this is out of this paper's scope.

We describe a novel way to build a simple information system from an OWL [13] ontology. The presented tool automatically generates a fully-functional data storage and retrieval type information system based on a given ontology's RDFS schema part. The result is a web-based platform for editing and overlooking ontology knowledge base. The main focus of our work is on user friendliness of information system creation process rather than on advanced features of the system.

Additional system configuration options allow one to add more complicated features later on, but that is outside the scope of this paper. The approach is called an *ontology-based information system* (OBIS).

W. Kim et al. (Eds.): JIST 2013, LNCS 8388, pp. 33–47, 2014.
DOI: 10.1007/978-3-319-06826-8_3, © Springer International Publishing Switzerland 2014

The core of our system is based on the factual part where a user edits information about individuals – their attribute values and classes they belong to. For now, OBIS uses the RDFS subset of ontologies – named classes, properties and associations between them, and subclass hierarchies. This subset corresponds to the inference capabilities of the underlying SPARQL endpoint that we use[1].

We envision that the proposed approach can be used for semantic system development for small organizations as they often describe information systems as expensive and hard to maintain and resort to, for example, large Excel files instead. Instead they would only need an ontology describing their domain, which would allow them to create the information system required.

For example, it would be great to access one's hairdresser's schedule to know when he or she is available. Providing hairdressers ontology one can create automatically the system. Thus, the hairdresser's workers edit information when they are available and when they are occupied. Clients (or client agent) access system (or SPARQL endpoint) in read-only access to see available spots or even register to the hairdresser. This is one step towards the Semantic Web original idea about agents [10].

The described OBIS prototype is available at http://obis.lumii.lv.

2   Use Case of the OBIS Methodology

In this section we describe the process how a typical user would create and use the ontology-based information system.

2.1   Selecting or Creating an Ontology

To build an OBIS information system one needs an OWL ontology that describes the conceptual structure of the data to be held in the information system. One can search ontology repositories (e.g., the DERI vocabulary catalogue[2]) or the Web for an appropriate ontology, or one can create an ontology from scratch.

For illustration purposes we will create an information system for a veterinary clinic. As we did not find a suitable ontology we created a simple ontology depicted in Fig. 1 using the OWLGrEd editor [3]. The ontology could have been created in any other ontology editor, for example, Protégé[3].

Ontology consists of all doctors that work in a clinic. Doctors hold appointments that can be classified in two subclasses: planned and past appointments. For each appointment there is an animal to be examined. For past appointments there are also details about performed examinations and their types.

---

[1] We use Virtuoso (http://virtuoso.openlinksw.com/) endpoint which implements RDFS reasoning and supports SPARQL 1.1. We considered StarDog endpoint with full OWL reasoning, but it does not support SPARQL 1.1 update.

[2] http://vocab.deri.ie/

[3] http://protege.stanford.edu/

2.2   Creating a New Ontology-Based Information System

After the ontology is selected or created we can proceed with generating a new information system. This task is very simple as we just need to open the system generator in a web browser and upload the ontology, see Fig. 2. The OBIS tool will generate a working information system that is available through web browser for data browsing and editing.

The generated OBIS information system uses a preselected (in the OBIS generation tool the creator enters the SPARQL endpoint address) SPARQL endpoint for data storage; the data in this endpoint is also available to other tools.

The generated user interface is quite simple – there is a header, a tree-like menu with a list of all pages and a content frame (see Fig. 3). The information system has a page for each class in the ontology, listed in the page menu. The content part allows users to view, edit and create new class instances.

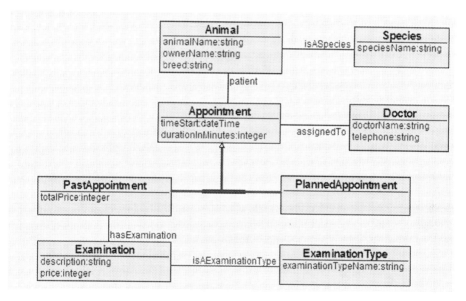

Fig. 1. Veterinary clinic ontology

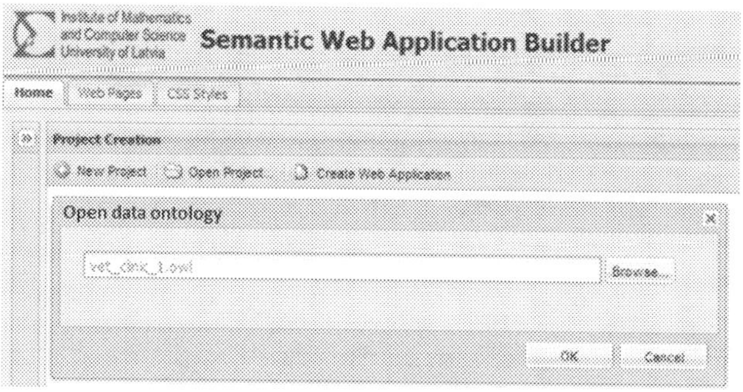

Fig. 2. The OBIS generation tool

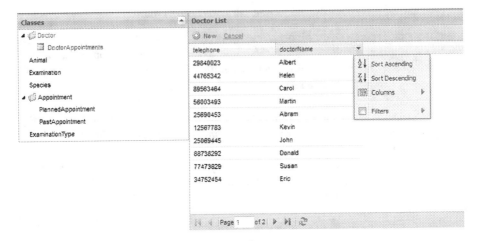

Fig. 3. A list of class instances in a table form

2.3   Using the Created Information System for Data Manipulation

The data manipulation takes place in the content frame of the information system's web interface. It is similar to a database entry manipulation in other information systems, for example, MS Access. Users can click on a class in the tree menu that they want to inspect at the instance level.

Figure 3 depicts the browsing view of a selected class. It displays all individuals of the selected class in a table format. A user can browse, sort or filter individuals by data property (attribute, column) values. If the user selects one individual (instance, row) then a detailed browsing/editing window for that individual (instance) is opened.

Figure 4 shows the individual (instance) data browser. Its upper part displays the individual's (instance) data properties (attributes) and their respective values. Below it lists all the outgoing and incoming[4] object properties (connections, associations) and individuals (instances) they connect to are displayed.

Thus, a user can easily add, update and delete instances for all classes in the ontology. This is done using the SPARQL 1.1 protocol and with data being stored in the underlying SPARQL endpoint[5].

2.4   Report Browsing

One of the most important parts of relational databases is the view definition mechanism. It allows to create various, more complex looks on the underlying data. This is important in any information system.

---

[4] Although inverse object properties are available in OWL, their inference is not supported by SPARQL endpoint such as Virtuoso. Therefore, it is more convenient to allow navigating object properties bi-directionally.

[5] We use Virtuoso endpoint - http://virtuoso.openlinksw.com/.

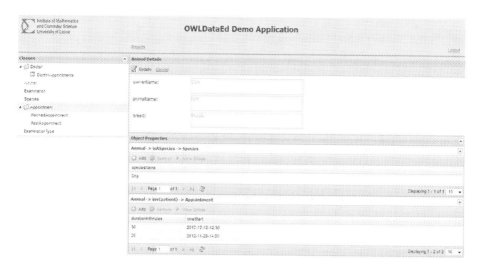

Fig. 4. The individual data browser

We consider it an important part of OBIS, as well. However, we want to extend this view mechanism to allow navigating from any individual description that is included in a view further to the individual's details.

To achieve this result, we define the requirements of view creation. Each view must have defined one central class. In OBIS, the view is a SPARQL query where a specific variable binding (?ID) is used to identify individuals of the central class. It is also permitted to include in the view additional attributes from connected individuals by connecting them in the SPARQL query. In order to include the defined view in the hierarchy of ontology classes offered by OBIS, one would create a named subclass of the view's central class and annotate it with the SPARQL query describing the view.

Alternatively, one could dynamically load the SPARQL queries corresponding to the views into OBIS system via its explicit report mechanism that offers the SPARQL query execution to obtain the result set that is enriched with navigation to every report's line central entity details' pages.

This feature is most useful for data browsing. If we run the OBIS in read-only version for enterprise clients then it would also be useful to define additional views to enable end users to understand the underlying data better.

For example, we would like to add a report view for Animal class to OBIS based on the veterinary ontology depicted in Fig. 1. The report contains all information about an animal including animal species name that is stored in a separate class "Species".

SPARQL query defining the view is depicted in Fig. 5 while the report view of the information system is depicted in Fig. 7. This report is named "AnimalSpecies" and is highlighted with a different icon in the tree list.

```
SELECT DISTINCT
  ?ID ?Animal__animalName ?Animal__breed
  ?Animal__ownerName ?Species__speciesName
WHERE {
  ?ID rdf:type :Animal.
  OPTIONAL { ?ID :animalName ?Animal__animalName.}
  OPTIONAL { ?ID :breed ?Animal__breed.}
  OPTIONAL { ?ID :ownerName ?Animal__ownerName.}
  ?ID :isASpiecies ?Species.
  ?Species rdf:type :Species
  OPTIONAL { ?Species :speciesName
?Species__speciesName.}}
```

Fig. 5. SPARQL code to define view

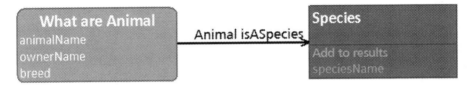

Fig. 6. ViziQuer lite query defining the report

2.5 SPARQL Endpoint Access

All data is stored in the SPARQL endpoint, thus it can be accessed by SPARQL queries or by other tools that visualize or gather data from a SPARQL endpoint. Automatic data agents could also access the SPARQL endpoint.

An example of an application that can connect to the SPARQL endpoint is the ViziQuer[6] graphical SPARQL language front-end [4]. The ViziQuer tool can also be used for report creation as it gives end-users a simple and visual way to create SPARQL queries. For example, the query in Fig. 5 corresponds to the ViziQuer graphic query shown in Fig. 6.

Finally, the OBIS graphical interface can be used in a read only mode for data browsing by enterprise clients via web interface.

3 OBIS Overview

We can perceive the ontology definition as two separate parts. First is the factual part (Abox or Assertion box) that includes known facts about our knowledge base, for example, a statement that some individual is Doctor. Second (Tbox or Terminology box) is the reasoning and constraint rules for the factual part, for example, a rule "if

---

[6] http://viziquer.lumii.lv

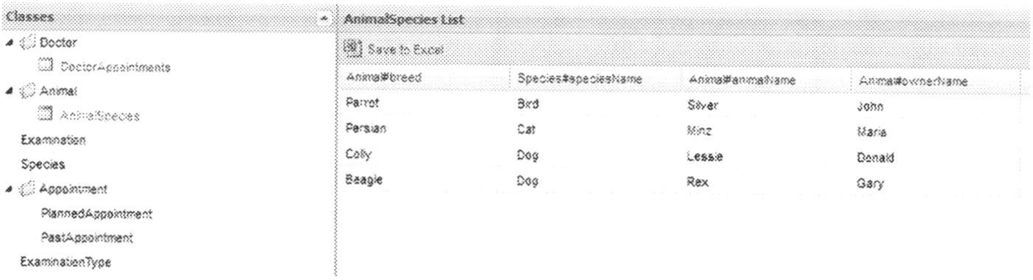

Fig. 7. Preview of report table view

time of an appointment is earlier that today then it must be past appointment" or a constraint that an appointment has just one animal to examine.

As the users at first need to enter the data into the system then our core focus now is based on the factual part where a user edits information about individuals. For now, OBIS uses the RDFS subset of ontologies – named classes, properties and associations between them, and subclass hierarchies as it covers all necessity to work with data input and editing.

Named class instances in the ontology are edited through OBIS interface. The class instance assertions for anonymous restriction classes are derived implicitly from the individual's properties and can be checked e.g. using SPARQL queries and non-editable views, similar to those described in Subsect. 2.4.

We see that reasoning and constraint rules are an important part of the ontology. It helps to maintain correct and consistent input, and also reveals which classes individuals belong to. We envision that a part of constraints could be enforced on client side (for example, checking of simple input constraints, cardinality) while another part could be performed in server side. Obvious implementation would be to translate OWL constraints to SPARQL queries that would allow monitoring data integrity regularly as is done in similar works [11].

An essential part of OBIS is its ability to show the instances of a class together with their data property values in a tabular form thus obtaining a compact overview of a list of class instances at the same time. The key assumption behind the possibility of the - OWL data property cardinality is restricted to 1, which allows data properties of individuals to be visualized as a table or a data input form that is typical for traditional information systems. We use this restriction only for now to make the system a bit simplistic and easier to explain and more performance-efficient. In the future we plan to use SPARQL 1.1 aggregate feature GROUP_CONCAT to show all values comma separated in one table cell or to adapt some other understandable display form.

The architecture of our solution is depicted in Fig. 8 We divide it in three main parts – (i) an ontology that represents a factual data structure, (ii) an OBIS information system that is generated from the data ontology and can be used by enterprise workers for data editing and (iii) SPARQL endpoint for enterprise clients that allows them to gather meaningful data and that can be accessed via different tools, for example, the OBIS information system in a read-only version.

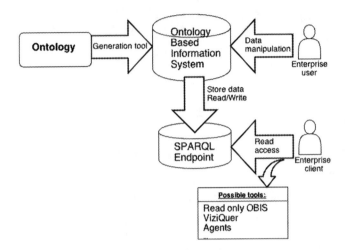

Fig. 8. System architecture

We note that the implementation described here is truly model driven – every OBIS information system is viewed as a set of instances of the universal OBIS meta-ontology, see Fig. 10.

The OBIS generation tool loads the data ontology and generates from it a set of instances of the OBIS meta-ontology. The resulting set of OBIS meta-ontology instances is afterwards executed by the interpreter that provides that every OBIS information system is presented as a composition of web-pages, see Fig. 9. The appearance of the automatically generated OBIS meta-model instance can later be customized for each enterprise.

The OBIS interpreter effectively presents the generated OBIS information system as composition of webpages. It is a web-based system where one can browse through

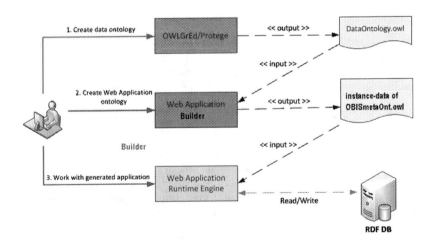

Fig. 9. High-level OBIS workflow architecture

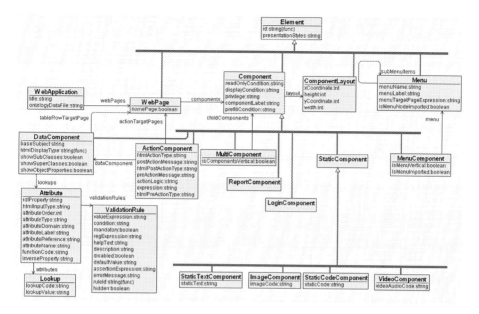

Fig. 10. OBIS meta-ontology

pages and see all class instances, edit them and add new ones. The data is stored and updated in an SPARQL 1.1 endpoint.

Every enterprise customer can query the SPARQL endpoint in a read-only mode and retrieve information that interests them. A user can access the SPARQL endpoint in a read-only mode by various tools, including read-only OBIS information systems, ViziQuer [4] or other data agents. In this paper we do not discuss the question of information privacy or sensitivity.

If we look back at one of the initial ideas of the Semantic Web [10] then a major role was meant for agents that could gather data and make suggestions to a user. By means of the OBIS system's SPARQL endpoints it could be one step closer to achieving the idea about agents gathering information.

Thus, we are effectively combining the mentioned Sowa's statement with Model Driven Architecture [2] approach that says that it is possible to develop universal platform independent meta-model (PIM or OBIS meta-ontology). Any specific instance of this platform independent meta-model can be converted to executable program (sometimes referred to as platform specific model or PSM or OBIS interpreter).

4   Technical Implementation

The described implementation is based on the model driven approach – every OBIS information system is viewed as a set of instances of the universal OBIS meta-ontology, see Fig. 10.

Web Application Builder (the OBIS generation tool) uses a data-ontology to generate the instance-data of the OBIS meta-ontology, see Fig. 9. Afterwards the *Web Application Runtime Engine* interprets this OBIS meta-ontology instance-data to provide the OBIS information system that is used by end-users through a web browser.

OBIS meta-ontology (Fig. 10) captures the common standard functionality of a typical web information system consisting of:

- Data representation and processing;
- Display of user interface components - web pages, data forms, tables, menus, images, etc.;
- Client and server side validation rules;
- Navigation between web pages and between web page components;
- Application user management (authentication and authorization);
- Reporting.

This ontology is used as the backbone of the OBIS and enables the truly model-driven design principle detailed here.

Web Application Builder is responsible for building an OBIS information system. In terms of model driven approach, this means to fill the OBIS meta-ontology with required instance-data based on user's data-ontology.

Besides fully automatic web application generation mode, this component provides also some configuration possibilities such as adjusting visual layout of components on the web page or other adjustments permitted within the OBIS meta-ontology.

The Web Application Builder is created as GWT[7]/GXT[8] web application with Java on server side. It reads data ontology file (RDF/XML syntax) using JENA parser[9].

Web Application Runtime Engine is the OBIS interpreter executing an OBIS information system during runtime. It:

- reads previously created OBIS meta-ontology instance-data file and interprets it;
- maps OBIS meta-ontology instance data to corresponding GWT/GXT user interface components;
- processes user input/output requests based on information in the OBIS meta-ontology instance data (processing includes client side form field validation, report creation, data representation and processing).

On a more detailed level Web Application Runtime Engine is a GWT/GXT web application with Java on server side. It can be deployed on Java compatible web application servers[10] such as Tomcat 7.0.

---

[7] GWT (Google Web Toolkit) - https://developers.google.com/web-toolkit/.

[8] GXT (Google Extended Toolkit) - http://www.sencha.com/products/gxt.

[9] http://jena.apache.org/

[10] http://tomcat.apache.org/

The main component of Web Application Runtime Engine is the *Web Application Ontology Reader*. Rather than using Jena API or similar tools, it reads and translates OWL file with OBIS meta-ontology instance data directly into Java class instances – OWL individuals are mapped directly into Java class instances. Thus, we actually make PIM to PSM transformation, an insight about PIM to PSM process can be gained in [14].

Such OWL/Java class and individual/instance mapping is made possible by pre-defined OBIS meta-ontology being a 1:1 copy of Java class hierarchy within the Web Application Ontology Reader component (This actually is platform independent to platform specific model transformation).

To facilitate such mapping, Java class structure is annotated with semantic types and reader automatically maps OWL XML (the reader reads OBIS meta-ontology as XML document) tags to annotated Java classes, see Table 1. Semantic annotations also include class or property names they correspond to in the OBIS meta-ontology. Currently semantic annotations cover only a small part of RDF/OWL but they can be extended if needed.

Table 1. OWL to Java mapping annotations

Java annotation	OWL XML tag	JAVA element
RdfOntClass	owl:Class	Class
RdfDataTypeProperty	owl:DatatypeProperty	Class internal variable with primitive data type (String, Integer, etc.)
RdfObjectProperty	owl:ObjectProperty	Class internal variable with reference data type (points to other class)

This OWL to Java mapping approach has been developed specifically for our system, but is not limited to web application ontology – it can actually be used with any meta-model ontology and corresponding Java class hierarchy.

We choose such architecture for performance reasons as this effectively creates a compiled runtime rather than interpreted runtime. This is important for interactive systems, as end-users want to see their action results immediately.

Other Web Application Runtime Engine software components are:

- *UI Generator* - maps OBIS meta-ontology instance-data to corresponding GWT/GXT user interface components;
- *Application Controller* - handles user requests (browser events) and executes appropriate services (save, search, login, etc.) and displays results;
- *RDF Repository Service* - provides CRUD operations with Virtuoso RDF server;
- *Rule Engine* - executes validation rules on form fields;
- *Report Engine* - generates report queries and displays results in GWT/GXT tables with possibility to save in MS Excel.

The components listed above interact in the following way. The main entry point is invoked by GWT/GXT when web application is opened from a web browser for the first time. This entry point calls server side method, which reads OBIS meta-ontology instance-data using Web Application Ontology Reader (described above).

As a result a Java class instance hierarchy is returned, populated with data from OBIS meta-ontology instance-data. App entry point then calls UI Generator which creates GWT/GXT widgets.

The UIGenerator is invoked on request, i.e. new web page is created only after user has requested it (e.g., a user pressed a link/button which navigates to another page).

Every web page consists of components which can be just display objects or can call some service. WebPage in OBIS meta-ontology is mapped to the GXT component LayoutContainer. WebPage invocation adds all subcomponents to this container as GWT/GXT widgets, which are automatically gathered from OBIS meta-ontology represented by the Java class hierarchy.

Web Application Runtime Engine logic is split into two modules communicating through GWT/GXT asynchronous calls:

- client services package (displays UI and handles UI input/output requests)
- server services package (calls other server-side services or third-party libraries, for example, Virtuoso RDF API).

All data is stored in a Virtuoso RDF database, which the Web Application Runtime Engine communicates with using SPARQL 1.1, as it provides both SPARQL select and update. Virtuoso RDF database also provides a public SPARQL endpoint that can be used by other systems.

5   Evaluation

The OBIS was evaluated as a part of a linguistic research project at the University of Latvia, where it was used for navigating and editing FrameNet[11] data while adapting it from English to Latvian language.

The need for the OBIS-like fast prototyping solution was dictated by frequent database structure changes requested by linguists. It would have been arduous to manually adapt the information system structure and interface every few days as linguists came up with new ideas for improved data structure (ontology).

Attempts to use a universal Protégé-like ontology individual editor interface for the frequently changing information system were not successful, as it turned out that linguists were thinking in terms of "tables" rather than RDF graphs - just like most relational SQL DB users and even programmers do. Thus creating good table-like views on ontology individuals became the focus of the OBIS development team.

It is relatively straightforward to create a table of individuals belonging to the selected class and to display data property attributes of each individual in additional columns of the table. But such simplistic approach turned out to be insufficient for the needs of linguists - they wanted to see in the same table not only data properties of the selected class, but also data properties of some individuals in neighboring classes, linked through object properties. This requirement lead to incorporation of arbitrary SPARQL queries as a way to generate arbitrary table-views on the data.

---

[11] http://framenet.icsi.berkeley.edu

But even this solution turned out to be insufficient for the needs of linguists - they wanted these table-views to be easy to edit in case some errors in the data were spotted (regular SPQRQL-generated reports are not editable). A solution was found how to make these SPARQL generated table-views editable - each SPARQL query must be associated with some "central class" individuals that this query is selecting (meanwhile the same query may also select individuals and attributes from other classes).

In this way each row in the SPARQL-generated answer table can be linked to the URI of the corresponding individual from the "central class" of the SPARQL query; by clicking on any row in the query answer table the system can automatically navigate to the corresponding individual of the "central class". From there the user can easily navigate to the attribute he wants to edit - it is either an attribute of this individual, or one of its neighboring individuals linked through object properties.

Since such "editable" SPARQL query generated table-views became integral part of the ontology based information system, it became natural to store them as annotations in the data ontology along with the "central class" they belong to. From there OBIS can automatically retrieve these queries and present them to the user as "editable" tables-views as part of the special leaves in the class-tree menu.

As a result of these improvements OBIS met the requirements of the linguistic research project and evolved into the universal approach described in this paper.

To test OBIS performance on larger data sets we loaded Latvian medical data into SPARQL endpoint and generated the OBIS information system from the Latvian medical ontology described in more details in [12]. Overall there are around 60 classes and more than 300 000 individuals in the ontology. The OBIS information system worked fine, although loading time for the Person class that has more than 70000 records took around 30 s.

Additional evaluation was conducted regarding convenience of the ViziQuer tool as a way to create SPARQL queries associated to some "main class". Original version of ViziQuer [4] did not use the concept of "main class" during SPARQL query construction. To adapt to OBIS needs a ViziQuer Lite version was created with "main class" concept at the center. ViziQuer Lite was further refined and usability improvements of these refinements were evaluated through blind tests conducted through Amazon Mechanical Turk.

6 Related Work

We position our platform as a fast way to deploy an information system from an ontology. A similar tool from the functionality perspective is WebProtégé [5].

Main difference is that WebProtégé focuses on collaborative ontology editing; our focus is on collaborative instance data editing. WebProtégé tool also allows editing on the instance level of the ontology, but it does not enforce data property cardinality restriction to one. Thus, it is hard to overview all the created individuals as they are listed by their identifier only, while the rest of properties are scattered.

WebProtégé presents a way to create individuals through custom forms, but individual creation form specification must be given in a XML specification file written by hand. Another drawback of WebProtégé is that it does not interface with a public SPARQL endpoint.

Also it does not separate ontology editing from instance editing, as it is important part in information systems. It is normal that data conceptual model can be edited only by system's administrator, but system data editing is done by enterprise common users. WebProtege does not separate browsing capabilities from editing capabilities. In contrast, our system provides an SPARQL endpoint access point where one can connect from outside in browsing mode or by read-only OBIS information system.

Tools like OntoViews [6], Longwell [7] allow users to browse semantic data in faceted manner. There are lot of options from browsing side point of view, but these tools do not provide capability to edit the underlying data.

Koodl[12] provides a user with collaborative ontology editing, connection to a SPARQL endpoint, creating and editing SPARQL queries, visualizing and analyzing created results. This approach also is based on data browsing and analyzing, not editing.

Another similar tool is OntoWiki [8] that is proposed as collaborative ontology data editor and it targets mostly instance data browsing and editing. It also provides access to data by SPARQL queries. New version is stated to incorporate a visual query builder [9]. As in wiki it does not strictly enforce ontology structure for instance data – ontology serves more like a dictionary. Individuals can have properties that are not defined in the ontology.

In relation database field there exist different solutions that ease web-based application building from a database (Oracle APEX[13], SQL Maestro[14]). Still, we did not encounter one that would fully automatically generate a working system from the given database.

7 Conclusions and Future Work

We have successfully implemented and tested the ontology-based information system (OBIS) development process. The first results are good, which is supported by our linguistic application test case in case of usability experience. The experiment with the medical data provides evidence that the system can scale up to larger data sets and is usable in real world applications.

We still see room for further development and technical realization. Important aspect that we have not targeted is how to move an individual (instance) from one class to another class or manipulate list of classes where the individual belongs. In our example of veterinary clinic it is natural to move planned appointments to past appointments. For information systems used over prolonged periods of time it is typical requirement. For example, a student becomes a professor; a prospect becomes a customer, etc.

---

[12] http://knoodl.com

[13] http://apex.oracle.com/

[14] http://www.sqlmaestro.com/

Other important part is to improve constraint and reasoning mechanism, as it is very important part to improve data quality and make system more useful. Other focus area is system configuration part as user require not only functionality, but also that system looks as they like it.

We see that the OBIS can be successfully used for web-based SPARQL endpoint data access and editing to promote the Semantic Web usage in wider user base.

Acknowledgments. Mārtiņš Zviedris has been supported within the European Social Fund Project "Support for Doctoral Studies at University of Latvia - 2". Aiga Romāne and Kārlis Čerāns have been supported by the European Union via European Regional Development Fund Project No. 2011/0009/2DP/2.1.1.1.0/10/APIA/VIAA/112. Guntis Bārzdiņš has been supported by Latvian 2010–2013. National Research Program Nr.2 project Nr.5.

References

1. Sowa, J.F.: Serious semantics, serious ontologies, panel. Presented at Semantic Technology Conference (SEMTEC 2011), San Francisco (2011)
2. Soley, R.: Model driven architecture. OMG White Paper (2000)
3. Barzdins, J., Barzdins, G., Cerans, K., Liepiņs, R., Sprogis, A. OWLGrEd: a UML style graphical notation and editor for OWL 2. In: Clark, K., Sirin, E. (eds.) Proceedings of the 7th International Workshop "OWL: Experience and Directions" (OWLED-2010) (2010)
4. Zviedris, M., Barzdins, G.: ViziQuer: a tool to explore and query SPARQL endpoints. In: Antoniou, G., Grobelnik, M., Simperl, E., Parsia, B., Plexousakis, D., De Leenheer, P., Pan, J. (eds.) ESWC 2011, Part II. LNCS, vol. 6644, pp. 441–445. Springer, Heidelberg (2011)
5. Tudorache, T., Nyulas, C., Noy, N., Musen, M.: WebProtégé: a collaborative ontology editor and knowledge acquisition tool for the web. Semant. Web J. **4**, 89–99 (2013)
6. Mäkelä, E., Hyvönen, E., Saarela, S., Viljanen, K.: ONTOVIEWS – a tool for creating semantic web portals. In: McIlraith, S.A., Plexousakis, D., van Harmelen, F. (eds.) ISWC 2004. LNCS, vol. 3298, pp. 797–811. Springer, Heidelberg (2004)
7. Longwell RDF Browser, SIMILE (2005). http://simile.mit.edu/longwell/
8. Auer, S., Dietzold, S., Riechert, T.: OntoWiki – a tool for social, semantic collaboration. In: Cruz, I., Decker, S., Allemang, D., Preist, C., Schwabe, D., Mika, P., Uschold, M., Aroyo, L.M. (eds.) ISWC 2006. LNCS, vol. 4273, pp. 736–749. Springer, Heidelberg (2006)
9. Riechert, T., Morgenstern, U., Auer, S., Tramp, S., Martin, M.: Knowledge engineering for historians on the example of the *Catalogus Professorum Lipsiensis*. In: Patel-Schneider, P.F., Pan, Y., Hitzler, P., Mika, P., Zhang, L., Pan, J.Z., Horrocks, I., Glimm, B. (eds.) ISWC 2010, Part II. LNCS, vol. 6497, pp. 225–240. Springer, Heidelberg (2010)
10. Berners-Lee, T., Hendler, J., Lassila, O.: The semantic web. Sci. Am. **284**, 29–37 (2001)
11. Tao, J., Sirin, E., Bao, J., McGuinness, D.L.: Integrity constraints in OWL. In: Fox, M., Poole, D. (eds.) Proceedings of the 24th AAAI Conference on Artificial Intelligence, AAAI 2010. AAAI Press (2010)
12. Barzdins, G., Liepins, E., Veilande, M., Zviedris, M.: Semantic Latvia approach in the medical domain. In: Proceedings of the 8th International Baltic Conference (Baltic DB&IS 2008), 2–5 June 2008, Tallin, Estonia, pp. 89–102. Tallinn University of Technology Press
13. Motik, B., Patel-Schneider P.F., Parsia B.: OWL 2 Web Ontology Language Structural Specification and Functional-Style Syntax (2009)
14. Escalona, M.J., Aragón, G.: NDT a model-driven approach for web requirements. IEEE Trans. Softw. Eng. **34**(3), 377–390 (2008)

Towards Sentiment Analysis on Parliamentary Debates in Hansard

Obinna Onyimadu[1,2], Keiichi Nakata[1(✉)], Tony Wilson[2],
David Macken[2], and Kecheng Liu[1]

[1] Informatics Research Centre, Henley Business School, University of Reading,
Whiteknights, UK
k.nakata@henley.reading.ac.uk
[2] System Associates Ltd, Maidenhead, UK
obinnao@systemassociates.co.uk

Abstract. This paper reports on our ongoing work on the analysis of sentiments, i.e., individual and collective stances, in Hansard (Hansard is a publicly
available transcript of UK Parliamentary debates). Significant work has been
carried out in the area of sentiment analysis particularly on reviews and social
media but less so on political transcripts and debates. Parliamentary transcripts
and debates are significantly different from blogs and reviews, e.g., the presence of sarcasm, interjections, irony and digression from the topic are com-
monplace increasing the complexity and difficulty in applying standard
sentiment analysis techniques. In this paper we present our sentiment analysis
methodology for parliamentary debate using known lexical and syntactic rules,
word associations for the creation of a heuristic classifier capable of identifying
sentiment carrying sentences and MP stance.

Keywords: Hansard · Sentiment analysis

1 Introduction

Further to our work on semantic search [1], our objective remains the evolution of
search from the retrieval of semantic content to addressing the hidden meaning and
expressions behind the speeches and views aired in the UK Houses of Parliament. This
is based on the assumption that sentiment, opinion and stance expressed by speakers
can be employed to achieve this objective. Sentiment analysis focuses on the task of
automatically identifying whether a piece of text expresses a positive or negative
opinion about a subject matter [2]. Stance refers to the overall position held by a
person towards an object, idea or proposition, for instance, whether a member of
parliament (MP) is for or against a subject of debate. Our objective is to determine an
MP's stance on a subject from analysis of the sentiments expressed in sentences and
utterances. This will enable queries that ask for a list of MPs with a particular stance,
e.g., all pro-EU MPs, based on their utterances. In addition, this analysis offers a
deeper understanding of MPs individually and collectively since it provides further
insight into MPs leanings and also has the potential to divulge trends in opinions when
combined with external content such as social media, external events etc.

W. Kim et al. (Eds.): JIST 2013, LNCS 8388, pp. 48–50, 2014.
DOI: 10.1007/978-3-319-06826-8_4, © Springer International Publishing Switzerland 2014

2   Methods

2.1   Representation of sentiments

Based on [3], we express opinion/sentiment as a sextuple featuring (1) the target entity or subject matter for which an opinion is expressed, (2) the sub-components of the subject matter, (3) the polarity of the sentiment/opinion (positive, negative, or neutral), (4) the intensity of the sentiment, (5) the opinion holder, and (6) the time the opinion was expressed For 2), in the example, *Conservative MPs aged over 50 voted overwhelmingly against the bill*, "Conservative MPs over 50" is a subcomponent of "Conservative MP". Hansard's document structure allows for identification of time and opinion holder. In cases where the opinion holder is expressed using his position (such as Health Secretary), the MP ontology [1] is used to disambiguate and resolve the name of MP. Typically, the subject matter is expressed clearly in the document structure and captures a broad explanation of the topic debated; however, during the debate, it is not uncommon for MPs to digress, expressing views on unrelated but significant topics or stay on point while expressing views on sub topics of the broad generalization of the overall topic in question. For instance, a debate might be labeled as "EU Vote" during which an MP might touch on related topics such as EU Economy, UK Independence or unrelated topics such as the performance of English football teams, etc. Identifying these sub-topics and their opinions is not trivial especially without a training corpus. Subjective expressions exists in various forms – allegations, desires, beliefs, speculations [4] – and since most opinionated sentences are subjective, we attempt to classify subjective and objective sentences before assigning subjective sentences as positive, negative or neutral. As such the unit of analysis is a sentence.

2.2   Sentiment Assignment

The proposed approach involves the use of heuristic classifiers based on the use of statistical and syntactic clues in the text to identify subjective sentences (e.g., "It is obvious to everyone that if the PM had arrived yesterday the strike would have been called off") and objective sentences (e.g. "15 MPs joined the demonstrators as the PM's motor cade arrived number 10"). Our approach involves a hybrid of syntactic and lexical clues in identifying subjective and objective sentences. The sentiment lexicon base – MPQA corpus [5] – is used for the identification of sentences bearing known positive and negative words. Sentences containing more positive or negative words are annotated initially as positive or negative before syntactic clues are applied to improve their classification. Syntactic clues such as the absence or presence of negations such as "not" or "never" affect sentence's polarity. Our initial classifier involved marking all adjective and adverb bearing sentence as subjective since subjectivity is associated with the presence of adjectives or adverbs [6]. We also observe the presence of conjunctions such as "in spite of" and "notwithstanding" as suitable heuristic subjective rules. Intensifying adverbs, e.g., "very disastrous", and comparative adjectives, e.g., "greater", provide the intensity of the sentiment allowing us detect levels of friendliness, anger and emotion. Having extracted a set of likely subjective/objective sentences, we improve precision by eliminating other non-sentiment bearing sentences such as questions. However, questions embedded between two subjective sentences are not filtered out as the sequence of statements is likely to have an assumed context. To cater for the variations to the rules, we take into account the proportion of adjectives and adverbs per sentence through which we set a threshold for the occurrence of adjectives or adverbs as three or more for subjective sentences, and less than three for objective.

3   Results, Conclusions and Future Work

The classifier was applied on thirteen debates and the results of annotations made by the classifier against the gold standard were compared. On average, 43 % of the sentences were correctly annotated (range: 40 %–45 %). Correctly annotated sentences were mostly sentences without compound opinions, sarcasm and comparative sentences featuring a comparison of several entities. In addition, we found that rhetorical questions which connote a positive or negative sentiment are incorrectly identified as neutral by our rule set. As seen in our results, the complexities inherent in debate speech styles means that the use of syntactic clues and a lexical base is quite insufficient to fully grasp the polarity and intensity of expressed sentiments. However, it still provides a foundation for the development of more advanced techniques since simple opinionated sentences can now be eliminated from the text. Having collected statistical and syntactic features such as term frequencies, part of speech tags, word positions, noun phrases and verb chunks from our initial analysis, our current objective is to implement supervised learning algorithms capable of learning and deriving the sentiments in complex sentences particularly ones featuring compound opinions.

References

1. Onyimadu, O., Nakata, K., Wang, Y., Wilson, T., Liu, K.: Entity-based semantic search on conversational transcripts: semantic search on Hansard. In: 2nd Joint International Conference, JIST 2012, 2–4 December 2012, Nara Japan (2012)
2. Melville, P., Gryc, W., Lawrence, R.D.: Sentiment analysis of blogs by combining lexical knowledge with text classification. In: KDD '09, 28 June–1 July 2009, Paris, France (2009)
3. Liu, B.: Sentiment analysis and subjectivity. In: Indurkhya, N., Damerau, F.J. (eds.) Handbook of Natural Language Processing, 2nd edn. CRC Press, Baco Raton (2010)
4. Wiebe, J.: Learning subjective adjectives from corpora. In: Proceedings of National Conference on Artificial Intelligence (AAAI-2000) (2000)
5. Wiebe, J., Riloff, E.: Creating subjective and objective sentence classifiers from unannotated texts. In: Gelbukh, A. (ed.) CICLing 2005. LNCS, vol. 3406, pp. 486–497. Springer, Heidelberg (2005)
6. Wiebe, J.M.: Learning subjective adjectives from corpora. In: Kautz, H.A., Porter, B.W. (eds.) American Association for Artificial Intelligence. AAAI Press, Menlo Park (2000)

Implementing Mobility Service
Based on Japanese Linked Data

Wataru Okawara[1], Takeshi Morita[2(⊠)],
and Takahira Yamaguchi[1(⊠)]

[1] Keio University, 3-14-1 Hiyoshi, Kohoku-ku, Yokohama-shi 223-8522, Japan
yamaguti@ae.keio.ac.jp
[2] Aoyama Gakuin University, 5-10-1 Fuchinobe, Chuo-ku,
Sagamihara-shi 252-5258, Japan
t_morita@si.aoyama.ac.jp

Abstract. This study aims at developing a web service with Japanese Linked Data and evaluating the service. In Japan, government sets Open Data as a new strategy in Information Technology field and focuses on "Linked Open Data (LOD)" [1] as a means to publish. However, the number of dataset as Japanese Linked Data is small, and the effect by introducing Japanese Linked Data has not been shown yet. Therefore, we created Linked Data in Japanese focused on geographical or positional data, and implemented a mobility service to reduce user's cognitive load. Moreover, we conducted verification experiment for using the service and compared with conventional services. As a result, the possibilities of Linked Data to respond to various queries easily and to apply for information services by crossing some domains were explored.

Keywords: Linked Open Data · Mobility service · Positional data

1 Introduction

Recently, in Japan government sets Open Data as a new strategy in the Information Technology field and aims to realize a society in which new industries and services will be created by publishing public data such as geospatial, disaster and statistical data and combine them with enterprise or personal data freely.

"Linked Open Data (LOD)" handles data of the smallest unit and connects data with links across domains. By publishing data to open place on the web and sharing, third parties can make use of data freely and add data to LOD. Therefore, the movement to open data as LOD by Japanese government is growing. Actually, some organizations and local government such as National Diet Library and Sabae City publish their data as LOD. Moreover, "Linked Open Data Challenge JAPAN" [2], in which people present the application using Japanese Linked Data, is held. However, this movement is behind the USA and the UK, and the number of datasets as Japanese Linked Data is small compared to them. In addition, the effect by introducing Linked Data has not been shown yet.

The purpose of this study is to develop a web service with Japanese Linked Data and to evaluate the service. Particularly, public data is closely related to the geospatial

W. Kim et al. (Eds.): JIST 2013, LNCS 8388, pp. 51–66, 2014.
DOI: 10.1007/978-3-319-06826-8_5, © Springer International Publishing Switzerland 2014

data such as geographic name, latitude/longitude and population. Therefore we created Linked Data in Japanese focused on geographical or positional data. The Japanese Linked Data we created has several domains such as highway, administrative district, traffic regulation, weather forecast and restaurant review. Then, we implemented a mobility service based on it to reduce user's cognitive load. Finally we evaluated the service through verification experiment and comparison with conventional services.

2   Related Work

Representative service using positional linked data is "DBpedia Mobile" [3]. DBpedia Mobile is an application to display position-centric data DBpedia [4] has on a Map for mobile devices. Based on the current GPS position of a mobile device, DBpedia Mobile renders a map containing information about nearby locations from the DBpedia dataset. Starting from the map, users can explore background information about locations and can navigate into DBpedia and other interlinked datasets such as Geonames, Revyu, EuroStat and Flickr. In addition, DBpedia mobile is easy to operate with an intuitive user interface and users can publish data as RDF. However, conventional application including DBpedia Mobile doesn't deal with high modifiability dataset. In addition, it is difficult to use them directly for Japanese information services.

"Telematics" [5] is cited as an example of mobility service. Telematics is a next generation information service for cars to receive information by using the device such as a mobile phone and connecting to the Internet, for example, reception of traffic information or sending and receiving e-mail and tracking service of stolen cars. In addition, it has a lot of functions like the text-to-speech and operating service. However, these services are individually provided, so it is difficult to combine them. On the other hand, our proposed system integrates datasets provided by various services and can provide services which support flexible user requirements by using standard techniques such as HTTP URI for reference to the data.

3   Proposed System

3.1   System Overview

This system develops the conventional system, "A Mobility Service based on Japanese Linked Data [6]". There are few datasets as Linked Data, so it is necessary to create Linked Data by our own. The conventional system extracted data in XML or HTML format on the Web and created Japanese Linked Data.

This system expanded the conventional Japanese Linked Data. In addition, we created Linked Closed Data by using enterprise data. Then, we developed more applications that meet the various conditions such as stormy weather or congestion by using expanded Linked Data and Linked Closed Data.

3.2   Methods for Creating and Publishing Linked Data

As mentioned above, our proposed system is divided into "Creating Japanese Linked Data" and "Application using Japanese Linked Data". Figure 1 shows a screenshot of system overview.

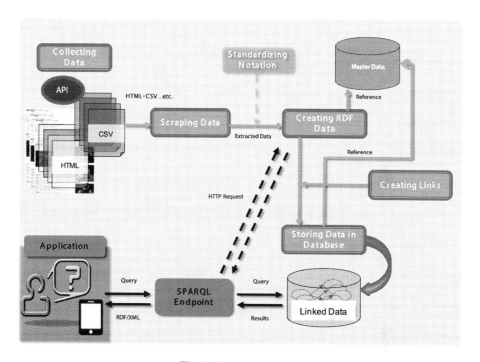

Fig. 1. System overview

(1) Collecting and Scraping Data
First, collect the data that is used to develop an application in various formats such as HTML, XML and CSV. The structures of Web sites are different, so develop the data scraping programs for each Web site and extract data. As appropriate, create master data to change it to RDF format and store it in a database.

(2) Creating RDF Data
Do data modeling for each data and convert it to RDF. In this method, RAP (RDF API for PHP V0.9.6) [7] that is use. It is a software package for parsing, querying, manipulating, serializing and serving RDF models. By this API, create RDF statements as model for representing data.
The modules for converting data to RDF are programs that return RDF presenting information about the resource when HTTP request is sent to URI for the resource. This is used to store RDF data in the database as API.

Additionally standardize date and time notation. In this study, standard data types that are defined in XML schema are used to use SPARQL query.

(3) Creating Links

Create links among resources in datasets and links to other external datasets.

In this study, if there is a proper name for the resource to identify the same resource that can be done by matching the name against the different dataset exactly. However, if the name of the resource about the spot is spelled in several different ways, for example, in the case of the datasets about shops, the same resource can be identified by checking whether the resources meet following two criteria. It's assumed that the resources have information about lat/long or address.

1. Coincidence in address/Distance between the resources is below the criterion value

If the resource has the information about the address of the spot, check whether both address information are same. If the resource has only the information about the lat/long of the spot, check whether the distance that calculated by the both lat/long data is below the criterion value (for example, within 300 m). Because sometimes the data about the same shop in different data sources do not have the same information about lat/long exactly.

2. Matching the name for the spot partially.

After removing spaces and special characters, such as "·", from the name for the spot, check whether both name for the spot match partially.

Furthermore, in this study, we create Linked Data in Japanese focused on geographical or positional data. In the case of the data about the spots, a link to the dataset about administrative district is created. The dataset about administrative district is RDF dataset that we created based on the data about postcode opened by Japan Post Service and represent trees of address (prefecture→city→small district). The objective of this is to function as the hub of the data and the dataset about local information.

(4) Storing the Data in the Database and Creating a SPARQL Endpoint

Store created RDF data in the database by using ARC2 (RDF Classes for PHP) [8].

In the case of low modifiability dataset, store the data in the database. In this study, ARC2 (RDF Classes for PHP) [8] is used. For this, acquire the list of instances by using the master data. Then acquire RDF data by using prepared API and sending HTTP request to URI for the instance, and store them by rotation. In the case of high modifiability dataset, acquire RDF data by using API each time.

If the external datasets are used, also store it by using dump data or opened SPARQL endpoint [9]/API.

Additionally create the SPARQL endpoint to search data in the database by SPARQL query. This is a program to return the results by SELECT query and CONSTRUCT query.

3.3   Linked Data for a Mobility Service

We created Linked Data based on the above-mentioned method for a mobility service. Figure 2 shows an overview of the Linked Data. We created highway including service area (SA), parking area (PA) and interchanges (IC), administrative district, traffic regulations, tourist sites and events, weather forecast, service area facilities and restaurant reviews dataset as Linked Data. Moreover, we linked Japanese Wikipedia Ontology [10] as an external dataset. Part of Traffic Regulations, Weather Forecast, Service Area Facilities consist of enterprise data and are regarded as Linked Closed Data. The total number of Linked Data we created are 52634 triples.

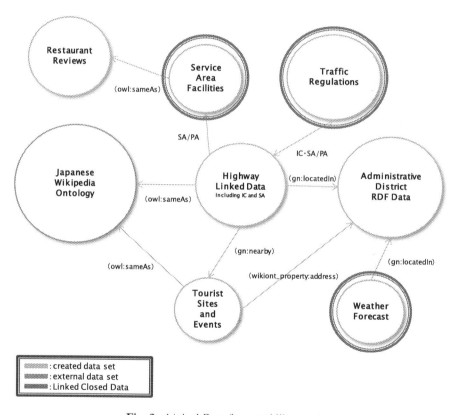

Fig. 2. Linked Data for a mobility service

Figure 3 shows links between resources. For example, in order to show SA weather information, this system acquires the district data from lat/long of SA and shows weather forecast. By linking resources over different domains, it is possible to correspond to various queries.

As for modifiability, datasets are classified as Table 1 shows. It depends on the frequency of update for the data on each website of the data collection data. We regard updated data within several hours such as weather forecast, parking or traffic regulations and congestion, and updated data within one day such as local community

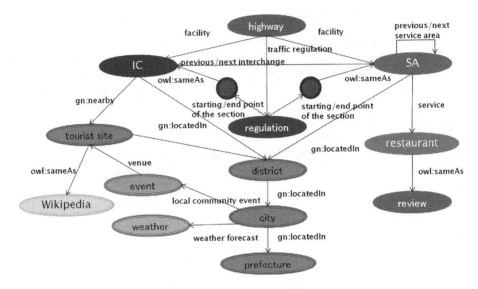

Fig. 3. Links between resources

events and gas station as high modifiability. We regard data that will not be updated for a long time such as highway, IC or SA as low modifiability data. With respect to restaurant reviews, update time depends on the timing of filling in reviews, so it is difficult to define modifiability exactly. Frequency of update reviews to each restaurant is low, so we regard restaurant reviews as low modifiability data.

Data sources such as Web Site and Web API are on the right side of the table. Enterprise Data is used about a part of Weather forecast, Parking and Traffic Regulations and congestion.

3.4   A Mobility Service Overview

We developed a mobility service using the Japanese Linked Data. It has the following five functions. Two functions are implemented by using only Linked Open Data. Three functions are implemented by using Linked Open Data and Linked Closed Data.

- Functions only with Linked Open Data

  1. Route Search
     This is a function to search a route with a place of departure and arrival entry. The text of detailed explanation about the route and a polyline of the route on the map are shown.
     On the interchange point on a route, a marker is shown on the map to present the information about tourist spots around there and a link to information about local information. If the link to dataset about local community events is clicked, the markers are shown on the venue to present abstract of the event information. In the case of weather forecast, the data of that is shown on the right side of the map.

Table 1. Modifiability of dataset

Modifiability	Information	Web site / WebAPI (Data Sources)	Update time
Low	Highway/Interchange (IC)	·kosoku.jp[a]	
		·Wikipedia[b]	
	Service Area(SA)	·drive plaza[c] ·SA information guide[d]	
	Restaurant Reviews	·tabelog API (gourmet site)[e]	
High	Weather Forecast	·tenki.jp[f] ·Enterprise Data	3 hours
	Local Community Events	·Yahoo!JAPAN local information[g]	1 day
	Gas Station	·Gas Price Information[h]	1 day
	Parking	·Enterprise Data	5 minutes
	Traffic Regulations and Congestion	·Enterprise Data	5 minutes

[a] http://kosoku.jp/
[b] http://ja.wikipedia.org/
[c] http://www.driveplaza.com/
[d] http://www.c-exis.co.jp/
[e] http://tabelog.com/
[f] http://weather.nifty.com/
[g] http://loco.yahoo.co.jp/
[h] http://api.gogo.gs

Dragging the current position marker or clicking the markers that show points on a route can specify the current position. The geographic coordinates of the current position are acquired by the position of that marker, and the road name that the current position is on is acquired by geocoding. If the road is a highway and there is information about it in the database, the markers are shown to present service areas.

2. Spot Search

This is a function to search tourist sites or landmarks around the highway with a highway name and a keyword entry such as castles, zoos and hot springs. Practically, it is used to search the class of the spots near the interchange as an instance on the highway. The class is defined by Japanese Wikipedia Ontology that is linked to the highway dataset. The information that the ontology has can be seen. And the location of the spot and the interchange is shown on the map.

Figure 4 shows a screenshot of Spot Search. The left figure shows the home page of this function. The Middle figure shows the entry of Highway name and a keyword of spot. And search button is clicked, search will start. The right figure shows the search result. Spot name and the nearest IC are shown. Moreover, the buttons linked to Wikipedia information and map around are shown on the right side. If the button is clicked, more detail information can be seen.

Fig. 4. A screenshot of spot search

- Functions with Linked Open Data and Linked Closed Data

1. Service Area Search

   This is a function to services/facilities in the service areas on the highway with a highway name, road direction and a facility category. If the GPS button is clicked, the lat/long at the current position is acquired with GPS. Since the lat/long at every facility is stored as RDF, facilities information can be seen in close order from current position by using them. If the checkbox that says "Search only the services/facilities in business at present" is checked, the scope of the search is limited to only the services/facilities in business at present in the service areas. The business hours of services/facilities are described every day of the week. Practically, it is used to check whether the current local time is within the business hours of the day of the week today with SPARQL FILTER function.

   If the facility is a restaurant and has a link to restaurant reviews dataset, the reviews are shown. Moreover, this system can recommend the nearest vacant SA by using positional data (Linked Open Data) and parking information (Linked Closed Data).

2. Calling a user's attention during stormy weather

   This is a function to call a user's attention and provide related information when the road condition is changed due to the stormy weather. If the road is closed, alert and the detour and detail information of the route are shown. If the icon of Snow Removal, Spraying antifreeze or Stormy Weather is clicked, current traffic regulation information on the highway (Linked Closed Data) is shown and that section of the road is shown on the map. By searching the data combined RDF data about the regulations of the highway that acquired by the API and the data about facilities (service areas, interchanges) on the highway in the database as Linked Data, geographical coordinates of service areas/interchanges that is starting or end point of the regulation section can be presented.

   Figure 3 shows a screenshot of attention during stormy weather. The left figure shows the home page of this function and explains the icons. The Middle figure shows entry of Highway names and road direction. And if search button is clicked, search will start. The right figure shows the search result. On a map, regulated section is shown with icons and polyline. Moreover, the detail information such as start and end point of regulation, cause, ending time and others is shown below a map (Fig. 5).

3. Support during congestion

This is a function to support decision-making whether dropping in near service area or pass congestion by showing the time of eliminating congestion and passing congestion. With a highway name, road direction, and a favorite gas station name, the congested section and the accident point are shown on the map and detail information is shown below. If the button of Information Update is clicked, the lat/long at current position and starting point of congestion are all acquired. Then calculate the distance between current position and starting point of congestion (Linked Closed Data).

If the distance is more than 30 km, "No congestion ahead" is shown.

If the distance is less than 30 km and there is more than one service area as far as the starting point of congestion, the list of service areas is shown and the map, gas station name, gasoline price and parking information of every service area are shown.

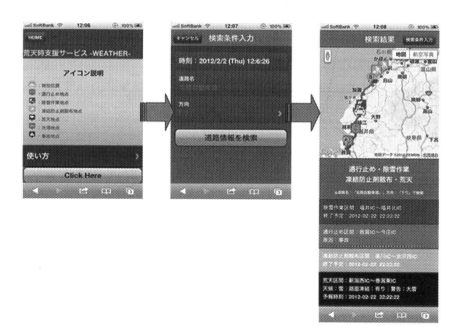

Fig. 5. A screenshot of attention during stormy weather

4   Discussion

4.1   Experiment for Using Proposed Service

We conducted a verification experiment for using proposed service. In the following, we explain how to use Linked Data we created by showing links between resources. Table 2 shows the contents of the experiment.

Table 2. Contents of verification experiment

Driving section	Yagami campus at Keio university $\sim$ Gotemba premium outlet
Highway	Tomei highway: Tomei Kawasaki IC $\sim$ Gotemba IC
Date	Departure: around 10:00, Tuesday, November 22, 2012
SA/PA to visit	Ebina SA, Kouhoku PA

1. Departure from Yagami Campus
   Before departure, we searched a route with a place of departure and arrival entry. On a map, passing route and IC are confirmed.
2. Search traffic regulation information around Kohoku PA (down)
   The list of regulation by construction and car accident was shown in real time.
3. Search service areas where to eat a snack
   We entered "bread" as a category and search, "Portugal" at Ebina SA (down) was recommended as the highest rated restaurant based on the restaurant reviews.
4. Search spot information around Gotemba IC
   We entered "zoo" as a keyword and researched, "Fuji Safari Park" around Gotemba IC was recommended and address, opening hours, and facilities in zoo were shown.
5. Search a rest area
   To take a break, we searched vacant area and Kohoku PA (up) was recommended.

Service Area Search recommended very popular bakery "Portugal" by following links and referring reviews of restaurant and current position. Figure 6 shows links between resources of experiment 3. First, near SA from the current position on the road is selected (Highway Domain). Then, information of facilities in the SA such as restaurant and parking is acquired (SA Facilities Domain). Restaurant matching user's condition is recommended considering to business hours and rating of restaurants (Restaurant Reviews Domain). We verified that it is possible for this system to perform such a complex process in a single query by crossing some domains.

Figure 7 shows links between resources of experiment 4. First, near IC from the current position on the road is selected (Highway Domain). Then weather information at the city where the IC is located is acquired (Weather Forecast Domain). In addition, tourist sites near the IC are shown (Tourist Site Domain). Since the tourist site is linked to Japanese Wikipedia Ontology, detail information of the tourist site is acquired (Japanese Wikipedia Ontology Domain).

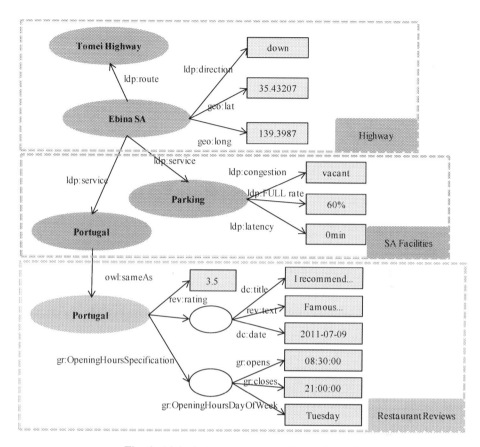

Fig. 6. Links between resources of experiment 3

Moreover, vacant SA was recommended at break time considering to congestion situation of Parking (Linked Closed Data). Traffic information as high modifiability data was provided at the appropriate time. In this way, we showed the usefulness of high modifiability data and the combination of Linked Open Data and Linked Closed Data.

On the other hand, links with external dataset weren't fully used. At present, we link only Japanese Wikipedia Ontology as external dataset, but information is only shown and not fully utilized. It is necessary to consider the service of using external dataset as well as to increase the number of those.

4.2   Comparison with Conventional Service

(1)   Reduce User's Cognitive Load

Our proposed system aims to reduce cognitive load by using Linked Data. Accordingly, we compared our proposed system with web service in two ways.

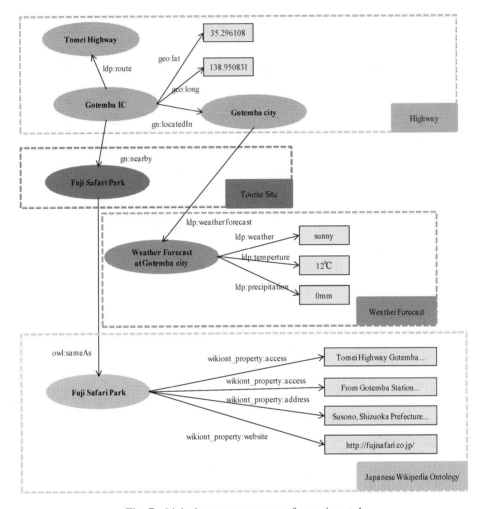

Fig. 7. Links between resources of experiment 4

First, we suppose to the condition that an end user looks for only restaurants under business in SA to examine user's cognitive load. Figure 8 shows comparison about user's search method.

If end user uses conventional services, he performs in the following procedure.

1. Search the web page of the SA.
2. Select some candidates for stores in the SA.
3. Search the candidates on the restaurant web page.
4. Confirm each review and business hours of the restaurants.

If he uses our proposed service, he performs in the following procedure.

1. Click icon of the SA.
2. Select displaying restaurant information. (Only restaurants under business are acquired.)

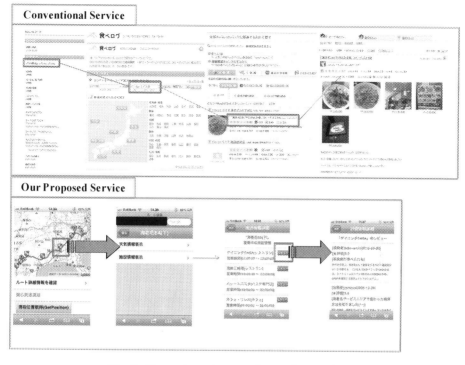

Fig. 8. Comparison about user's search procedure 1

3. Click review button.

   Secondly, we suppose to the condition that an end user looks for tourist site near current position to examine user's cognitive load. Figure 9 shows comparison about user's search method.

   If end user uses conventional services, he performs in the following procedure.

1. Search the near IC from current position.
2. Search the administrative district of the IC.
3. Search the tourist sites on the web page.
4. Narrow the list with keyword.

   If he uses our proposed service, he performs in the following procedure

1. Search with road name and keyword.
2. Involved tourist sites are shown.
3. If necessary, refer to information of Japanese Wikipedia Ontology

   In case of conventional services, he has to perform those processes for all the selected candidates. There are heavy loads to transition pages many times.

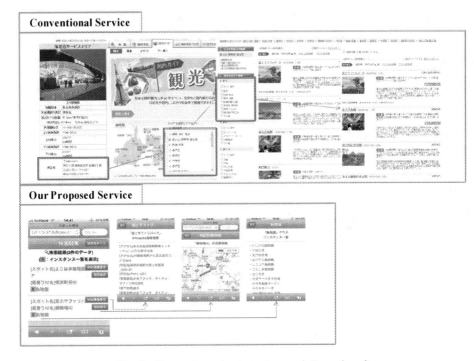

Fig. 9. Comparison about user's search Procedure 2

Our proposed system makes it possible to search with simple operation by accessing different domains with one query. In this way, it was verified that our proposed system reduce user's cognitive load.

(2) Comparison with Telematics
We compared our proposed system with telematics. Telematic services individually provide a route search service, a store search service, traffic regulation information, and others. On the other hand, our proposed system implements various services by crossing domains of LOD.

Figure 10 shows comparison with "G-BOOK" [11], which is one of the telematics. In the left figure, G-BOOK only shows positional and congestion information of the parking on the map. In the right figure, our proposed system recommends the restaurant which matches user's condition by using positional data, business hours, evaluation point, parking information.

Moreover, our proposed system predicts dissolution time of congestion from current position and congestion onset point and induces user to SA or general way. With this comparison, telematics is superior to our system in terms of a single application in limited domain, but it is difficult to expand an application and to cooperate other application. On the other hand, our system can develop an application across wide domains by using Linked Data. If dataset is expanded, it will be easier than conventional database to expand an application.

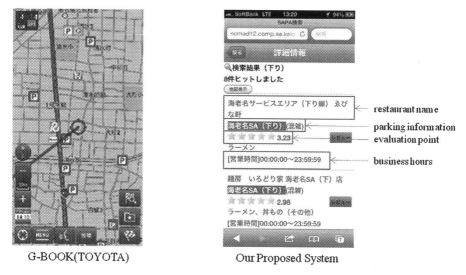

G-BOOK(TOYOTA)   Our Proposed System

Fig. 10. Comparison with telematics

5 Conclusions

In this paper, we created Japanese Linked Data focused on geographical or positional data and developed a mobility service with it. We conducted experiment for verifying the usefulness of cooperating various data such as Linked Open Data, Linked Closed Data and high modifiability data. Moreover, compared with conventional development methods, the possibilities of Linked Data to respond to various queries easily and to apply for information services by crossing some domains were explored.

For the future work, we will utilize not only conventional dataset on the web, but also various external dataset which government and local government will be expected to release. If transportation census which government has is linked to Linked Data, we will be able to improve a proposed mobility service. We also would like to extract user context and preference from Social Network Service (SNS) and cooperate high modifiability dataset with user schedule. It will be possible to provide services specific to each user.

References

1. Bizer, C., Heath, T., Berners-Lee, T.: Linked data—the story so far. Special issue on linked data. Int. J. Semant. Web Inf. Syst. (IJSWIS) **5**(3), 1–22 (2009)
2. Linked Open Data Challenge JAPAN: http://lod.sfc.keio.ac.jp/
3. Becker, C., Bizer, C.: DBpedia mobile: a location-enabled linked data browser. Web Semant. Sci. Serv. Agents World Wide Web **7**(4), 278–286 (2008)
4. DBpedia: http://dbpedia.org/
5. Telematics: http://analysis.telematicsupdate.com/

6. Iijima, C., Morita, T., Enomoto, Y., Yamaguchi, T.: A mobility service based on japanese linked data. In: König, A., Dengel, A., Hinkelmann, K., Kise, K., Howlett, R.J., Jain, L.C. (eds.) KES 2011, Part III. LNCS, vol. 6883, pp. 246–255. Springer, Heidelberg (2011)
7. RAP - RDF API for PHP V0.9.6: http://www4.wiwiss.fu-berlin.de/bizer/rdfapi/
8. ARC2: https://github.com/semsol/arc2/wiki
9. SPARQL endpoint - semanticweb.org: http://semanticweb.org/wiki/SPARQL_endpoint
10. Tamagawa, S., Sakurai, S., Tejima, T., Morita, T., Izumi, N., Yamaguchi, T: Learning a large scale of ontology from Japanese Wikipedia. In: 2010 IEEE/WIC/ACM International Conference on Web Intelligence and Intelligent Agent Technology, WI-IAT, pp. 279-286 (2010)
11. G-BOOK: http://g-book.com/

Multilingual Issues

Toward Interlinking Asian Resources Effectively: Chinese to Korean Frequency-Based Machine Translation System

Eun Ji Kim and Mun Yong Yi[✉]

Department of Knowledge Service Engineering, KAIST, Daejeon,
Republic of Korea
{eunjik,munyi}@kaist.ac.kr

Abstract. Interlinking Asian resources on the Web is a significant, but mostly unexplored and undeveloped task. Toward the goal of interlinking Asian online resources effectively, we propose a novel method that links Chinese and Korean resources together on the basis of a new machine translation system, which is built upon a frequency-based model operated through the Google Ngram Viewer. The study results show that Chinese characters can be mapped to corresponding Korean characters with the average accuracy of 73.1 %. This research is differentiated from the extant research by focusing on the Chinese pronunciation system called Pinyin. The proposed approach is directly applicable to voice translation applications as well as textual translations applications.

Keywords: LOD · Asian resources · Machine translation · Google Ngram Viewer · Multilingual resources

1 Introduction

The current Web provides seriously limited support for sharing and interlinking online resources at the data level. Linked Open Data (LOD) is an international endeavor to overcome this limitation of the current Web and create the Web of Data on a global level. The Web of Data is an absolute prerequisite for the realization semantic Web. Since Web 2.0 emerged, a vast amount of data has been released to the LOD cloud in the structured-data format so that computers can understand and interlink them. Many tools and frameworks have been developed to support that transition and successfully deployed in a wide number of areas. However, as the proportion of non-Western data is comparatively small, a couple of projects have been only recently initiated to extend the boundary of LOD over non-Western language resources [1].

Toward the goal of interlinking Asian resources effectively, we pay our first attention to interlinking Chinese and Korean resources and propose a novel method that links Chinese and Korean resources together. Due to the two countries' historical backgrounds, almost all of the Chinese characters can be converted to one or several Korean characters and these Korean characters often reflect the pronunciation of the Chinese characters (e.g., for ' 夢(dream)' in Chinese, whose pronunciation is 'meng',

W. Kim et al. (Eds.): JIST 2013, LNCS 8388, pp. 69–74, 2014.
DOI: 10.1007/978-3-319-06826-8_6, © Springer International Publishing Switzerland 2014

it can be uniquely converted to Korean character ' 몽' pronounced as 'mong') [2]. According to a study on the frequency in Korean vocabulary use [3], about 70 % of Korean words are common in Chinese and Korean.

In this research, we propose a new method that recognizes the identical or similar nature of the words even though their pronunciations are different, and test its effectiveness using a sample data set consisting of 33,451 words, which is more than enough to cover most of the words commonly used by two countries.

2 Background

Pinyin is an official phonetic system for transcribing the sound of Chinese characters into Latin script in China, Taiwan, and Singapore. It has been used to teach standard Chinese and spell Chinese names in foreign publications and used as an input method to enter Chinese characters into computers [4].

Besides, the usage of Pinyin has come into the spotlight recently in light of the developments in voice recognition technology. It is a good means to express Chinese pronunciation in comparison with Chinese characters, which are not based on phonogram. Furthermore, the total number of Pinyin is about 400 and it is much fewer than the number of Chinese characters.

However, mapping one Pinyin to one Korean pronunciation is not a simple problem. One Chinese character can be read by various Chinese pronunciations and different Chinese characters can have same Pinyin. For example, a Chinese character ' 的' has various Pinyin such as 'de' and 'di'. Also, Chinese characters ' 十' and ' 事' have same Pinyin 'shi', but the ' 十' is read as ' 십'(sib), and ' 事' is read as ' 사'(sa) in Korean.

Therefore, mapping Pinyin to Korean cannot be characterized as 1:1 relationships. As shown in Table 1, we can see that even the same Pinyin can be corresponded to different Korean Characters.

3 Methods

3.1 Resource Used in the Machine Translation System

In this research, we used 3,500 Chinese characters in common use. According to a statistical analysis, 2,400 Chinese characters cover 99 %, 3,800 characters cover 99.9 % of documents written in Chinese [5]. For this reason, we chose 3,500 Chinese characters in common use sets as our standard Chinese characters, covering over 99.7 % (when it was estimated with simple extrapolation).

Table 1. Pinyin to Korean pronunciation relations

Chinese Pronunciation	Korean Pronunciation
Kua	과(gua)
Kuai	괴(gue), 쾌(que), 회(hue)
Kuang	광(gwang), 황(hwang)

3.2   Frequency-Based Table

Utilizing a Pinyin to Korean relation table is unlikely successful because one Pinyin can be mapped to many Korean. Thus, to identify a matching Korean character, we devised a frequency based recommendation system. According to the Chinese Ministry of Education, just 581 Chinese characters can cover more than 80 % of commonly used Chinese language. Exploiting this distribution information, we assigned high points to frequently used Chinese characters so as to properly determine the ranking of corresponding Korean Characters.

For the measurement of frequency in language, we took the advantage of Google Ngram Viewer. Google Ngram Viewer shows statistical data that describe how frequently the queried word was used in books. We queried all 3,500 common Chinese characters and got each character's frequency. A part of the results is shown in the Table 2.

In setting up the frequency measurement, we had to consider when one Chinese character is pronounced in many different ways. In this case, we divided the measured frequency f into the number of ways pronounced differently n and applied the normalized frequency f/n to each pronunciation. For example, 的 can be pronounced as 'de' and 'di' in Chinese. Thus, n is equal to 2, so the normalized frequency 0.0629/2 was applied to each.

3.3   Pinyin-to-Korean Mapping Table

Through the previous steps, we recorded each Chinese character's frequency. In this section, we explain how to build a Pinyin-to-Korean frequency based translation system. To find their mapping relation, calculating each Korean character's possibility to be mapped to each Pinyin is required. We performed this process using the Eq. (1).

$$Frequency(KP|Pinyin) = \sum_{i=1}^{n} Frequency(kp_i|Pinyin) \tag{1}$$

$Frequency(KP|Pinyin)$ means how frequently a certain Pinyin is chosen by a certain Korean pronunciation KP in case of Chinese-Korean common words. Also, kp_i means all of the same Korean pronunciations that came from different Chinese characters. Based on the result of calculating total sum points, Table 3 shows Pinyin to Korean frequency based mapping table.

The way to use this table as a translator is as follows. When the Pinyin input 'a' is given, we can find Korean characters corresponding to 'a' in the table. In this case, both '아' and '가' are candidates. However, the table shows that each candidate's frequency value is different. According to the table, the pronunciation 'a' in Chinese is

Table 2. Frequency of Chinese characters

Chinese Character	Frequency
的	0.0629
是	0.0109
在	0.0106

Table 3. Frequency-based Pinyin to Korean relation

	Pinyin	Mapped Korean Pronunciation			
	a	아(ah)	가(ga)		
Frequency(KP|Pinyin)		2.06E-5	1.78E-7		
	ai	애(eh)	왜(wai)		
Frequency(KP|Pinyin)		1.24E-5	3.73E-8		
	an	안(an)	엄(um)	전(jeon)	암(am)
Frequency(KP|Pinyin)		6.33E-5	4.35E-5	9.69E-9	1.78E-8

mapped to Korean ' 아' with 2.06E-5 frequency while 'a' is mapped to Korean ' 가' with 1.78E-7 frequency. Therefore, this system recommends ' 아' as the first choice and recommends ' 가' as the next choice.

4   Experiment

4.1   Experimental Setup

To measure our system's accuracy, we collected Chinese-Korean common 33,451 words. When Pinyin is given as an input, our Pinyin to Korean translator is executed and it recommends its Korean counterpart. With each recommendation, if it is exactly the same with the original answer we have, it stops to recommend other Korean character and we measure its accuracy according to the number of times it has recommended.

4.2   Performance Evaluation

In this experiment, relying on an existing well known evaluation method was inappropriate because we had to evaluate the accuracy of Korean characters that has not been actively researched and far different from Roman characters. Therefore, we devised our own evaluation method based on a mathematical model.

Our evaluation method measures the recommended Korean's accuracy compared with the original Korean answer with the score from 0 to 1. In consideration of each word's length difference, when the length of the word is m, we distributed $1/m$ as a total score to each character. To sum up, each character can take $1/m$ score as its maximum, and the total score for each word cannot exceed 1. In addition, each character's score is measured by $(1/m) \times (1/j)$ when j means the total number that a new Korean character was recommended. For example, when a firstly recommended character was exactly the same with the compared answer character, j is equal to 1 and the total score for the character becomes $1/m$. However, according to our translation system, if the recommended character is not equal to the answer, it recommends the next ordered character continuously until it can find exactly the same one. For this reason, when the recommended character is low-ranked in its recommendation list, the corresponding character score becomes low. Besides, when there is no recommended character in the recommendation list in the end, the character takes the score of 0.

Table 4. Sample program results for Chinese to Korean machine translation

Chinese	Pinyin	Korean	Recommended Korean	Accuracy
加工	jia gong	가공	가|공|	1
加工工場	jia gong gong chang	가공공장	가|공|공|장|	1
加工順序	jia gong shun xu	가공순서	가|공|순|수{허서}|	0.833333
加工業	jia gong ye	가공업	가|공|야{업}|	0.833333
加工特性	jia gong te xing	가공특성	가|공|특|성|	1

4.3 Evaluation Results

According to the experiment result obtained by using the Chinese-Korean 33,451 common words, a Pinyin input was able to be mapped to a corresponding Korean character with the average accuracy of 73.1 %. Also, the number of words which got 1 in accuracy score was 7,926 and its proportion was about 23.7 %. The machine translation program offered many good translations but still makes errors in recommendation order of priority. For example, the Chinese word ' 加工業'(jia gong ye) could not get the score of 1 because the program regarded ' 야'(ya) as a corresponding Korean character to ' 業'(ye) instead of the correct answer ' 업'(up) based on statistical data analysis. The sample results of the translation program are shown in Table 4.

5 Conclusion

In this study, we proposed a new method to translate Chinese to Korean. Different from the extant approaches, we focused on Chinese phonetic system Pinyin. In analyzing 3,500 Chinese characters in common use, we noticed the problem of one to many mapping between the two character systems and consequently realized the necessity of priority in translated results. Thus, we adopted a frequency value which was obtained through Google Ngram Viewer as its ranking criteria. To verify this translator's performance, we employed about 33,000 words which are commonly used in both Chinese and Korean and evaluated the translated results. Consequently, our Pinyin to Korean frequency based translator showed 73.1 % accuracy on average.

The results of this study have practical implications for linking Asian resources on Web. There are a number of well-developed translator systems for Western languages, whereas translator systems for Asian languages are almost non-existent. However, according to the statistical data of Miniwatts Marketing Group, the percentage of Internet users who are from Asian is bigger than any other continent and its rate is increasing rapidly [6]. Therefore, studies for linking Asian resources will become more important in the near future as there is great necessity for resource sharing among Asians. With this new machine translation system, we can link Chinese resources to Korean resources available in the LOD cloud.

Furthermore, voice translation has become a promising field recently. Google has collected a massive amount of voice data through voice search service. The amount of data Google receives in a day is the same with the amount that a person speaks continuously in two years [7]. In this situation, a demand for research about Pinyin will be naturally increased. Furthermore, different from most existing machine translation systems, we did not use dictionary definition at all. We devised a rule that exploits the two languages' similarity and applied the rule to automatic translation system. Therefore, the proposed approach requires less time and less memory.

On the other hand, there still remain some improvements that need to be made. Above all, we did not consider the five tones in Chinese language in this research. When we cover five pitches in Chinese, it is expected to have higher accuracy. Besides, the proposed approach can be used for only Chinese-Korean common words. Although a high percentage of Korean words are commonly used in Chinese and Korean, the existence of uncovered words needs to be addressed in the following researches by complementing the proposed approach with other solutions. Notwithstanding these limitations, however, it should be noticed that the proposed approach is a promising first attempt toward interlinking Asian resources effectively.

Acknowledgements. This work was supported by the IT R&D program of MSIP/KEIT. [10044494, WiseKB: Big data based self-evolving knowledge base and reasoning platform]

References

1. Hong, S.G., Jang, S., Chung, Y.H., Yi, M.Y.: Interlinking Korean resources on the web. In: Takeda, H., Qu, Y., Mizoguchi, R., Kitamura, Y. (eds.) Semantic Technology. LNCS, vol. 7774, pp. 382–387. Springer, Heidelberg (2013)
2. Huang, J.-X., Choi, K.-S.: Chinese-Korean word alignment based on linguistic comparison. In: ACL'00 Proceedings of the 38th Annual Meeting on Association for Computational Linguistics, pp. 392–399 (2000)
3. Lee, Y.: Research About Korean Chinese Common Words. Samyoung-sa, Seoul (1974)
4. Chen, Z., Lee, K.-F.: A new statistical approach to chinese pinyin input. In: ACL'00 Proceedings of 38th Annual Meeting on Association for Computational Linguistics, pp. 241–247 (2000)
5. Gillam, R.: Unicode Demystified : A Practical Programmer's Guide to the Encoding Standard. Addison-Wesley, Boston (2003)
6. Internet World Stats. http://www.internetworldstats.com/
7. Na, S.H., Jung, H.Y., Yang, S.I., Kim, C.H., Kim, Y.K.: Big data for speech and language processing. Electronics and Telecommunications Trends, pp 52–61 (2013)

CASIA-KB: A Multi-source Chinese Semantic Knowledge Base Built from Structured and Unstructured Web Data

Yi Zeng[✉], Dongsheng Wang, Tielin Zhang, Hao Wang,
Hongwei Hao, and Bo Xu

Institute of Automation, Chinese Academy of Sciences, Beijing, China
{yi.zeng,dongsheng.wang,hongwei.hao,xubo}@ia.ac.cn

Abstract. Knowledge bases play a crucial role in intelligent systems, especially in the Web age. Many domain dependent and general purpose knowledge
bases have been developed to support various kinds of applications. In this
paper, we propose the CASIA-KB, a Chinese semantic knowledge base built
from various Web resources. CASIA-KB utilizes Semantic Web and Natural
Language Processing techniques and mainly focuses on declarative knowledge.
Most of the knowledge is textual knowledge extracted from structured and
unstructured sources, such as Web-based Encyclopedias (where more formal
and static knowledge comes from), Microblog posts and News (where most
updated factual knowledge comes from). CASIA-KB also aims at bringing in
images and videos (which serve as non-textual knowledge) as relevant
knowledge for specific instances and concepts since they bring additional
interpretation and understanding of textual knowledge. For knowledge base
organization, we briefly discussed the current ontology of CASIA-KB and the
entity linking efforts for linking semantically equivalent entities together. In
addition, we build up a SPARQL endpoint with visualization functionality for
query processing and result presentation, which can produce query output in
different formats and with result visualization supports. Analysis on the entity
degree distributions of each individual knowledge source and the whole
CASIA-KB shows that each of the branch knowledge base follows power law
distribution and when entities from different resources are linked together to
build a merged knowledge base, the whole knowledge base still keeps this
structural property.

Keywords: Chinese semantic knowledge base · Web of data · Semantic web ·
Information extraction

1 Introduction

Large-scale knowledge bases are essential to support various knowledge driven
applications, especially in the context of the Web. Many efforts have been made such
as CYC [1], DBPedia [2], YAGO [3], ConceptNet [4], NELL [5], etc. Most of them
are either based on Semantic Web techniques or Natural Language Processing. In
addition, most of them are English or at least English centric semantic knowledge
bases.

W. Kim et al. (Eds.): JIST 2013, LNCS 8388, pp. 75–88, 2014
DOI: 10.1007/978-3-319-06826-8_7, © Springer International Publishing Switzerland 2014

In this paper, we introduce the construction of CASIA-KB developed at Institute of Automation, Chinese Academy of Sciences (CASIA), which is a large-scale semantic knowledge base built from multiple structured and unstructured sources, ranging from textual contents (e.g. Web-based Encyclopedia, microblog posts, news pages) to non-textual contents (e.g. images). The whole knowledge base is constructed based on a combination of Semantic Web, Natural Language Processing, statistical analysis and network visualization principles and techniques.

Firstly, we crawl the large volume of Web pages and extract structured facts in infoboxes from the three sources, including Baidu Encyclopedia, Hudong Encyclopedia and Chinese Wikipedia. The structured facts are represented in RDF N3 format and stored in Jena TDB[1], resulting in a large scale triple-based semantic knowledge base. In order to enrich the structured knowledge from Web-based Encyclopedia infoboxes, we obtained a large volume of News Web pages and social media contents (e.g. microblog posts) as potential knowledge sources. Then we use pattern-based methods to automatically extract knowledge triples from different resources (Encyclopedias, News Web pages, and microblog posts), and enrich the knowledge base.

Secondly, we extract the "sub class of" and "instance of" relation from the Web-based Encyclopedias category information, and construct a hierarchical ontology. In order to merge knowledge from different resources, we embed an entity linking algorithm into the knowledge base where entities from different sources can be dynamically linked together. Finally, we build up an SPARQL and a visualization interface for query results presentation.

The rest of this paper is organized as following. Section 2 provides a brief introduction of related works on developing knowledge bases. Section 3 details the construction of CASIA-KB. Section 4 presents the conclusion and future works.

2   Related Works

After successful development of state-of-the-art knowledge bases such as CYC [1], the efforts for building large-scale knowledge bases have been shifted to automatic construction of knowledge bases based on Web contents. DBpedia [2] is based on structured knowledge extraction from Wikipedia structured sources such as infoboxes. It assigns a URL for each entity, with hierarchical ontologies managing all these triple data together. In addition, based on the multiple language versions of Wikipedia, DBPedia knowledge base also provides multiple language versions. Another well accepted semantic knowledge base YAGO [3], which is built from Wikipedia and WordNet provides temporal and spatial extensions for existing knowledge extracted from Wikipedia. Besides extraction of knowledge from structured sources, recent efforts on knowledge base construction based on information extraction as well as rule learning from unstructured texts are with great potential (e.g. NELL [5]).

As for large-scale Chinese knowledge bases, National Knowledge Infrastructure (NKI) is a pioneer work and has been applied to several domains (e.g. call center

---

[1] http://jena.apache.org/documentation/tdb/

support) [6]. Recent efforts for building Chinese semantic knowledge bases have been focusing on extracting structured knowledge from Web-based encyclopedias and representing the knowledge using Semantic Web techniques. Zhishi.me [7], which is developed by Shanghai Jiaotong University crawl facts embedded in information tables (known as infobox) from three sources including Baidu Encyclopedia, Hudong Encyclopedia and Chinese Wikipedia. Another semantic knowledge base maintained by Tsinghua University extracts and refines hierarchical ontologies for managing entities and infobox knowledge from Baidu Encyclopedia and Hudong Encyclopedia [8, 9]. Nevertheless, for the above knowledge bases which are based on semantic Web techniques, knowledge extraction from unstructured texts has not been included, while most knowledge on the Web is embedded in unstructured Web contents.

Early versions of CASIA-KB contains declarative knowledge from structured sources such as infobox triples and ontology from Web-based Encyclopedias [10], and extensions of representing uncertain knowledge with typicality was introduced in [11]. In this paper, we introduce the whole road map, current status and implementation details of CASIA-KB.

3   Knowledge Base Construction

This section introduces the design and implementation of the CASIA-KB in details. Section 3.1 provides the roadmap and current status of CASIA-KB. Section 3.2 illustrates the large RDF triple data construction from structured sources. Section 3.3 discusses the efforts on enriching the knowledge base from unstructured sources. Section 3.4 introduces the OWL ontologies of the three resources and how to adopt entity linking algorithm for data level entity linking. Section 3.5 introduces and demonstrates the SPARQL endpoint as the interface for accessing the CASIA-KB and the visualization as well as structural analysis of the knowledge base.

3.1   The Roadmap and Current Status of CASIA-KB

CASIA-KB is a general purpose and multi-source semantic knowledge base built from various resources. CASIA-KB mainly focuses on declarative knowledge. Most of the knowledge is textual knowledge extracted from structured and unstructured sources, such as Web-based Encyclopedias (where more formal and static knowledge comes from), Microblog posts and News (where most updated knowledge comes from). CASIA-KB also aims at bringing in images and videos (which serve as non-textual knowledge) as relevant knowledge for specific instances and concepts since they bring additional interpretation and sometimes better understanding of textual knowledge [12]. The road map of CASIA-KB is illustrated as Fig. 1.

The first version of CASIA-KB only contains knowledge from structured sources such as infobox triples from Web-based Encyclopedias [10]. In this paper, we provide detailed introduction of how the CASIA-KB is been built and the extraction of knowledge (including triples with Is-A, Part-Whole, and Has-Property relation) from unstructured sources (e.g. free texts in Web-based Encyclopedias, microblog posts,

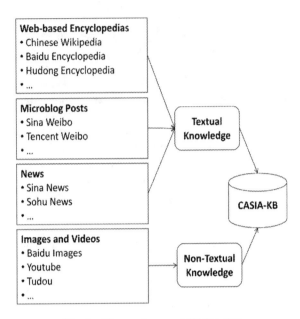

Fig. 1. The road map of CASIA-KB

and news pages) will also be discussed. For non-textual knowledge, currently, CASIA-KB contains images crawled from Baidu Images and Web-based Encyclopedias to support various applications, such as the CASIA tour recommendation system[2]. In this paper, we will mainly focus on the construction of textual knowledge base. The main resources that CASIA-KB contains are listed in Table 1.

Table 1. Current status of CASIA-KB

Sources	Entities	Triples from Structured Sources	Triples from Un-structured Sources	Triples with Typicality
Baidu Encyclopedia	1,967,259	9,752,018	276,676	15,926,018
Hudong Encyclopedia	1,581,055	11,216,471	253,588	
Chinese Wikipedia	460,439	4,569,236	13,154	
Sina Weibo	53,172		920,823	
Sogou News	10,963		31,140	

---

[2] CASIA Tour Recommendation System: http://www.linked-neuron-data.org/CASIAPOIR.

All the above literal triple knowledge is transformed to RDF triples and stored in Jena TDB. Since the knowledge in CASIA-KB are all from the Web contributed by various users and organizations with different understanding of the world knowledge. It is obvious that the extracted knowledge may be with different uncertainties. Hence, CASIA-KB is designed to be with uncertainties. In [11], we introduced a typicality approach for describing knowledge with uncertainty. How the generated typicality values are used for entity conceptualization and knowledge validation are also discussed [11]. In this paper, we focus on the knowledge extraction and ontology organization of CASIA-KB.

3.2 Knowledge Extraction from Structured Data and Resource Organization

In this paper, we consider information tables such as infoboxes, and category information on the Web-based Encyclopedia pages as structured data sources. Firstly, we crawl the entire Web pages from Baidu Encyclopedia, Hudong Encyclopedia and Chinese Wikipedia. Secondly, we extract the entity names, their properties and property values, as well as category information to obtain the literal knowledge triple. Thirdly, all the literal triples are transformed to RDF triples.

As shown in Fig. 2, the title "中国科学院" is extracted as the entity name, and the infobox information is represented as the properties and property values of the entity (e.g. <中国科学院, 总部地址, 北京>). The category information in the bottom of the page is transformed to triples with the "rdf:type" relation. In addition, "Related Terms" in the pages are stored as triples with a general binary relation "RelatedTo".

Since the knowledge is contributed by different users in various sources, and they may mention the same entity using different names, synonyms are very important for better knowledge organization in the knowledge base in the way of linking semantically equivalent entities together. In our study, the synonym set includes possible resources from synonym labels, redirect labels, nick names, previous names, etc. from the three encyclopedia sources (e.g. as shown in Fig. 2, "中科院" is a synonym of the full title "中国科学院", and this piece of information is extracted to build up the synonym set). 476,086 pairs of synonyms are added in the synonym set. In order to have better supports for users to retrieve resources, we utilize the synonym set to build labels for resources to support different input alternatives.

Triples from the same knowledge source are more consistent, and well managed by the ontology designed for this specific knowledge source. Hence, triples from the same knowledge source are stored in the same named graph. Users can choose which knowledge source(s) they would like to use.

3.3 Knowledge Extraction from Unstructured Data Sources

Web-based encyclopedia Infoboxes and tables are direct sources for extracting structured knowledge, while there are a lot more knowledge which are represented in unstructured text resources, such as free texts in Web-based encyclopedia and microblog posts. In this paper, we try to use rule based methods to extract knowledge

← Synset

中国科学院 ←---- Title

🗒 编辑词条

百科名片

中国科学院院徽

中国科学院（Chinese Academy of Sciences, CAS）是中国在科学技术方面的最高学术机构和中国自然科学与高新技术的综合研究与发展中心，于1949年11月在北京成立。中国科学院把"唯实、求真、协力、创新"作为院风，秉持"科学、民主、爱国、奉献"的优良传统。以中国富强、人民幸福为己任，为中国科技进步、经济社会发展和国家安全做出了不可替代的重要贡献。中国科学院服务国家战略需求和经济社会发展，始终围绕现代化建设需要开展科学研究，产生了许多科技成果，奠定了新中国的主要学科基础，自主发展了一系列战略高技术领域，形成了具有中国特色的科研体系，带动和支持了中国工业技术体系、国防科技体系和区域创新体系建设。中国科学院与中国工程院在中国并称"两院"。

Infobox ⟍

中文名：	中国科学院	总部地址：	北京
外文名：	Chinese Academy of Sciences	机构地位：	中国科学技术方面的最高学术机构
中文简称：	中科院	成立时间：	1949年
英文简称：	CAS		

开放分类： Category

| 科学 | 机构 | 中国 | 科研机构 | 中国科学院 | 中国网站 | 互联网 | 分子生物学 | 国家机关 | 学术组织 | 技术 | 冶金组织 |

| 智库 | 科学组织 | 北京大学 | 中国科学院院士 数学物理学部 | 中国科学院院士 化学部 | 中国科学院院士 生命科学和医学学部 |

| 中国科学院院士 地学部 | 中国科学院院士 信息技术科学学部 | 中国科学院院士 技术科学部 | 中华人民共和国国务院下属机构 |

基础科学

"中国科学院"相关词条：   ✎ 我来完善

清华大学 中国工程院 山东农业大学 中央美术学院 联想 中国科学技术大学 美国国家科学院 中国地质科学院
中国科学院首届高等学校

Fig. 2. Screenshot of a page in Baidu Encyclopedia

triples to enrich the original knowledge base. We mainly focus on the extraction of three types of declarative knowledge, namely, Is-A relation, Part-Whole relation and Has-Property relation.

There are two types of binary relations which can be used to enrich the organization of the knowledge base ontology, namely, Is-A relation and Part-Whole Relation. Is-A relation represents a hierarchical binary relation either between two concepts or between an instance and a concept. Unlike extracting Is-A relation from unstructured English texts [13], there are various ways to represent Is-A relation in Chinese. Some possible (but not complete) candidates are listed in Table 2. In the extraction phase, we find the subject of the triple before the possible candidate terms which represent the Is-A relation, and the object should appear after the Is-A relation. In this paper, we focus on extracting Is-A relation whose subjects and objects are all nouns.

Part-Whole relation represents the Part-Whole binary relation among two resources, in which one resource is a part of another. The most important English pattern for extracting Part-Whole relation are "part of" and "is composed of", while in Chinese, some possible patterns are listed in Table 2.

Table 2. A list of possible Chinese terms relevant to specific relations

Relation Type	Possible Chinese Terms Relevant to This Relation
Is-A	是/为一(个/只/群/把/条/种/件/名/门/款/颗/根/头/尾/片/次/座/块/本/代/张/双/篇/堆/批/章/节/步/副/缕/位/台/侏), 属于, ...
Part-Whole	是/为... 的一部分, 由...组成/构成
Has-Property	...的...是/为

Properties are essential for describing concepts and instances in knowledge bases. In this paper, we focus on extracting property names and construct declarative triples with Has-Property relation. In English, properties can be extracted from the pattern "the X of Y..." while "X" represents a possible property of the concept or instance "Y". The relevant pattern in Chinese is "Y的X是/为", while "X" is the property here, and the triple is represented as <Y, Has-Property, X>.

By using the extraction strategies discussed above, Table 3 provides a list on the number of triples for each type of extraction from different sources.

Table 3. Unstructured knowledge from different sources

Source	Is-A	Part-Whole	Has-Property
Baidu Encyclopedia	9,586	115,149	151,941
Hudong Encyclopedia	129,417	98,658	25,513
Chinese Wikipedia	3,809	6,737	2,608
Sina Weibo	40,315	29,625	850,883
Sogou News	13,962	9,806	7,372

It is noticed that although microblog posts and News pages are always considered as events related resources, through our experimental studies, it shows that there are also many common declarative knowledge which can be extracted from these two types of sources.

Although the pattern based extraction strategy is simple, it is straightforward and practical for large-scale information extraction. Table 4 gives some illustrative examples on the three types of extracted triples from different sources.

3.4 Conceptual Ontology and Entity Linking

We extract the category level information from Baidu Encyclopedia and Hudong Encyclopedia. We extract all the "*subClassOf*" relationship among classes and create an OWL file for them. We observe the file by protégé and find that the Baidu Encyclopedia has a relatively clean hierarchical classification system officially

Table 4. Some examples on extracted triples from different sources

Extracted Triples	Source
<表皮层, part-whole, 皮肤> <每一个幸福的今天, part-whole, 人生> <实践部, part-whole, 团总支>	Sina Weibo
<摩托车, Is-A, 机动车> <胆碱, Is-A, 麻醉剂> <纪念碑, Is-A, 文物>	Sogou News
<郁金香, Has-Property, 原产地> <出版社, Has-Property, 负责人> <影片, Has-Property, 主角>	Baidu Encyclopedia

maintained by Baidu Inc., while the Concept-Instance relations are contributed by common users. As shown in Fig. 3, there are 13 very broad categories which are with many sub-categories under each of them, and generally there are three to four hierarchical layers. While the hierarchy of Hudong Encyclopedia is with more detailed classification, but the organization of the hierarchy needs improvements (such as self-loops in the ontology [9]).

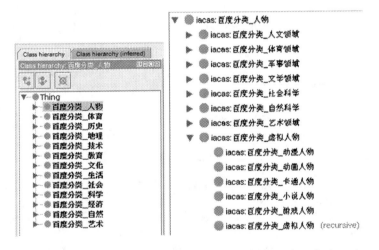

Fig. 3. Hiearchical ontology of CASIA-KB Baidu Encyclopedia branch

However, the OWL ontology only contains a well-structured hierarchy with "*rdf:subClassOf*" relationship among classes (concept). It is not enriched with object property, data type property, or any constraint information. Therefore, we can use the Concept-Instance relations to perform some statistical analysis and conclude some constraints on the OWL ontology, such as generating Top-K attributes for a given concept or generating domain and range for a specific relation [11].

Although entities are locally managed by their own ontology, many of them appeared in different sources might be semantically equivalent. Hence, entity linking is required for better organization of CASIA-KB. The entity linking is a dynamic process, with entity disambiguation as one of the key step. For example, given a search input "*Apple*", the Baidu Encyclopedia returns several possible candidate entities (such as "Apple" in the fruit category and "Apple" in the company category), and the Hudong Encyclopedia also returns several possible candidate entities which share the same semantics with the ones in Baidu Encyclopedia. So the entity linking algorithm measures the similarity of each two entities and if the similarity value is greater than a threshold, the two are regarded to be the same and linked together. In [14], we developed the CASIA-EL subsystem and a Stepwise Bag-of-Words based entity disambiguation algorithm for CASIA-KB. The linking precision of CASIA-EL is 88.5 % based on the Chinese microblog entity linking contest organized by the 2nd Conference on Natural Language Processing and Chinese Computing (CASIA-EL ranked as the 2nd team).

3.5   Knowledge Base Interface and Knowledge Network Visualization

We build a Web-based interface to support the visualization of SPARQL query. The highlighted function is the online graph visualization based on JavaScript, as well as

Fig. 4. Screenshot of the CASIA-KB SPARQL endpoint

Table 5. An example of a query results

Subject	Predicate	Object
Hudong:互动资源_龙	Hudong:名称	"龙"^^<http://www.w3.org/2001/XMLSchema#string>
Hudong:互动资源_龙	Hudong:拼音	"lóng"^^<http://www.w3.org/2001/XMLSchema#string>
Hudong:互动资源_龙	Hudong:繁体	"龍"^^<http://www.w3.org/2001/XMLSchema#string>
Hudong:互动资源_龙	Hudong:英文	"dragon"^^<http://www.w3.org/2001/XMLSchema#string>
Hudong:互动资源_龙	Hudong:笔画	"5"^^<http://www.w3.org/2001/XMLSchema#string>
Hudong:互动资源_龙	Hudong:部首	"龙"^^<http://www.w3.org/2001/XMLSchema#string>
Hudong:互动资源_龙	Hudong:笔顺	"横撇那勾撇点 "^^<http://www.w3.org/2001/XMLSchema#string>
⋯⋯⋯		

different forms of text outputs. When we just want a text form output, we are supposed to input "*SELECT * FROM*" SPARQL format query. If we want a graph based visualized result, we are required to input "*CONSTRUCT {*} FROM*" SPARQL format query since *CONSTRUCT* can return a sub-model of the graph model.

Figure 4 is a snapshot of the Web interface of the SPARQL Endpoint. By using the default query as an example, if we select the "*互动百种*" and click "*执行搜索*" (run query), the triple format query result is listed as in Table 5.

The textual output may be helpful for developers but not intuitive for common users. In order to improve user experience, we take advantage of Javascript to demonstrate the result as visualized nodes and edges. To generate a subgraph of the knowledge base, SPARQL "*CONSTRUCT*" query is required instead of "*SELECT*" query. For instance, when we input the query as shown in Table 6:

Then, the corresponding subgraph is generated and presented to the end user, as shown in Fig. 5.

Table 6. An example of construct query on CASIA-KB

```
prefix baidubaike:<http://www.ia.cas.cn/baike_baidu/resource/>
CONSTRUCT {?s ?p ?o}
WHERE {baidubaike:百度资源_北海公园 ?p ?o}
```

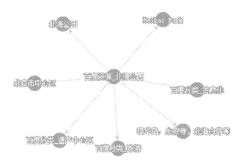

Fig. 5. An example of a graph visualization

Besides a SPARQL Construct visualization tool, we also utilize complex network visualization tools such as NodeXL to visualize and analyze the complex knowledge network of CASIA-KB. Figure 6 is a visualization of 10 thousand triples from the Hudong Branch of CASIA-KB. In this figure, we select one normal node and its two-degree adjacency nodes are shown. We can observe that a normal node in the knowledge network does not have many connections since it is only connected to a few entities through a few triples. Figure 7 shows a pivotal node and its two-degree adjacency nodes in the knowledge network, which indicates that this node connects to many other entities, and only by two degrees, most of the entities are connected together.

In [10], we analyzed the structural properties on the Baidu branch of the CASIA-KB. From the degree distribution perspective, it yields a power law distribution. As a step forward, it is not surprising to see that the Hudong branch and the Chinese Wikipedia Branch also follow power law distributions, as shown in Fig. 8. When we merge these branches together as one, the whole CASIA-KB also follows a power law distribution (This is consistent with previous findings for structural properties

Fig. 6. Knowledge network visualization and a normal node's two-degree adjacency nodes

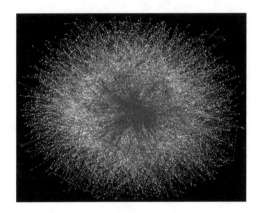

Fig. 7. Knowledge network visualization and a pivotal node's two-degree adjacency nodes

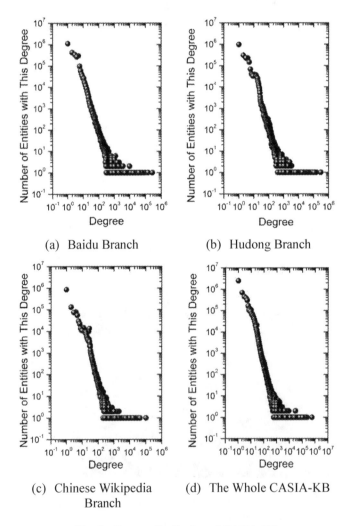

(a) Baidu Branch

(b) Hudong Branch

(c) Chinese Wikipedia
      Branch

(d) The Whole CASIA-KB

Fig. 8. Degree distribution of CASIA-KB

preservation in the network merged from multiple complex networks [15]). This observations show that from a global point of view, although the entities and relevant relations are from different resources on the Web, there are always several pivotal entities (but not many) which serve as the hub of the whole knowledge base and connect the rest of the entities together. This structure keeps the entire knowledge base as an interconnected Web of knowledge.

4   Conclusion and Future Works

Many semantic knowledge bases are being developed in a variety of areas and multiple languages such as YAGO and Zhishi.me. Although the Chinese knowledge base constructed in previous efforts based on Semantic Web techniques are from multiple sources, they just extract structured information from Encyclopedia pages.

This paper introduces CASIA-KB, a general purpose Chinese semantic knowledge base built from structured and unstructured resources, including Web-based Encyclopedias, microblog posts and Web News pages, etc. CASIA-KB mainly focuses on declarative knowledge and most of the knowledge is textual knowledge extracted from structured and unstructured sources, such as Web-based Encyclopedias (where more formal and static knowledge comes from), Microblog posts and News (where most updated knowledge comes from). CASIA-KB also aims at bringing in images and videos (which serve as non-textual knowledge) as relevant knowledge for specific instances and concepts since they bring additional interpretation and understanding of textual knowledge. The structured facts are represented in RDF N3 format and stored in Jena TDB, resulting in a large scale triple-based semantic knowledge base. For triple extraction from unstructured texts, we focus on three types of declarative knowledge, namely, those with Is-A, Part-Whole, or Has-Property relations. The extractions are based on patterns, and further investigations on the extraction quality need to be improved. For this reason, currently, we store knowledge triples from unstructured texts as independent resources. After quality evaluation and improvements, they should be linked to knowledge from structured resources such as Baidu, Hudong, Wikipedia infobox knowledge and corresponding ontologies.

As a step forward, events related knowledge base should be considered as an important branch for CASIA-KB since they can be considered as an important source where most updated knowledge is from. Inspired by YAGO's efforts on extraction of time and location from Wikipedia, each event in CASIA-KB can be with more provenance relation extracted from News contents and microblog posts, such as event time, location, source of the News, etc.

Acknowledgement. The set of news Web pages are from Sogou Lab[3]. Peng Zhou is involved in crawling the Baidu and Hudong Encyclopedia Web pages. The set of Sina Weibo microblog posts are crawled by Bo Xu and Feng Wang from the Computational Brain Group at Institute of Automation, Chinese Academy of Sciences. This study was supported by the Young Scientists Fund of the National Natural Science Foundation of China (61100128).

---

[3] Sogou Lab News Data <http://www.sogou.com/labs/dl/ca.html>.

References

1. Lenat, D.B.: CYC: a large-scale investment in knowledge infrastructure. Commun. ACM **38**(11), 32–38 (1995)
2. Bizer, C., Lehmann, J., Kobilarov, G., Auer, S., Becker, C., Cyganiak, R., Hellmann, S.: DBpedia - a crystallization point for the web of data. Web Semant.: Sci. Serv. Agents World Wide Web **7**(3), 154–165 (2009)
3. Suchanek, F.M., Kasneci, G., Weikum, G.: YAGO: a large ontology from Wikipedia and WordNet. Web Semant.: Sci. Serv. Agents World Wide Web **6**(3), 203–217 (2008)
4. Havasi, C., Speer, R., Alonso, J.: ConceptNet: a lexical resource for common sense knowledge. In: Nicolov, N., Angelova, G., Mitkov, R. (eds.) Recent Advances in Natural Language Processing, vol. 5. John Bernjamins Publishers, Amsterdam (2009)
5. Carlson, A., Betteridge, J., Kisiel, B., Settles, B., Hruschka, E.R., Jr., Mitchell, T.M.: Toward an architecture for never-ending language learning. In: Proceedings of the Twenty-Fourth AAAI Conference on Artificial Intelligence (AAAI 2010), Georgia, USA, 11–15 July 2010
6. Cao, C., Feng, Q., Gao, Y., Gu, F., Si, J., Sui, Y., Tian, W., Wang, H., Wang, L., Zeng, Q., Zhang, C., Zheng, Y., Zhou, X.: Progress in the development of national knowledge infrastructure. J. Comput. Sci. Technol. **17**(5), 523–534 (2002)
7. Niu, X., Sun, X., Wang, H., Rong, S., Qi, G., Yu, Y.: Zhishi.me - weaving Chinese linking open data. In: Aroyo, L., Welty, C., Alani, H., Taylor, J., Bernstein, A., Kagal, L., Noy, N., Blomqvist, E. (eds.) ISWC 2011, Part II. LNCS, vol. 7032, pp. 205–220. Springer, Heidelberg (2011)
8. Wang, Z., Wang, Z., Li, J., Pan, J.: Knowledge extraction from Chinese Wiki Encyclopedias. J. Zhejiang Univ.-SCI. C **13**(4), 268–280 (2012)
9. Wang, Z., Li, J., Wang, Z., Tang, J.: Cross-lingual knowledge linking across Wiki knowledge bases. In: Proceedings of the 21st World Wide Web Conference (WWW 2012), Lyon, France, 16–20 April 2012, pp. 459–468 (2012)
10. Zeng, Y., Wang, H., Hao, H., Xu, B.: Statistical and structural analysis of web-based collaborative knowledge bases generated from Wiki Encyclopedia. In: Proceedings of the 2012 IEEE/WIC/ACM International Conference on Web Intelligence (WI 2012), Macau, China, 4–7 December 2012, pp. 553–557 (2012)
11. Zeng, Y., Hao, H., Xu, B.: Entity conceptualization and understanding based on web-scale knowledge bases. In: Proceedings of the 2013 IEEE International Conference on System, Man, and Cybernetics (SMC 2013), Manchester, UK (2013)
12. Larkin, J., Simon, H.: Why a diagram is (Sometimes) worth ten thousand words. Cogn. Sci. **11**, 65–99 (1987)
13. Song, Y., Wang, H., Wang, Z., Li, H., Chen, W.: Short text conceptualization using a probabilistic knowledgebase. In: Proceedings of the 22nd International Joint Conference on Artificial Intelligence (IJCAI 2011), Barcelona, Spain, 16–22 July 2011, pp. 2330–2336 (2011)
14. Zeng, Y., Wang, D., Zhang, T., Wang, H., Hao, H.: Linking entities in short texts based on a Chinese semantic knowledge base. In: Zhou, G., Li, J., Zhao, D., Feng, Y. (eds.) NLPCC 2013. CCIS, vol. 400, pp. 266–276. Springer, Heidelberg (2013)
15. Sun, X., Zhuge, H.: Merging complex networks. In: Proceedings of the Seventh International Conference on Semantics Knowledge and Grid (SKG 2011), Beijing, China, 24–26 October 2011, pp. 233–236 (2011)

TANLION – TAmil Natural Language Interface for Querying ONtologies

Vivek Anandan Ramachandran[⊠] and Ilango Krishnamurthi

Department of Computer Science and Engineering, Sri Krishna College
of Engineering and Technology, Coimbatore 641 008 Tamilnadu, India
{rvivekanandan, ilango.krishnamurthi}@gmail.com

Abstract. An ontology contains numerous information in a formal way which
cannot be easily understood by casual users. Rendering a natural language
interface to ontologies will be more useful for such users, as it allows them to
retrieve the necessary information without knowing about the formal specifications existing in the ontologies. Until now, most such interfaces have only
been built for accepting user queries in English. Besides English, providing
interface to ontologies in other native and regional languages should also be
explored, as it enables the casual users to gather information from an ontology
without any language barrier. One of the most popular languages in South Asia
is Tamil, a classical language. In this paper, we present our research experience
in developing TANLION, a Tamil interface for querying ontologies. It accepts
a Tamil query from an end-user and tries to recognize the information that the
user needs. If that information is available in the ontology, our system retrieves
it and presents to the user in Tamil.

Keywords: Natural Language Interface · Ontology · Tamil

1 Introduction

Knowledge management systems that utilize semantic models for representing a
consensual knowledge about a domain are widely in use. In such systems, knowledge
mostly exists in the form of ontologies[1]. Ontology is a graph consisting of a set of
concepts, a set of relationships connecting those concepts and a set of instances.
Usually ontologies are developed in formal languages such as Resource Description
Framework (RDF) [1], Web Ontology Language (OWL) [2] etc. In order to acquire
information from the knowledge available in an ontology, a casual user should know
about:

- The syntax of the formal languages used in modeling the ontology.
- The formal expressions and vocabularies used in representing the knowledge.

Providing a Natural Language Interface (NLI) to ontologies will help them to
retrieve the necessary information without knowing about the formal specifications

---

[1] In this paper we refer ontology as a knowledge base that includes concepts, relations and instances
existing in a domain.

W. Kim et al. (Eds.): JIST 2013, LNCS 8388, pp. 89–100, 2014.
DOI: 10.1007/978-3-319-06826-8_8, © Springer International Publishing Switzerland 2014

existing in the ontologies. In addition to English, investigations in providing NLIs should also be carried out in other native and regional languages, as it enables the casual users to acquire information from an ontology without any language hindrance. Further, such NLIs decrease the gap of ontology utilization between the professional and the casual users [3]. As the gap of usage diminishes both ontologies and Semantic web spread widely.

The need for regional language NLIs to ontologies can be explained with a scenario. A farmer who knows only Tamil[2] [4] wants to acquire the answer for the query, *nelvaaRpuuchchiyai aJikka e_n_na uram iTa veeNTum?* (*What fertilizer can be used to destroy threadworms in Rice-plant?*), from a Rice-plant ontology developed in English. In this scenario the farmer faces the following problems:

- He might not know English which is used in developing the ontology.
- He might not know the syntax of the formal language used in modelling the ontology.
- He might not be able to follow the formal expressions used in the ontology.

Considering this as a potential research problem, we have developed a Tamil NLI for querying ontologies (TANLION).

In the above scenario, we have considered using Tamil NLI for querying the Rice-plant ontology. But addressing the factor of customizing NLIs to other ontologies is also a crucial issue. Portable or Transportable NLIs are those that can be customized to new ontologies covering the same or different domain. Portability is an important feature of an NLI because it provides an option for an end user to move NLIs to different domains. TANLION is a portable NLI that can accept a Tamil Natural Language Query (NLQ) and a given ontology as input, and returns the result retrieved from the ontology in Tamil.

This paper is organized as follows: In Sect. 2 we discuss the studies related to TANLION. In Sect. 3 we brief about the design issues considered for developing TANLION. In Sect. 4 we describe about the System Architecture. In Sect. 5 we present the System Evaluation. In Sects. 6 and 7 we give the limitations and the concluding remarks respectively.

2   Related Work

Researches in NLIs have been reported since 1970s [5, 6]. Extensive studies have been conducted to provide NLIs to a database [7–9]. As a result of such research, good NLIs to database have emerged [10]. But the major constraint of database is that they are not easily shareable and reusable. Hence the usage of ontologies became increasingly common, as it can be easily reused and shared.

Thus the increased utilization of ontologies inspired researches for providing NLIs to ontologies [11, 12]. The main goal of such NLIs is to recognize the semantics of the

---

[2] Tamil is a Dravidian language spoken predominantly by the Tamil people of the Southern India. It is also a classical language. It has official status in India, Sri Lanka and Singapore. Tamil is also spoken by substantial minorities in Malaysia, Mauritius and Vietnam.

input NLQ and use it to generate the target SPARQL[3] [13]. Further, those NLIs should either assure a correct output or indicate that it cannot process the NLQ. Along with the efforts on rendering NLIs to ontologies many systems such as Semantic Crystal [14], Ginseng [15], FREyA [16], NLP-Reduce [17], ORAKEL [18], e-Librarian [19], Querix [20], AquaLog [21], PANTO [22], QuestIO [23], NLION [24] and SWAT [25] have evolved.

Semantic Crystal displays the ontology to an end-user in a graphical user interface. The end-user clicks on the needed portion in the ontology, and the system will display the answer accordingly. Comparatively, Ginseng allows the user to query ontology with the help of structures such as pop-up menus, sentence completion choices and suggestion boxes. FREyA system is termed after Feedback, Refinement and Extended Vocabulary Aggregation. It displays ontology contents to the end user in a tree structure. The end-user clicks on the suitable portions in the tree, and the system will display the answer accordingly. Remaining systems are functionally similar, but the difference is to recognize the semantics of the input NLQ:

- Querix uses Stanford parser output and a set of heuristic rules.
- PANTO utilizes the information in the Noun phrases of the NLQ.
- e-Librarian uses normal string matching techniques.
- NLP-Reduce employ stemming, WordNet and string metric techniques.
- AquaLog uses a shallow parser and hand-crafted grammar.
- NLION utilizes the semantic relation between the words in the NLQ and the ontology.
- ORAKEL uses tree structure.
- QuestIO employs shallow language processing and pattern-matching.

SWAT totally differs from the other systems as it transforms the input NLQ in Attempto Controlled English (ACE) to N3. The above systems render NLIs to ontologies by accepting the input NLQ in English. But providing Tamil NLIs to ontologies is not explored as the supporting technologies are required to a great extent and also the availability is scant. Among all the approaches NLION provides an inherent support to develop an NLI for ontology in an easy and a fast way for languages whereas the supporting technologies are availability is scant. So we decided to extend NLION approach to Tamil. In the next Section, we discuss about the technical issues needed to be addressed for developing a Tamil NLI.

3 TANLION Computing Issues

In this Section, we elaborate about the issues to be dealt for developing a standard Tamil NLI and how we handle those issues in TANLION.

---

[3] SPARQL is a Query Language used to retrieve and manipulate the information from the ontologies that is stored in the RDF/OWL format.

3.1 Splitting

Splitting is the process of separating two or more lexemes present in a single word. For example, the word ain-taaNTuttiTTam (five-year-plan) contains three lexeme ain-tu (five), aaNTu (year) and tiTTam (plan). When ain-taaNTuttiTTam (five-year-plan) is matched directly with 'ain-tu aaNTu tiTTam' (five year plan) it will be erroneous. So splitting should be done on such NLQ words for effective Information Retrieval (IR).

3.2 Stemming

Stemming is the process of reducing derived words to their root form. For example, reducing the word Mothers to the word Mother is stemming. The reason for using stemming in IR systems is that, most of the words exist in the target database in their root form. For example, to represent a concept 'Mother' in any ontology it will be mostly labeled as Mother and not as Mothers. So each NLQ word should be stemmed for effective IR. In TANLION we have used a stripping stemmer for improving the system retrieval ability [26].

3.3 Synonym Expansion

Synonym Expansion (SE) is a technique where variants of each word in a NLQ are used to improve the retrieval in an IR system. Say, an ontology contains a concept 'Mother' with the label Mother and an IR system is looking in the same ontology for the concept 'Mother' with the keyword mom. In such case, the IR system should use either Thesaurus or WordNet and make use of the fact contained in them that Mother is the synonym of mom [27].

3.4 Translator

Providing a Tamil NLI to English ontologies require a bi-directional translation service between English and Tamil. Say, a user query contains a word ammaa, it should be translated to Mother for mapping it with the concept Mother in the ontology. Currently, complete automized translation software does not exists for English to Tamil translation, as it is in the research stage. In TANLION, we handle the translation issue in a simple way. We annotate all the ontology elements with its Tamil equivalence. For the entity Mother we annotate its Tamil equivalent as ammaa. If a system needs the Tamil equivalence of Mother, it can refer to its Tamil annotation value and acquire the result ammaa. By dealing only with the translation of the ontology elements, it is conclusive that our system answering ability will be restricted to Factoid[4] and List[5] NLQs. This can be easily inferred by tracing TANLION working principle. In the next Section, we explain about TANLION working with its architecture.

---

[4] Factoid query focuses on questions whose answers are entities.
[5] List query focuses on questions whose answers are list of entities.

4 TANLION Computing Issues

In this Section we provide an overview about our System architecture, depicted in the Fig. 1, with its working principle. The system consists of four main parts viz. user interface, query expansion processor, triple extractor and SPARQL convertor. Each component function is explained in the following subsections as follows.

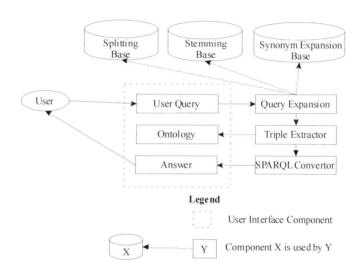

Fig. 1. System Architecture

4.1 User Interface

TANLION UI contains a query field, an answer field and an ontology selection field. TANLION UI is portrayed in the Fig. 2.

4.2 Query Expansion

Query Expansion (QE) is the process of reformulating the input query to improve the performance of the IR system. There are many QE techniques. We use three of them viz. splitting, stemming and synonym expansion. They are briefed as follows:

4.2.1 Splitting

In the example[6], the query token *varuTattiTTatti_n* (*yearly-plan*) is split to *varuTam* (*year*) and *tiTTam* (*plan*) for improving the system retrieval. So it becomes necessary to separate multiple lexemes existing in each input NLQ token for effective IR.

---

[6] Throughout this Section the term example refers to the query, *mutalaavatu aintu varuTatti_n poJutu yaar piratamaraaka iruntaar*, that is represented in the Fig. 3.

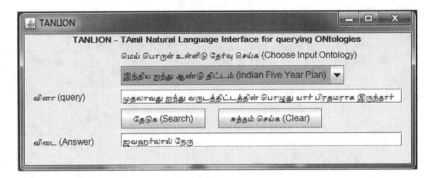

Fig. 2. TANLION User Interface

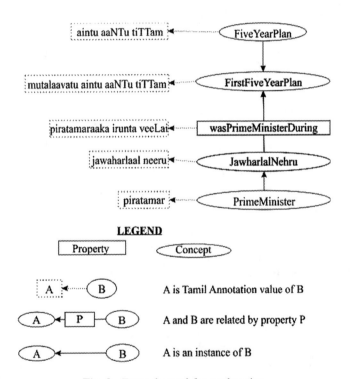

Fig. 3. Example used for explanation

4.2.2   Stemming

In the example, the token in the user query *mutalaavatu* (*The first*) is stemmed to *mutal* (*first*), to make it possible for the system to compare with '*mutal aintu aaNTu tiTTam*' (*first five year plan*). Hence, without stemming it will be difficult for TAN-LION to interpret the user requirement.

4.2.3 Synonyms Expansion

In the example, the synonyms of the token *varuTam* (*year*) viz. *aaNTu* (*year*) and *varusham* (*year*) should be used by the system to compare it with '*mutal ain-tu aaNTu tiTTam*' (*first five year plan*). Therefore without synonyms expansion it will be hard for TANLION to recognize the user requisite.

4.3 Triple Extractor

Until now basic NLP operations are performed on the user query. But the most important phase is to interpret from the input query what the user needs. In this sub-section, we explain the methodology adopted in TANLION to construct the user requirement.

4.3.1 Probable Properties and Resources Extractor

If all the lexicons of a property in an ontology exist in the user query, then there is always a probability that the corresponding property might be the one that the user requires. In the ontology, applying stemming on the lexicons of the annotation value '*piratamaraaka irunta veeLai*' (*was Prime minister during*) yields the result '*piratamar iru veeLai*'(*was Prime minister during*). All the lexicons in the resultant are present in the example stemmed user query. With this inference we assume that the property corresponding to '*piratamaraaka irunta veeLai*' is '*wasPrimeMinsiterDuring*' and it is the probable property that the user might need. So TANLION treats '*wasPrimeMinsiterDuring*' as the Probable Property Element (PPE). In this scenario there is only one PPE but in other cases there can also be more than one PPE. Similarly, we deduce that the Probable Resource Elements (PRE) which the user needs are *PrimeMinister*, *FiveYearPlan* and *FirstFiveYearPlan*.

4.3.2 Ambiguity Fixing

In NLP, there is always a possibility of encountering ambiguities while processing a NLQ. In our example user query too there is an ambiguity whether the user requires *FiveYearPlan* or *FirstFiveYearPlan*. After obtaining PPEs and PREs to resolve the ambiguities, our system follows the following hand-crafted rules:

Rule-1: If a PRE, say PREx, subsumes another PRE, say PREy, then remove the PREyfrom the PREs.

Rule-2: If a PRE, say PREx, is an instance of another PRE, say PREy, then remove the PREy from the PREs.

Rule-3: If a PPE subsumes a PRE, say PREx, then remove the PREx from the PREs.

Rule-4: If a PPE, say PPEx, is a sub-property of a PPE, say PPEy, then remove the PPEy from the PPEs.

In the example, as *FiveYearPlan* is subsumed in *FirstFiveYearPlan* our system removes *FiveYearPlan* from the PREs by applying *Rule-1*. Now, the system will be able to solve the ambiguity that the user intended *FirstFiveYearPlan* and not

FiveYearPlan. Similarly, as *PrimeMinister* is subsumed in *wasPrimeMinisterDuring* our system removes *PrimeMinister* from PREs by applying *Rule-3*. Finally, PREs list will contain only *FirstFiveYearPlan*.

4.3.3 Probable Triple extractor

Consider the fact, '*Jawaharlal Nehru was the Prime Minister during First five year plan*'. It is represented in a Triple as, '*:JawaharlalNehru :wasPrimeMinisterDuring :FirstFiveYearPlan.*'. After resolving the ambiguity, our approach tries to find whether any Triple exists in the ontology of the following forms:

?ar :PPEy :PREx . (or) *:PREx :PPEy ?ar.* (or) *:PREx ?ap :PREx.*

where '*?ar*' and '*?ap*' means any resource and any property in the ontology respectively. Notice that '*?*' is used in the Triple form to satisfy the standard notation requirement. If any such Triple exists, it is treated as a Probable Triple (PT).In our example, after resolving the ambiguities the PRE is *FirstFiveYearPlan* and the PPE is *wasPrimeMinisterDuring*. A Triple '*?ar :wasPrimeMinisterDuring :FirstFiveYearPlan.*', exists in the ontology. So it is treated as PT. In this example there is only one PT, in other cases there can be more than one PT too. In the next sub-section we explain how the PTs are converted to SPARQL for retrieving the user requested information from an ontology.

4.4 SPARQL Convertor

In general, any query language is used for retrieving and manipulating the information from a database. Similarly, a RDF Query Language (RQL) is used to retrieve and manipulate the information from the ontologies that is stored in the RDF/OWL format. SPARQL is a RQL and it is standardized by the RDF Data Access Working Group of the World Wide Web Consortium. After extracting all the PTs we convert it to SPARQL using the template:

*SELECT distinct * WHERE {PT1 . PT2 . ……….. PTn .}.*

It is basically extracting all PTs and placing them in a correct position of a formal SPARQL query to satisfy the syntax constraint. For the example user query, SPARQL generated by TANLION is:

*SELECT distinct * WHERE {?ar :wasPrimeMinisterDuring :FirstFiveYearPlan.}*

The above SPARQL is executed using the Jena SPARQL engine and the result is acquired [28]. Finally, the acquired result's Tamil annotation value is presented to the user. In the example, the result of SPARQL is *JawaharlalNehru*. Its Tamil annotation value *javaharlaal neeru* is displayed to the user.

5 Evaluation

In general, effectiveness means the ability to bring out the result that the user intended. In this Section, we present the parameters used in calculating the effectiveness of our system. To evaluate the performance of TANLION, we implemented a prototype in

Java with the help of the Jena framework. Also, we developed an Indian Five year plan ontology based and used it for evaluation. We analyzed the system effectiveness using two parameters; System ability and Portability.

5.1   System Effectiveness

After developing any NLI system, it is important to analyze the system effectiveness, i.e., to assess the range of the standard questions that the system is able to answer. Unfortunately, there are no standard Tamil query sets. So we requested 7 students, none of whom are not directly or indirectly involved in the TANLION project, to generate questions. They generated totally 108 queries[7]. A deeper analysis on executing these questions over the Indian five year plan ontology (which contains 102 concepts and 46 properties) revealed the following:

- 74.1 % (80 of 108) of them were correctly answered by TANLION.
- 7.4 % (8 of 108) of them were incorrectly answered.
- 10.2 % (11 of 108) of them were not answered due to the failure in Probable Triple generation.
- 8.3 % (9 of 108) of them were not answered due to the failure in query expansion.

After adding the user requested information that is not found in the ontology, we calculated the overall System Effectiveness (SE) using the formula (1).

$$System\ Effectiveness(SE) = \frac{Number\ of\ queries\ correctly\ answered}{Total\ Number\ of\ queries} \tag{1}$$

SE was found to be 74.10 %, which is considerably a good value. Yet, the reason behind achieving a moderate effectiveness value is that our system is in an earlier stage. Still a lot of functionalities, such as increasing the rules for resolving ambiguities in the user query are to be incorporated in our system.

5.2   Portability

We have addressed earlier that TANLION is portable. It could be inferred from our system working principle that construction of answer for the user query depends on the fact that whether the words in the user query exists in the ontologies as resources and properties. So extraction of answer is not dependent on the ontology rather it is dependent on the content of the ontology. The answer extraction phase is dependent on whether the words in the user query exist as resource and property in the ontology. In order to evaluate the portability factor, we interfaced TANLION with a Rice-plant

---

[7] We have given the information to the students that our system will work for only factoid and list based generic queries.

ontology[8] (which contains 72 concepts and 29 properties) and found the SE to be 70.8 %. This calculation was based on testing the system with 79 queries generated by the same students as mentioned in the previous sub-section. So far we have described about the analysis carried out by us to evaluate our approach. We now proceed to brief about our possible future enhancements in the next Section.

6 Future Work

The limitation with the current version of TANLION includes the following:

6.1 Restriction in Query Handling

Consider the user query, *aintaavatu aintaaNTu tiTTatti_n poJutu etta_nai pirata-markaL naaTTai aaTchi cheytaarkaL* (*How many Prime Minister governed the country during the Fifth five year plan*). The TANLION output is: '*intiraa kaanti, moraaji teechaay*'(*Indira Gandhi, Moraji Desai*), while the user required answer is '*2*'. This issue will be handled in the near future by classifying the questions and providing the result accordingly.

6.2 Scalability

The ontologies used for evaluation are relatively much smaller. Investigation on system performance with the larger ontologies is a part of our future work.

7 Conclusion

In this paper, we hypothesize an approach for providing a Tamil NLI to ontologies that allows an end-user to access the ontologies using Tamil NLQ. We process the user queries as a group of words and do not use complicated semantic or NLP techniques as a normal NLI systems does. This drawback is also TANLION's major strength as it is robust to ungrammatical user queries. Our approach is highly dependent on the quality of vocabulary used in the ontology. Yet this fact is also TANLION's big strength, as it does not need any changes for adapting the system to new ontologies. Evaluation results have shown that our approach is simple, portable and efficient. Further evaluation of correctness of TANLION in terms of precision and recall is a part of our future work. To conclude, in this paper we have extended NLION approach to Tamil with splitting as an additional supporting technology. Explorations on extending NLION approach to other languages will be done in the near future.

---

[8] Rice-plant ontology is developed by us based on the information given at the Tamilnadu Agricultural University Website.

Appendix: Tamil Transliteration Scheme Used in this Paper

a	aa	i	ii	u	uu	e	ee	ai	o	oo	au	q				
அ	ஆ	இ	ஈ	உ	ஊ	எ	ஏ	ஐ	ஒ	ஓ	ஔ	ஃ				

k	-N	ch	-n	n	T	N	t	n	p	m	y	r	l	v	J	L	R
க்	ங்	ச்	ஞ்	ன்	ட்	ண்	த்	ந்	ப்	ம்	ய்	ர்	ல்	வ்	ழ்	ள்	ற்

References

1. http://www.w3.org/RDF/
2. http://www.w3.org/TR/2003/CR-owl-guide-0030818/
3. Spink, A., Dietmar, W., Major, J., Tefko, S.: Searching the web: the public and their queries. J. Am. Soc. Inform. Sci. Technol. **52**(3), 226–234 (2001)
4. http://en.wikipedia.org/wiki/Tamil_language
5. Burger, J., Cardie, C., Chaudhri, V., et al.: Tasks and program structures to roadmap research in question & answering (Q&A). NIST Technical Report (2001)
6. Hirschman, L., Gaizauskas, R.: Natural language question answering: the view from here. Nat. Lang. Eng. Special Issue on Question Answering **7**(4), 275–300 (2001)
7. Copestake, A., Jones, K.S.: Natural language interfaces to databases. Knowl. Eng. Rev. **5**(4), 225–249 (1990)
8. Popescu, A.-M., Etzioni, O., Kautz, H.: Towards a theory of natural language interfaces to databases. In: Proceedings of the 8th International conference on Intelligent user interfaces, 12–15 January 2003, Miami, Florida, USA (2003)
9. Jerrold Kaplan, S.: Designing a portable natural language database query system. ACM Trans. Database Syst. (TODS) **9**(1), 1–19 (1984)
10. Androutsopoulos, I., Ritchie, G.D., Thanisch, P.: MASQUE/SQL: an efficient and portable natural language query interface for relational databases. In: Proceedings of the 6th International Conference on Industrial and Engineering Applications of Artificial Intelligence and Expert Systems, 1–4 June 1993, Edinburgh, Scotland, pp. 327–330 (1993)
11. Thompson, C.W., Pazandak, P., Tennant, H.R.: Talk to your semantic web. IEEE Internet Comput. **9**(6), 75–78 (2005)
12. McGuinness, D.L.: Question answering on the semantic web. IEEE Intell. Syst. **19**(1), 82–85 (2004)
13. http://www.w3.org/TR/rdf-sparql-query/
14. Kaufmann, E., Bernstein, A.: How useful are natural language interfaces to the semantic web for casual end-users? In: Aberer, K., et al. (eds.) ASWC 2007 and ISWC 2007. LNCS, vol. 4825, pp. 281–294. Springer, Heidelberg (2007)
15. Bernstein, A., Kaufmann, E., Kaiser, C., Kiefer, C.: Ginseng: a guided input natural language search engine for querying ontologies. In: 2006 Jena User Conference, Bristol, UK, May 2006
16. Damljanovic, D., Agatonovic, M., Cunningham, H.: Natural language interfaces to ontologies: combining syntactic analysis and ontology-based lookup through the user

interaction. In: Aroyo, L., Antoniou, G., Hyvönen, E., ten Teije, A., Stuckenschmidt, H., Cabral, L., Tudorache, T. (eds.) ESWC 2010, Part I. LNCS, vol. 6088, pp. 106–120. Springer, Heidelberg (2010)

17. Kaufmann, E., Bernstein, A., Fischer, L.: NLP-Reduce: a naïve but domain independent natural language interface for querying ontologies. In: 4th ESWC, Innsbruck, Austria (2007)

18. Cimiano, P., Haase, P., Heizmann, J. Porting natural language interfaces between domains – an experimental user study with the orakel system. In Proceedings of the International Conference on Intelligent User Interfaces (2007)

19. Stojanovic, N.: On Analysing Query Ambiguity for Query Refinement: The Librarian Agent Approach. In: Song, I.-Y., Liddle, S.W., Ling, T.-W., Scheuermann, P. (eds.) ER 2003. LNCS, vol. 2813, pp. 490–505. Springer, Heidelberg (2003)

20. Kaufmann, E., Bernstein, A., Zumstein, R., Querix: a natural language interface to query ontologies based on clarification dialogs. In: 5th International Semantic Web Conference (ISWC 2006), Springer (2006)

21. Lopez, V., Uren, V., Pasin, M., Motta, E.: AquaLog: an ontology-driven question answering system for organizational semantic intranets. J. Web Semant. 5(2), 72–105 (2007)

22. Wang, C., Xiong, M., Zhou, Q., Yu, Y.: PANTO: a portable natural language interface to ontologies. In: Franconi, E., Kifer, M., May, W. (eds.) ESWC 2007. LNCS, vol. 4519, pp. 473–487. Springer, Heidelberg (2007)

23. Tablan, V., Damljanovic, D., Bontcheva, K.: A natural language query interface to structured information. In: Bechhofer, S., Hauswirth, M., Hoffmann, J., Koubarakis, M. (eds.) ESWC 2008. LNCS, vol. 5021, pp. 361–375. Springer, Heidelberg (2008)

24. Ramachandran, V.K., Krishnamurthi, I.: NLION: Natural Language Interface for querying ONtologies. In: Proceedings of 2nd ACM International Conference on Applied research in contemporary computing, Compute 2009, Bangalore, India, 9–10 January, (2009)

25. Bernstein, A., Kaufmann, E., Göhring, A., Kiefer, C.: Querying ontologies: a controlled english interface for end-users. In: Gil, Y., Motta, E., Benjamins, V., Musen, M.A. (eds.) ISWC 2005. LNCS, vol. 3729, pp. 112–126. Springer, Heidelberg (2005)

26. Porter, M.F.: An algorithm for suffix stripping. Program 14(3), 130–137 (1980)

27. Fellbaum, C. (ed.): WordNet: An Electronic Lexical Database. MIT Press, Cambridge (1998)

28. http://www.jena.sourceforge.net/ontology/index.html

Biomedical Applications

Federating Heterogeneous Biological Resources on the Web: A Case Study on TRP Channel Ontology Construction

Se-Jin Nam[1], Jinhyun Ahn[1], Jin-Muk Lim[1], Jae-Hong Eom[1,2(✉)],
Ju-Hong Jeon[3], and Hong-Gee Kim[1,2(✉)]

[1] Biomedical Knowledge Engineering Lab. (BIKE),
Seoul National University, Seoul, Korea
{jordse,jhahncs,bikeljm}@gmail.com
[2] Dental Research Institute, School of Dentistry,
Seoul National University, Seoul 110-749, Korea
[3] Department of Physiology, College of Medicine,
Seoul National University, Seoul 110-799, Korea
{jhjeon2,zpage,hgkim}@snu.ac.kr

Abstract. TRP (Transient receptor potential) channel is a biological component which could be of factors in severe diseases such as heart attack and cancer. In order for researchers to easily search for protein-protein interactions for mammalian TRP channel, TRIP Database was created. However, TRIP Database does not contain information about proteins in details, making researchers in turn visit other database services such as UniProt and PDB. In this paper, we propose a semantic TRP Ontology made from TRIP Database, allowing users to be given the collected contents from other relevant databases as well using a single request. As a practical scenario, we generate RDF triples from TRIP Database by referring to TRP Ontology designed to have links with UniProt. A federated way of collecting contents from two different services is proposed.

Keywords: Ontology · SPARQL · TRP channel

1 Introduction

In biomedical domain, ontology has been widely used to describe domain specific data and to share and integrate with the other data. Furthermore, technical improvements in semantic web technology make it feasible to semantically integrate various data distributed world-wide, especially through SPARQL endpoint providing machine accessible means to access local data with a standard ways. In this paper, we provide ontology and SPARQL endpoint for TRIP database which maintains protein-protein interactions for TRP channel, and then show how this SPARQL endpoint could be dynamically connected to the other biomedical data such as UniProt[1] and RCSB Protein Data Bank[2] (PDB).

---

[1] http://www.uniprot.org/

[2] http://www.rcsb.org/

W. Kim et al. (Eds.): JIST 2013, LNCS 8388, pp. 103–109, 2014
DOI: 10.1007/978-3-319-06826-8_9, © Springer International Publishing Switzerland 2014

2 TRIP Database

TRIP Database [1, 2] is a database that maintains protein-protein interactions for mammalian TRP channel. The protein-protein interaction (PPI) pairs were manually curated by human experts reading peer-reviewed papers. As of August 2013, 668 of PPIs are accumulated in the database with 410 cellular proteins and 28 TRP channels. That was from 385 peer-reviewed papers. TRIP Database maintains PPI information in terms of screening, validation, characterization and functional consequence regarding experiments that had identified the PPI. It provides a web interface that help users search for the information like conventional web search services. In addition, TRP Channels in the databases are tagged with corresponding IDs in external databases such as Entrez Gene, UniProt, KEGG. TRIP Database is under continuous updates to allow users to search for up-to-date TRP channel related information.

3 TRP Ontology

TRP ontology represents PPIs related to TRP channels and experiment methods used to identify other related PPIs. We build TRP ontology from TRP database. The reason why we transform the TRP Database into ontology is that we do not want to care about schemas of external database to which TRP Database could be linked. TRP Database is maintained in relational database which is a structured form that can be easily processed by a program. It means that one might want to implement an automatic tool to mine useful information from the database. If we were able to put several other databases into a single database, more useful information could be discovered by analyzing heterogeneous data at a time. However, in general, it is not easy to put different relational database into a single relational database.

To address the problem, we use OWL as our ontology language in this paper. If data is written in OWL/RDF, it would be automatically integrated with the other OWL/RDF resources, without taking into account schemas. Moreover, since OWL/RDF is designed for the web, one can deal with the different resources just like with a single resource.

3.1 TRP Ontology Classes

We model TRP channel as a class. Since TRP channels are grouped into a hierarchy, we create super classes to represent part-of relationships for TRP channels such as *TRP super family* and *TRP sub family*. Note that a property called *OBO_REL:part_of* in OBO Relation Ontology are used to represent the part-of relationship. In Fig. 1, we depict how TRP ontology is associated with Protein ontology (PRO ontology) and external databases. The *TRP super family* class is especially linked to *Protein* class in PRO ontology. Establishing the link makes sense because TRP channel is a kind of protein. By doing so, we can enrich TRIP ontology with contents in PRO ontology. *TRP channel* class is linked to external databases as well with other additional information about protein.

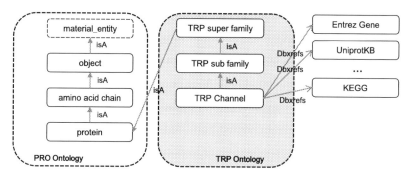

Fig. 1. TRP ontology classes

In order to represent knowledge acquired from experiments that have identified PPI, we create *identification, validation, characterization* and *consequences*. The *identification* class represents the way of identifying PPIs. Since screening is of an identification method, we put *screening* class as a child class of *identification* class. The *screening* class represents the experiment method and samples they used. We have two kinds of validation methods, say, *in vitro* and *in vivo*. We create *validation* class as a parent class of the classes, *in vitro* and *in vivo*. The class *analysisResult* is used for aggregating all of these information regarding experiments. The *documentSource* class represents source papers from which PPI information is extracted by human experts.

3.2   TRP Ontology Properties

The *interaction* class and *analysisResult* class are the main class, to which the other classes are associated. In Fig. 2, relationships between classes are depicted, drawing *interaction* and *anaysisResult* in the center. Since TRP channel and interactor are involved in a PPI interaction, two properties, *hasTRPChannel* and *hasInteractor*, are created to make an association within an *interaction* instance. Interaction is identified by an experiment, which leads us to make *hasTargetInteraction* whose domain is *analysisResult* class and range is *interaction* class. Since *analysisResult* is extracted from a research paper, *hasAnalysisResult* property links *documentSource* and *analysisResult*. The other classes are linked from *analysisResult* using properties show in Fig. 2.

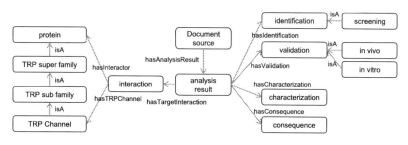

Fig. 2. A class diagram of TRP ontology

3.3   Example

In this section, we give an example data to help readers understand TRP ontology. Figure 3 is a screen-shot captured from TRP database web site, showing interactions between protein TRPC1 and I-MFA (i.e. two bottom lines). That has been extracted from a paper with PubMed ID 14530267. We can see that the interaction was identified by screening using "Yeast two-hybrid" method. That TRP channel is located in the region of 639-750 in DNA for mouse species. The interactor, I-MFA, had been taken out from a mouse embryo.

In Fig. 4, we depict instances of TRP ontology made from the data in Fig. 3 (i.e., for two bottom lines in Fig. 3). The *document source1* instance represents the PubMed ID of source paper. Since a research paper could introduce several PPIs, *document source 1* instance would be linked to multiple *analysisResult* instances, which is not shown here. The interaction instance has associations with I-MFA and TRPC1. TRPC1 is in particular linked to 7220 ID in Entrez Gene database and P48995 in UniProtKB. Since it is identified screening method, we create *screening* instance and make an association with *analysisResult* instance. In the *screening* instance, we assign appropriate data.

			Screening ✱						
			Experimental screening					Non-experimental screening	Reference
			TRP channel construct		Interactor source				
TRP channel	Interactor	Method	Species	Region	Species	Organ/tissue	Sample type		
TRPC1	↔ D2R	Yeast two-hybrid	Not used as a bait		Human	Fetal brain	cDNA library		18261457
TRPC1	↔ BMPR-2	Affinity purification-mass spectrometry	Not used as a bait		Mouse	Leg muscle	C2C12 lysates		15186402
TRPC1	↔ I-MFA	Yeast two-hybrid	Human	639-750	Mouse	Embryonic	cDNA library		14530267
TRPC1	↔ I-MFA	Yeast two-hybrid	Human	639-750	Human	Fetal kidney	cDNA library		14530267

Fig. 3. An example of TRIP database screening table

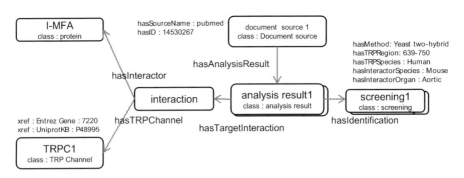

Fig. 4. TRP ontology based representation of example screening table in Fig. 3 (In this figure, ontologies only for the protein TRPC1 and I-MFA are shown as an example diagram)

4 TRP Ontology In-Use

In this section, we will show an application that makes use of TRP ontology. Implementing TRP Ontology in OWL allows other relevant resources on the web to be automatically integrated with TRP Ontology as long as valid URIs are matched. Acquiring knowledge from ontology is achieved by using SPARQL. SPARQL is a query language for RDF triples like SQL for relational database [3]. Recently, in SPARQL 1.1, federated query constructs were introduced so that one can specify in a single query multiple target endpoints even resided in remote servers. A SPARQL engine would collect data from multiple remote servers and then aggregate it.

Suppose that TRP Ontology with instances is stored in `sparql.trip.org` server which is a SPARQL endpoint. In addition, UniProt's SPARQL endpoint is http://beta.sparql.UniProt.org which is an actual public server. In order to make a query to both endpoints, we can specify two sub queries like Fig. 5.

```
prefix fn:    <http://www.w3.org/2005/xpath-functions#>
prefix trp:   <http://bike.snu.ac.kr/trp#>

SELECT ?trp ?UniProtKB ?p ?o
{
  {
    SELECT  DISTINCT ?trp ?UniProtKB
    {
      SERVICE <http://sparql.trip.org/>
      {
        ?iteraction trp:hasInteractor trp:S100A10 .
        ?interaction trp:hasTRPChannel ?trp .
        ?trp trp:hasUniProtKBID ?UniProtKB .
      }
    }
    LIMIT 1
  }
  {
    SELECT *
    {
      LET(?protein := URI(fn:concat("http://purl.UniProt.org/UniProt/", ?UniProtKB)))

      SERVICE <http://beta.sparql.UniProt.org/>
      {
        ?protein ?p ?o
      }
    }
    LIMIT 4
  }
}
```

Fig. 5. An example federated SPARQL query for TRP ontology and UniProt.

The SPARQL query is intended to acquire detailed information about a TRP Channel which had been identified being interacted with a protein S100A10 given by user. From TRP ontology, we can search for the TRP Channels. However, no detailed information about the TRP Channel would exist in TRP ontology. Fortunately, in TRP Ontology, there exists corresponding UniProt ID for the found TRP Channel. Using that ID we can further search for detailed information in UniProt database. In a SPAQL engine, the query is performed in the following two steps:

First Query. It sends the first query to the TRP Ontology server. It requests for TRP Channels interacted with protein interactor S100A10. Corresponding UniProt ID of the found TRP Channel is then looked up. Since we want to get only one TRP Channel, we restrict the number results to be one. The UniProt ID is stored in the variable called ? UniProtKB.

Second Query. The retrieved data in variables in the first query, ?UniProtKB, is passed to the variable with the same name. Note that UniProt ID stored originally in TRP Ontology only has postfix string, e.g. "Q13563," we have to append a namespace to make it a fully qualified URI that can be interpreted by the UniProt server. Now there is no string manipulation construct in SPARQL 1.1 specification. We make use of Jena ARQ engine [4], a SPARQL engine with some built-in functions. The fn:concat function takes a list of strings as input and returns a string by concatenating the input strings. The fully qualified string is converted into an actual URI string through the URI function. Using LET function, the URI is assigned as ?protein variable. In responding to the query with the URI, the UniProt server will return RDF triples with subject matched with the URI.

In Table 1, result data from the query is shown. The first two columns are from TRP ontology and the last two columns are from UniProt database. We can know that the given protein S100A10 is interacted with TRPP1 and PKD2_HUMAN is mnemonic of TRPP1 and so on.

Table 1. A real data acquired from TRP ontology and UniProt database.

trp	UniProtKB	P	O
trp:TRPP1	"Q13563"	<http://www.w3.org/1999/02/22-rdf-syntax-ns#type>	<http://purl.UniProt.org/core/Protein>
trp:TRPP1	"Q13563"	<http://purl.UniProt.org/core/mnemonic>	"PKD2_HUMAN"
trp:TRPP1	"Q13563"	<http://purl.UniProt.org/core/alternativeName>	<http://purl.UniProt.org/SHA-...400447B>
trp:TRPP1	"Q13563"	<http://purl.UniProt.org/core/alternativeName>	<http://purl.UniProt.org/SHA-...A5B5E90>

5 Conclusion

In this paper, we defined TRP Ontology made out of TRP Database schema. We transformed contents in TRP Database into instances in TRP Ontology, in order to make it be linked to external resources on the web to enrich knowledge. An application is shown to prove usefulness, where we propose a federated SPARQL query to collect contents from both TRP Ontology and UniProt, which are even stored in different servers.

In the future work, we will extend TRP Ontology to be linked with natural language resources such as PubMed. Based on the linked resources, one can make an automatic tool to add more knowledge regarding TRP channel.

Acknowledgement(s). This work was supported by the Industrial Strategic Technology Development Program (10044494, WiseKB: Big data based self-evolving knowledge base and reasoning platform) funded by the Ministry of Science, ICT & Future Planning (MSIP, Korea).

References

1. Shin, Y.-C., Shin, S.-Y., So, I., Kwon, D., Jeon, J.-H.: TRIP Database: a manually curated database of protein–protein interactions for mammalian TRP channels. Nucleic Acids Re. **39**(Suppl. 1), D356–D361 (2011)
2. Shin, Y.-C., Shin, S.-Y., Chun, J.N., Cho, H.S., Lim, J.M., Kim, H.-G., So, I., Kwon, D., Jeon, J.-H.: TRIP Database 2.0: a manually curated information hub for accessing TRP channel interaction network. PloS ONE **7**(10), e47165 (2012)
3. SPARQL 1.1, http://www.w3.org/TR/sparql11-query/
4. Jena ARQ, http://jena.apache.org/documentation/query/
5. UniProt, http://www.UniProt.org/
6. TRIP Database, http://trpchannel.org/
7. OBO Relation Ontology, http://obofoundry.org/ro/

Publishing a Disease Ontologies
as Linked Data

Kouji Kozaki[1(✉)], Yuki Yamagata[1], Takeshi Imai[2], Kazuhiko Ohe[3],
and Riichiro Mizoguchi[4]

[1] The Institute of Scientific and Industrial Research, Osaka University,
8-1 Mihogaoka, Ibaraki, Osaka 567-0047, Japan
{kozaki,yamagata}@ei.sanken.osaka-u.ac.jp
[2] Department of Planning Information and Management,
The University of Tokyo Hospital, 7-3-1 Hongo, Bunkyo-ku,
Tokyo 113-0033, Japan
ken@hcc.h.u-tokyo.ac.jp
[3] Department of Medical Informatics Graduate School of Medicine,
The University of Tokyo, 7-3-1 Hongo, Bunkyo-ku, Tokyo 113-0033, Japan
ohe@hcc.h.u-tokyo.ac.jp
[4] Japan Advanced Institute of Science and Technology,
1-1 Asahidai, Nomi, Ishikawa 923-1292, Japan
mizo@jaist.ac.jp

Abstract. Publishing open data as linked data is a significant trend in not only
the Semantic Web community but also other domains such as life science,
government, media, geographic research and publication. One feature of linked
data is the instance-centric approach, which assumes that considerable linked
instances can result in valuable knowledge. In the context of linked data,
ontologies offer a common vocabulary and schema for RDF graphs. However,
from an ontological engineering viewpoint, some ontologies offer systematized
knowledge, developed under close cooperation between domain experts and
ontology engineers. Such ontologies could be a valuable knowledge base for
advanced information systems. Although ontologies in RDF formats using
OWL or RDF(S) can be published as linked data, it is not always convenient to
use other applications because of the complicated graph structures. Consequently, this paper discusses RDF data models for publishing ontologies as
linked data. As a case study, we focus on a disease ontology in which diseases
are defined as causal chains.

Keywords: Ontology · Linked data · Disease ontology · Ontological
engineering

1 Introduction

Linked data has been adopted by numerous data providers as best practices for
publishing and connecting structured data on the Web. It is based on Semantic Web
technologies, such as RDF and SPARQL, and contributes to the publication of data as
part of a single global data space on the Web. Currently, many datasets are published

W. Kim et al. (Eds.): JIST 2013, LNCS 8388, pp. 110–128, 2014.
DOI: 10.1007/978-3-319-06826-8_10, © Springer International Publishing Switzerland 2014

as linked data in not only the Semantic Web community but also many other domains such as life science, government, media, geographic research and publication [1]. Linked data is based on the idea that considerable data are required to realize the Semantic Web. Therefore, it focuses on publishing more instances (data) instead of constructing more ontologies, which provide common vocabulary and schema to describe data. In other words, linked data is an instance-centric approach that aims to produce valuable knowledge by connecting considerable instances.

However, from an ontological engineering viewpoint, ontologies could be valuable knowledge bases for advanced information systems when developed as systematized knowledge under close cooperation between domain experts and ontology engineers. This paper discusses publishing such ontologies as linked data.

Because the standard format for linked data is RDF, it may be regarded an easy task to publish ontologies in RDF formats using OWL or RDF(S) as linked data. However, an ontology language such as OWL is designed for mainly class descriptions, and the assumption is that the language will be used for reasoning based on logic; yet finding and tracing connections between instances are main tasks in linked data. Therefore, OWL/RDF and RDF(S) are not always convenient and efficient for linked data because of their complicated graph structures. This is problematic, especially when we want to use an ontology's conceptual structures as a knowledge base with rich semantics.

Consequently, this paper discusses RDF data models for publishing ontologies as linked data. As a case study, we use a disease ontology built as part of a clinical medical ontology [2] with support from Japan's Ministry of Health, Labour and Welfare. In the disease ontology, a disease is defined as a causal chain of clinical disorders, and the ontology is based on a computational model called the *River Flow Model of Disease* [3]. That is, the ontology includes rich information about causal chains related to each disease, which represent how the diseases appear and advance in a human body. Publishing such an ontology as linked data can contribute to the development of advanced medical information systems. At the same time, considerations about how to publish the ontology as linked data are informative.

The rest of this paper is organized as follows. The next section provides an overview of the current state of ontologies in the context of linked data. Section 3 summarizes the disease ontology with its ontological definitions. Section 4 discusses how to publish the disease ontology as linked data and presents a browsing system for the published linked data, showing its effectiveness. Section 5 discusses the result of the case study for publishing ontologies as linked data with some related works. Finally, we present concluding remarks and discuss future work.

2   Publishing Ontologies as Linked Data

2.1   Uses of Ontologies in Linked Data

Gruber defines "an ontology" as "an explicit specification of a conceptualization" in the knowledge systems area [4]. In the Semantic Web, an ontology should provide vocabulary and schemas to represent metadata. Another important role of an ontology is to give data semantics. These roles remain unchanged, although some people think that publishing and linking instances (data) are more important than defining the ontologies in the context of linked data. It is true that considerable data linked with other data are valuable, even if it is not based on an ontology, but the common vocabulary and schema provided by ontologies are also very important in linking various data efficiently and consistently.

In fact, many datasets in the Linking Open Data Cloud have an ontology[1]. For example, DBpedia[2] English, famous for its linked data, describes 3.77 million things, 2.35 million of which are classified in a consistent ontology[3] that includes definitions of about 500 classes and 1,900 properties. This is a typical example of how the ontology plays a role in providing vocabulary and schema for linked data. Many ontologies are published for this purpose in life science domains. On the NCBO BioPortal [5], which is a portal site for publishing ontologies on bio informatics, 355 ontologies are available[4]. They are linked to each other, resulting in a significant amount of linked data on the Web.

However, some ontologies can be used as knowledge infrastructures that provide meanings of concepts for semantic processing. Such ontologies are classified as domain-independent general ontologies such as Cyc[5] or WordNet[6] or as domain-specific ontologies such as SNOMED-CT[7] or Galen[8]. They are considered linked data because they are connected to other linked data. Also, they tend to have several class definitions compared to the ontologies for vocabulary and schema discussed above. For example, Japanese Wikipedia Ontology (JWO)[9] was developed on the basis of the Japanese Wikipedia to provide general knowledge in Japanese [6], and includes 160,000 class definitions. In contrast, DBPedia, also based on Wikipedia, has only 500 classes. This difference is caused by a difference in their uses related to the linked data. That is, using an ontology itself as a structured knowledge infrastructure requires defining a sufficient amount of class definitions with rich semantics, while instances are not necessarily needed.

2.2   Ontology Representation for Linked Data

In general, RDF(S) and OWL are used as ontology representation languages in the Semantic Web. Recently, SKOS has also been used to publish knowledge organization systems such as thesauri, taxonomies, and classification schemes on the Web. Since

---

[1] When we searched for "ontology" as a keyword on the data catalog, 43 datasets were found at http://datahub.io/group/lodcloud. Also, 402 datasets were found at http://datahub.io/ and 365 of them have "lod" tag. These searches were conducted on August 17th, 2013.

[2] http://dbpedia.org/

[3] http://mappings.dbpedia.org/server/ontology/

[4] http://bioportal.bioontology.org/. The number was checked on August 17th, 2013.

[5] http://www.cyc.com/

[6] https://wordnet.princeton.edu/

[7] http://www.ihtsdo.org/snomed-ct/

[8] http://www.opengalen.org/

[9] http://www.wikipediaontology.org/query/

they are based on RDF, we can use them as representation languages to publish ontologies as linked data. Selecting a language that is good for publishing linked data depends on how to use the published ontology.

Because RDF(S) is a simple schema language, it is good for representing ontologies that do not have many complicated class definitions. However, when we want to describe more detailed class definitions to represent a conceptual structure (e.g., stating which properties a class must have), we have to use OWL. In particular, ontologies are used for reasoning (e.g., classifications of instances) by using semantics defined in these ontologies, most of which are represented in OWL DL. In OWL version 2, we can flexibly choose which specifications of OWL2 are used in our ontology. That is, OWL is very good for publishing ontologies that provide a detailed schema or semantic reasoning for linked data.

On the other hand, OWL ontologies represented as an RDF graph include many blank nodes when some OWL properties such as owl:restriction are used. Therefore we need complicated SPARQL queries to get class definitions in such ontologies. That is, we can say that OWL is not always good for representing ontologies published as linked data when the purpose is to share class definitions on the Web. Following approaches have been proposed for avoiding this problem. One is to avoid using OWL properties, which cause blank nodes. For example, the ontology in DBPedia is represented in OWL, but does not use owl:restriction. Another approach is to use another representation language designed for providing knowledge infrastructures on the Web. SKOS is designed as a common data model for sharing and linking knowledge organization systems via the Web. If the main purpose of publishing ontologies is to provide a common vocabulary with some stricter knowledge, such as thesauri, taxonomies, and classification schemes, then SKOS is a very good candidate to be used as the ontology representation language.

However, SKOS only supports primitive properties when representing classifications of concepts with mappings to other datasets. Therefore, to represent more detailed conceptual structures in ontologies, we have to design other representation languages. Of course, we have to avoid reinventing the wheel. However, it is a reasonable approach to design an RDF model for publishing an ontology as linked data, considering the efficiency and convenience this would provide.

3   A Case Study of a Disease Ontology

These days, medical information systems store considerable data. Semantic technologies are expected to contribute to the effective use of such systems, and many medical ontologies such as OGMS [7], DOID [8], and IDO [9] have been developed for realizing sophisticated medical information systems. They mainly focus on the ontological definition of a disease with related properties. On the other hand, we propose a definition of a disease that captures it as a causal chain of clinical disorders and a computational model called the River Flow Model of Disease [3]. Our disease ontology consists of rich information about causal chains related to each disease. The causal chains provide domain-specific knowledge about diseases, answering questions such as "What disorder causes a disease?" and "How might the disease advance, and what symptoms may appear?" This section provides an overview of the disease ontology.

3.1 Definitions

A typical disease, as a dependent continuant, enacts extending, branching, and fading processes before it disappears. As a result of these processes, a disease can be identified as a continuant that is an enactor of those processes. Such an entity (a disease) can change according to its phase while maintaining its identity. On the basis of this observation, we defined a disease and related concepts as follows [3].

Definition 1: A disease is a dependent continuant constituted of one or more causal chains of clinical disorders appearing in a human body and initiated by at least one disorder.

Note that, although all diseases have dynamic flows of causality propagation as its internal processes, it is also the enactor of its external processes, such as branching and extending its causal chain of disorders. When we collect individual causal chains belonging to a particular disease type (class), we can find a common causal chain (partial chain) that appears in all instance chains. By generalizing such a partial chain, we obtain the notion of a core causal chain of a disease as follows.

Definition 2: A core causal chain of a disease is a sub-chain of the causal chain of a disease, whose instances are included in all the individual chains of all instances of a particular disease type. It corresponds to the essential property of a disease type.

Definition 2 provides a necessary and sufficient condition for determining the disease type to which a given causal chain of clinical disorders belongs. That is, when an individual causal chain of clinical disorders includes instances of the core causal chain of a particular disease type, it belongs to that disease type. We can thus define such a disease type, which includes all possible variations of physical chains of clinical disorders observed for patients who contract the disease. According to a standard definition of subsumption, we can introduce an *is-a* relationship between diseases using the chain-inclusion relationship between causal chains.

Definition 3: *Is-a* **relationship between diseases.** Disease A is a supertype of disease B if the core causal chain of disease A is included in that of disease B. The inclusion of nodes (clinical disorders) is judged by taking an *is-a* relationship between the nodes, as well as sameness of the nodes, into account.

Definition 3 helps us systematically capture the necessary and sufficient conditions of a particular disease, which roughly corresponds to the so-called "main pathological conditions." Assume, for example, that (non-latent) diabetes and type-I diabetes are, respectively, defined as *<deficiency of insulin → elevated level of glucose in the blood>* and *<destruction of pancreatic beta cells → lack of insulin I in the blood → deficiency of insulin → elevated level of glucose in the blood>*. Then, we get *<type-I diabetes is-a (non-latent) diabetes>* according to Definition 3.

3.2   Considering the Types of Causal Chains

As discussed in the previous section, we define a disease as causal chains of clinical disorder. In this paper, we call causal chains that appear in the disease definition *disease chains*. In theory, we can consider three types of causal chains that appear in the disease definition, which are called *disease chains*, when we define a disease:

General Disease Chains are all possible causal chains of (abnormal) states in a human body. They are referred to by all disease definitions.

Core Causal Chain of a disease is a causal chain that is shown in all patients of the disease (see Definition 2).

Derived Causal Chains of a disease are causal chains obtained by tracing general disease chains upstream or downstream from the core causal chain. The upstream chains imply possible cause of the disease, and the downstream ones imply possible symptoms in a patient suffering from the disease.

Figure 1 shows the main types of diabetes constituted by the corresponding types of causal chains. The figure shows that subtypes of diabetes are defined by extending its core causal chain according to its derived causal chains upstream or downstream.

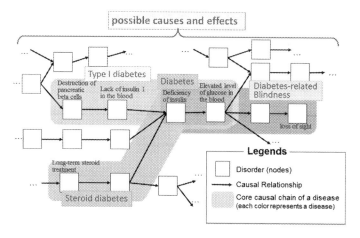

Fig. 1. Types of diabetes constituted of causal chains.

However, it is obviously difficult to define all general causal chains in advance, because it is impossible to know all possible states in the human body and the causal relationships among them. To avoid this problem, we take a bottom-up approach to building the disease ontology by focusing on the definition of core/derived causal chains. The core causal chain of a disease can be obtained to observe many instances of the disease. Similarly, typical parts of derived causal chains of the disease can be obtained through their observations. That is, these causal chains can be defined on the basis of the existing medical knowledge of clinicians. Therefore, we first define these two types of causal chains by clinicians, and then define the general disease chains by generalizing them. In practice, a core causal chain of a disease is easier to define than its derived causal chains, because its range seems to be limited to some extent, as a result of the restriction that it must appear in all patients of the disease. On the other hand, clinicians can define various derived causal chains, depending on their knowledge. Therefore, we ask clinicians to define only core causal chains and typical derived causal chains of each disease, according to their knowledge and the textbooks on the disease. Despite these limitations on the range for definitions, the defined causal chains overlap each other after many disease concepts are defined in several areas. As a result, we can get general disease chains that cover all diseases defined in the ontology. They do not cover all possible causal chains in human bodies, but represent all causal chains in which clinicians are interested when they consider all defined diseases. Therefore, we can get not only derived causal chains, defined by the clinician directly, but also causal chains, derived by tracing the general disease chains through all clinical areas. Note here that each clinician defined disease concepts in his or her special field without knowing how other diseases were defined in other fields by others. After they finished defining disease concepts, we collected all causal relationships from all disease concepts defined in the 13 special fields. As of 11 May 2013, the disease ontology has about 6,302 disease concepts and about 21,669 disorder (abnormal state) concepts with causal relationships among them.

3.3   Implementing the Disease Ontology

We developed the disease ontology using Hozo. Although Hozo is based on an ontological theory of roles and has its original ontology representation model, we show an OWL representation of the ontology to aid reader understandability[10]. Figure 2 shows an OWL representation of *angina pectoris*, whose causal chain is shown in Fig. 3. Abnormal states that appear in the disease are listed using *owl:Restriction* properties on *hasCoreState/hasDerivesState* properties[11]. The former represents abnormal states in its core causal chain, and the later represents the ones in its derived causal chain defined by a clinician. Causal relationships among them are represented by *hasCause/hasResult* properties[12]. Because causal chains (states and causal relationships among them) in core causal chains are necessary (Definition 3), *owl:someValuesFrom* properties are used. On the other hand, because causal chains in derived causal chains are possible, *owl:allValuesFrom* properties are used to represent possible causes/results[13]. Definitions of diseases refer to definitions of abnormal

---

[10] Note that we used a simplified OWL representation of the disease ontology to show its overview while it does not support full semantics of Hozo. The detailed semantics of Hozo are discussed in [10].

[11] Note that property names such as *hasCause* and *hasCoreState* represent *owl:Restriction* on them in Figs. 3 and 4.

[12] If probability of the causal relationship is high, *hasProbableCause/hasProbableResult* properties are used instead. We do not discuss how its probability is decided since it is beyond the scope of this paper.

[13] If there are more than two possible causes/results, *owl:unionOf* is used to list them.

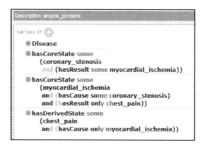

Fig. 2. Class definition of *angina pectoris* in OWL

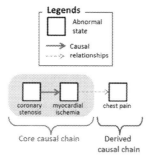

Fig. 3. Causal chain of *angina pectoris.*

Fig. 4. Class definition of *myocardial_ischemia* in OWL

states, which represent possible causes and results, as shown in Fig. 4. General disease chains are represented as an aggregation of definitions of abnormal states.

4   Publishing the Disease Ontology as Linked Open Data

4.1   Basic Policy

As discussed in Sect. 2.2, we can publish the disease ontology in the OWL format as linked data. However, it is not convenient and efficient to use the ontology for linked data because it has rich class definitions compared to other ontologies for providing data schema. For example, when we obtain a general disease chain, which is probably caused by *myocardial_ischemia*, we have to repeat SPARQL queries to obtain RDF graphs, which include blank nodes such as those shown in Fig. 5. Furthermore, when

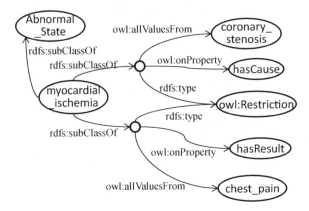

Fig. 5. RDF graph of a general disease chain in OWL.

we obtain definitions of a disease, we have to make more complicated queries to obtain graphs which correspond to OWL descriptions such as those shown in Fig. 3, with restrictions inherited from the disease's super classes. These queries and patterns of graphs are intuitively very different from the disease chains we want to produce. They are not good for publishing the disease ontology as linked data because they are not easy to use through SPRAQL queries.

If we want to use the ontology to provide schema for describing instances of diseases, we should publish it in OWL. However, our goal in publishing it as linked data is to share definition of diseases and disease chains as structured domain knowledge for medical information systems. We propose semantic processing using the ontology to obtain three types of disease chains, as discussed in Sect. 3.3; however, the ontology does not need reasoning based on OWL semantics if we use simple procedural reasoning.

Given these considerations, we designed an RDF model to publish the disease ontology as linked data, because we thought a simple RDF model was more efficient than a complicated OWL model for representing the ontology in RDF. In this case study, we focus on publishing causal chains, which consists of definitions of disease as a primary trial, because these chains are core of the disease ontology.

4.2   RDF Model for Causal Chains of a Disease

According to the basic policy discussed in the previous section, we extracted information about causal chains of diseases from the disease ontology and concerted them into RDF formats as a linked data. We call the dataset the Disease Chain LOD. It represents definitions of diseases based on causal chains among abnormal states as instances (individuals), while these definitions are classes in the original ontology.

The Disease Chain LOD *consists* of diseases, abnormal states, and the relationships among them. Abnormal states are represented by instances of *Abnormal_State* type. Causal relationships between them are represented by describing *hasCause* and *hasResult* properties as bidirectional links. Abnormal states connected by these properties are a possible cause/result. Therefore, general disease chains can be obtained by collecting all abnormal states according to these connections.

Diseases are represented by instances of *Disease* type. Abnormal states that constitute a core causal chain and a derived causal chain of a disease are represented by *hasCoreState* and *hasDereivedState* properties, respectively. Is-a (sub-class-of) relationships between diseases and abnormal states are represented by *subDiseaseOf/subStateOf* properties instead of *rdfs:subClassOf* because diseases and abnormal states are represented as instances, while *rdfs:subClassOf* is a property between classes.

Figure 6 shows an example of RDF representation of diseases. It represents *disease A* and its sub-disease *disease B*, whose causal chains are shown in Fig. 7. Note that causal chains consist of abnormal states and causal relationships between them. Therefore, when we obtain a disease's core causal chain or derived causal chain, we have to obtain not only abnormal states connected to the disease by *hasCoreState/hasDereivedState* properties but also causal relationships between them. Although causal relationships are described without determining whether they are included in the causal chains of certain diseases, we can identify the difference by whether abnormal states at both ends of *hasCause/hasResult* properties are connected to the same disease by *hasCoreState/hasDereivedState* properties. In the case of disease A, shown in Fig. 6, although abnormal state 4 is described as a result of abnormal state 2, abnormal state 4 is not included in the core causal chain/derived causal chain because it does not have *hasCoreState/hasDerivedState* properties with disease A.

Furthermore, when we obtain the causal chain of a disease that has a super disease, such as disease B in Fig. 5, we have to obtain causal chains of its super disease in addition to the causal chain directly connected with it, and aggregate them. The processing is not complicated; it just requires simple procedural reasoning.

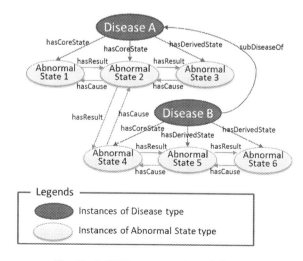

Fig. 6. A RDF representation of diseases.

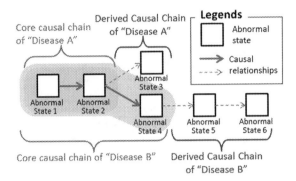

Fig. 7. Causal chains of diseases shown in Fig. 6.

In summary, we can obtain a disease's causal chains, which define the disease through the following queries to the Disease Chain LOD:

(1) Obtain all super diseases of the selected disease using repetitive queries to identify the super disease connected with a disease through the *subDiseaseOf* property.

(2) Obtain abnormal states, which constitute each disease identified in step 1, using queries to obtain abnormal states connected with the disease by *hasCoreState/ hasDerivedState* properties.

(3) Identify *hasCause/hasResult* properties for each abnormal state obtained in step 2. If both ends of the properties are included in the abnormal states, then the properties should be regarded as causal relationships that constitute the selected disease.

As discussed above, the proposed RDF model can represent causal chains of diseases more intuitively than OWL representation. Therefore, the users can obtain necessary information about disease chains through simple SPARQL queries.

4.3   Publication of Disease Chain LOD and Its Trial Use

4.3.1   Overview of Dataset and Systems

We published the disease ontology as linked data based on our RDF model. It includes definitions of 2,103 diseases and 13,910 abnormal states in six major clinical areas[14] extracted from the disease ontology on May 11, 2013. The dataset contained 71,573 triples. Then, we developed two prototype systems. One is a SPARQL endpoint with a supporting function to help beginners input queries. The other is a visualization system for disease chains, called the Disease Chain LOD Viewer.

The SPARQL endpoint provides the following functions to support users, even if they do not have much knowledge about SPARQL (Fig. 8):

---

[14] Although the disease ontology includes definitions diseases in 13 clinical areas, we published parts of them that were well reviewed by clinicians. The rest of them will be published after reviews.

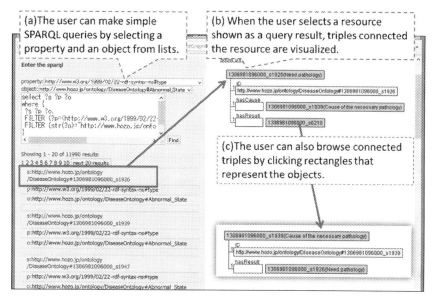

Fig. 8. The SPARQL endpoint with supporting functions.

Query making support: The user can make simple SPARQL queries by selecting a property and an object from lists (Fig. 8(a)). For example, by selecting *rdfs:type* in the property lists and *Disease (type)* in object list, the user can obtain a list of diseases defined in the dataset.

Visualization of triples: When the user selects a resource shown as a query result, triples connected the resource are visualized as shown in Fig. 8(b). The user can also browse connected triples by clicking rectangles that represent the objects (Fig. 8(c)). For example, the user can browse general disease chains by selecting objects with *hasCause/hasResult* properties.

Links to the Disease Chain LOD Viewer: When the user browses triples connected with a disease, the system shows a link to visualize the disease's causal chains in the Disease Chain LOD Viewer.

The Disease Chain LOD Viewer can visualize diseases' causal chains in a user friendly representation, as shown in Fig. 9. The user can select the target disease by the following two methods:

Performing a keyword search for the disease name or abnormal state name, which is included in the definitions of diseases and

Browsing the is-a hierarchy of diseases by selecting lists of diseases, as shown in Fig. 9(a).

The system is designed as a web service that supports not only PCs but also tablets or smartphones. It is implemented using AllegroGraph 4.10 for its RDF database and HTML 5 for visualizations of causal chains.

When a user selects a disease, the system makes SPARQL queries according to the procedures discussed in Sect. 4.2, and it obtains the necessary information about the

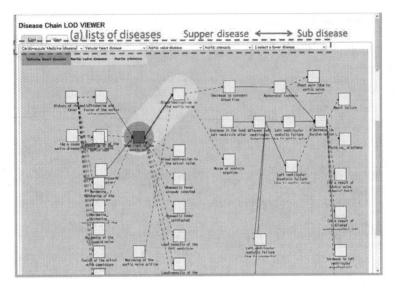

Fig. 9. Disease Chain LOD Viewer.

disease definition from the results. Then, the obtained disease definition is visualized on the fly. That is, if the system can visualize disease definitions correctly, it can obtain the necessary information using SPARQL queries. In fact, the average time it takes to visualize a disease definition is around 2–3 s, depending on the computer environment, such as the client type and network speed. This performance should be sufficient for practical use.

4.3.2   Example Queries

In this section, we show typical example queries to the Disease Chain LOD. Figure 10 shows example queries to get abnormal state in the dataset. Because all abnormal states are defined as individual resources *of Abnormal_State* type, we can get them by

```
(a1) Get all abnormal states.
select ?abn
where {?abn rdf:type dont:Abnormal_State}
(a2) Get all cause of abnormal state <abn_id>.
select ?o
where {<abn_id> dont:hasCause* ?o }
(a3) Get general disease chain which includes ab-
      normal state <abn_id>.
select ?o
where {{<abn_id> dont:hasCause* ?o}
UNION{<abn_id> dont:hasPosibleCause* ?o}
```

Fig. 10. Example queries to get abnormal states. In this figure "dont:" represents a prefix of the Disease Chain LOD and <abn_id> represents id of a selected abnormal state.

the query shown in (a1). When we want to get causes/result of a selected abnormal state, we can follow *hasCasue/hasResult* properties. For example, (a2) is a query to get all causes of the selected abnormal state. Furthermore, we can get a general disease chain which includes the abnormal state by the query shown in (a3). This query means to follow all that *hasCasue/hasResult* properties recursively from the selected abnormal state.

On the other hand, Fig. 11 shows example queries to get definitions of diseases. We can get all diseases in the dataset by query (d1) which is almost the same with (a1). When we want to get all super disease (super class) of a selected disease, we can use the query show in (d2). In order to get core causal chains or derived causal chains of a selected disease, we can use query (d3) or (d4) respectively. Combining query (d2), (d3) and (d4), we can get all causal chains which appear in definitions of disease by the query shown in (d5). Furthermore, when we want to get not a list of abnormal state but a list of causal relationships which appear in causal chain of definition of the diseases, we can use the query shown in (d6). This query finds all properties among all abnormal states which appear in the definition of the selected disease.

```
(d1) Get all disease.
select ?dis
where {?dis rdf:type dont:Disease}
(d2) Get all super diseases of disease <dis_id>.
select ?o
where {<dis_id> dont:subDiseaseOf* ?o }
(d3) Get core causal chains of disease <dis_id>.
select ?o
where {<dis_id> dont: hasCoreState ?o}
(d4) Get core derived chains of disease <dis_id>.
select ?o
where {<dis_id> dont: hasDerivedState ?o}
(d5) Get all causal chains which appear in defini-
     tions of disease <dis_id> as a list of abnormal
     state.
select ?o
where { <dis_id> dont:subDiseaseOf* ?dis .
 {?dis dont:hasCoreState ?o }
 UNION {?dis dont:hasDerivedState ?o }}
(d6) Get all causal chains which appear in defini-
     tions of disease <dis_id> as list of causal rela-
     tionships.
select DISTINCT ?abn1 ?p ?abn2
where {<dis_id> dont:subDiseaseOf* ?dis .
 ?abn1 ?p ?abn2 .
 {?dis dont:hasCoreState ?abn1}
 UNION {?dis dont:hasDerivedState ?abn1}
 {?dis dont:hasCoreState ?abn2}
 UNION {?dis dont:hasDerivedState ?abn2}}
```

Fig. 11. Example queries to get definitions of diseases. In this figure "dont:" represents a prefix of the Disease Chain LOD and <dis_id> represents id of a selected disease.

We believe all queries shown the above are easy to understand and intuitive for many people. This is a big advantage of our RDF model in compare to the case we use our original disease ontology in OWL.

5   Related Works

As discussed in Sect. 2.1, many ontologies are published as linked data. Most of them use OWL, RDF(S) or SKOS for representation language because these languages are widely used as standard language of semantic web. On the other hand, some of them design original representation language (RDF model) considering to their usability as linked data. We took the later approach, especially, considering designing our RDF model so that it can represent detailed conceptual structure of class definition in our ontology while existing approaches mainly aim to represent simple schema to represent term definitions in their ontologies. For example, Assem discusses RDF representation of WordNet [11, 12] and published it as linked data[15]. A similar work is discussed in [13]. They suppose mainly to represent synset which is a set of synonyms for a word while they do not consider gathering information through many numbers of definitions such like our disease ontology. Nuzzolese and et.al. also discuss patterns to publish lexical resource as Linked Data [14]. They present a method to convert the 1.5 XML version of FrameNet into RDF and OWL. A main contribution of this paper is that we show a good case study for publishing an ontology which includes rich class definitions as linked data with comparison of existing approaches. As the result, we published the disease ontology as linked data and showed that it works well in practice through developments of prototype system to use it.

The feature of our approach is that we design an original RDF model to represent class definitions in our disease ontology without using OWL. This is why we aim to support simple SPARQL queries to get disease chains. There are some portal site to publish ontology on Web such as BioPortal [5] and Spatial Decision Support[16] [15]. They support OWL ontology with ontology visualization tool. They visualize ontologies not as simple RDF graph by as some user-friendly graph formant. They also support ontology search and browsing using the visualization tool. Although the Disease Chain LOD Viewer is designed based on the same approach with them, we suppose other systems access the LOD through SPARQL endpoint. That is, ontology visualization is not enough to use the ontology as a LOD.

At the same time, the disease ontology published as linked data is also valuable in compare to existing ones. The main feature of the disease ontology is the definition on a disease as causal chain based on ontological consideration of causal chains. It can capture characteristics of diseases appropriately, and it is also very friendly to clinicians since it is similar to their understanding of disease in practice. Moreover, it includes richer information about causal relationships in disease than other disease ontologies or medical terminologies such as SNOMED-CT. Disease Chain LOD got

---

[15] http://semanticweb.cs.vu.nl/lod/wn30/

[16] http://www.spatial.redlands.edu/sds/

"Life science prize" in Linked Open Data Challenge Japan 2012 because it received recognition as linked data which includes valuable dataset based on medical knowledge.

6 Discussions

6.1 Generalization of the Proposed Approach

We designed a RDF model for publishing the disease ontology as linked data instead of use OWL. As discussed in Sect. 4.1, it is mainly because we considered that efficiency and convenience as linked data is more important than strict representation of class definition of ontologies. The proposed RDF model for the disease ontology can represent class definitions of disease enough to obtain necessary information to visualizing disease chains discussed in Sect. 3 while it omits some axioms such as *owl:someValuesFrom* and *owl:cardinality* restrictions. This suggests that we can apply the same approach to other ontologies if we allow omitting such axioms for publishing them as linked data. That is, the proposed approach could be generalized into a method to publish an ontology as linked data in an ontology representation which allows the users to make queries in intuitive way as linked data. We have a plan to publish other parts of the medical ontology developed in our research project as additional case studies. For example, it includes class definitions which represent connections among all major blood vessels in a human body. We think it would be very valuable to publish as linked data in a purpose to provide an infrastructure medical knowledge.

Furthermore, we suppose that we can generalize our approach as a pattern of RDF model for publishing ontologies as linked data. Its basic idea consists of two convert methods; (1) To omit blank nodes for representing owl:restriction and convert them to triples (e.g. hasCause in our case study) and (2) To convert rdfs:subClassOf to property between individuals (e.g. subDiseaseOf in the case study). These methods can be applied into other ontologies easily.

6.2 Queries to Obtain Class Definitions of an Ontology

We discussed how we can obtain necessary information about a disease's causal chains through SPARQL queries to the disease ontology published as linked data in Sect. 4.3. Then, we showed that it works in enough performance in practical use while we have not compared it with performance in a case we represent the ontology in OWL. If we represent causal relationships between abnormal states in OWL as shown in Fig. 5, it needs three times triples for a *hasCause/hasResult* property. We can say the same thing about class definitions. It would obviously cause complicated queries to obtain class definitions. However, it is an important future work to make a quantitative evaluation in order to compare performances for queries to linked data based on the proposed RDF model and OWL.

On the other hand, we can consider another approach for publishing an ontology as linked data. It is to design query patterns to obtain class definition in OWL. For example, DL Query is designed for the user to search concepts whose definitions match a search condition described by the user in Manchester OWL Syntax [16]. The authors also are considering this approach focusing on concept search [17]. If the user can use such query patterns and it work in practical performance, it would be a good method to use ontology in OWL as linked data. Note here that, even if it works in good performance, one advantage of the proposed approach in this paper is that the user can obtain necessary information through combination of simple SPARQL queries in intuitive ways.

6.3   Remaining Issues

The prosed method fully ignores to use the original ontology as schema because it focuses on intuitive understanding and explore ontologies. Therefore, in order to make instances using ontologies transformed based on the proposed method, we have to use not semantics provided by RDFS or OWL but a special mechanism for the ontologies. Although it is an ad hoc approach, we suppose it is not difficult to prepare for an instance management mechanism using the transformed LOD at least in the case of our disease ontology. It is because we already have confirmed that we can get necessary semantics of the ontology through simple SPAQRL queries. Even so, there are room to consider that we use some major properties such as rdfs:SubClsssOf in the proposed method so that the user easy to understand the transformed LOD.

Another approach is to provide both of original ontologies and LOD transformed from them. Then, the users to choose them according to their purpose of use them. However, it might be somehow confusing that there are two formats for an ontology.

7   Concluding Remarks

This paper discussed how to design RDF data models for publishing ontologies as linked data. As a case study, we focused on a disease ontology in which diseases are defined as causal chains and considered an RDF data model for the disease ontology, offering the convenience of LOD. Then, we developed the resulting disease-chain LOD with two prototype systems for its trail use. As the result, we showed that we can obtain the necessary information of the disease ontology using SPARQL queries to the resulting LOD in sufficient performance for practical use. We think our experience is very informative when you consider publishing ontologies as linked data.

Future work includes applications of the proposed approach to other ontologies and generalization of them, quantitative evaluation to compare alternative approaches. We plan to consider publishing other parts of the clinical medical ontology we are developing. Developments of more practical applications using the Disease Chain LOD are also important future works. For example, we have a plan to use it with some other linked data in bioinformatics domain.

Disease Chain LOD is available at the URL http://lodc.med-ontology.jp/.

Acknowledgement. A part of this research is supported by the Japan Society for the Promotion of Science (JSPS) through its "FIRST Program" and the Ministry of Health, Labour and Welfare, Japan. The authors are deeply grateful to medical doctors (Natsuko Ohtomo, Aki Hayashi, Takayoshi Matsumura, Ryota Sakurai, Satomi Terada, Kayo Waki, et.al.), of The University of Tokyo Hospital for describing disease ontology and providing us with broad clinical knowledge. The authors also would like to thank Hiroko Kou for describing the primary version of disease ontology, Nobutake Kato for implementation the proposed systems and Enago (www.enago.jp) for the English language review.

References

1. Bizer, C., Heath, T., Berners-Lee, T.: Linked Data - The Story So Far. Int. J. Semant. Web Inf. Syst. Spec. Issue Linked Data **5**(3), 1–22 (2009)
2. Mizoguchi, R., Kou, H., Zhou, J., Kozaki, K., Imai, T., Ohe, K.: An advanced clinical ontology. In: Proceedings of ICBO, pp. 119–122. Buffalo, NY (2009)
3. Mizoguchi, R., Kozakil, K., et al.: River flow model of diseases. In: Proceedings of ICBO2011, pp. 63–70. Buffalo, USA (2011)
4. Gruber, T.R.: A translation approach to portable ontologies. Knowl. Acquis. **5**(2), 199–220 (1993)
5. Whetzel, P.L., Noy, N.F., Shah, N.H., Alexander, P.R., Nyulas, C., Tudorache, T., Musen, M.A.: BioPortal: enhanced functionality via new Web services from the National Center for Biomedical Ontology to access and use ontologies in software applications, Nucleic Acids Res. 39 (Web Server issue):W541-5 (2011)
6. Tamagawa, S., Morita, T., Yamaguchi, T.: Extracting property semantics from Japanese wikipedia. In: Huang, R., Ghorbani, A.A., Pasi, G., Yamaguchi, T., Yen, N.Y., Jin, B. (eds.) AMT 2012. LNCS, vol. 7669, pp. 357–368. Springer, Heidelberg (2012)
7. Scheuermann, R.H., Ceusters, W., Smith, B.: Toward an ontological treatment of disease and diagnosis. In: Proceedings of the 2009 AMIA Summit on Translational Bioinformatics, pp. 116–120, San Francisco, CA (2009)
8. Osborne, J.D., et al.: Annotating the human genome with Disease Ontology. BMC Genomics **10**(1), S6 (2009)
9. Cowell, L.G., Smith, B.: Infectious disease ontology. In: Sintchenko V. (ed.) Infectious Disease Informatics, ch. 19, pp. 373–395. Springer, New York (2010)
10. Kozaki, K., Sunagawa, E., Kitamura, Y., Mizoguchi, R.: Role representation model using OWL and SWRL. In: Proceedings of 2nd Workshop on Roles and Relationships in Object Oriented Programming, Multiagent Systems, and Ontologies, pp. 39–46. Berlin (2007)
11. van Assem, M., Gangemi, A., Schreiber, G. (eds.): RDF/OWL Representation of WordNet, W3C Working Draft 19 June 2006, http://www.w3.org/TR/2006/WD-wordnet-rdf-20060619/
12. van Assem, M., Gangemi, A., Schreiber, G.: Conversion of WordNet to a standard RDF/OWL representation. In: Proceedings of LREC (2006)
13. Koide, S., Morita, T., Yamaguchi, T., Muljadi, H., Takeda, H.: RDF/OWL Representation of WordNet 2.1 and Japanese EDR Electric Dictionary. In: 5th International Semantic Web Conference (ISWC2006), Poster (2006)
14. Nuzzolese, A.G., Gangemi, A., Presutti, V.: Gathering lexical linked data and knowledge patterns from FrameNet. In: Proceedings of the 6th International Conference on Knowledge Capture (K-CAP), pp. 41–48. Ban, Alberta, Canada (2011)

15. Li, N., Raskin, R., Goodchild, M., Janowicz, K.: An ontology-driven framework and web portal for spatial decision support. Trans. GIS **16**(3), 313–329 (2012)
16. Koutsomitropoulos, D.A., Borillo Domenech, R., Solomou, G.D.: A structured semantic query interface for reasoning-based search and retrieval. In: Antoniou, G., Grobelnik, M., Simperl, E., Parsia, B., Plexousakis, D., De Leenheer, P., Pan, J. (eds.) ESWC 2011, Part I. LNCS, vol. 6643, pp. 17–31. Springer, Heidelberg (2011)
17. Kozaki, K., Kitagawa, Y.: Multistep expansion based concept search for intelligent exploration of ontologies. In: Proceedings of IESD 2013. http://imash.leeds.ac.uk/event/2013/program.html (2011)

Adapting Gloss Vector Semantic Relatedness Measure for Semantic Similarity Estimation: An Evaluation in the Biomedical Domain

Ahmad Pesaranghader[1](✉), Azadeh Rezaei[1], and Ali Pesaranghader[2]

[1] Multimedia University (MMU), Jalan Multimedia,
63100 Cyberjaya, Malaysia
{ahmad.pgh,azadeh.rezaei}@sfmd.ir
[2] Universiti Putra Malaysia (UPM), 43400 UPM Serdang, Selangor, Malaysia
ali.pgh@sfmd.ir

Abstract. Automatic methods of ontology alignment are essential for estab-lishing interoperability across web services. These methods are needed to measure semantic similarity between two ontologies' entities to discover reli-able correspondences. While existing similarity measures suffer from some difficulties, semantic relatedness measures tend to yield better results; even though they are not completely appropriate for the 'equivalence' relationship (e.g. "*blood*" and "*bleeding*" related but not similar). We attempt to adapt Gloss Vector relatedness measure for similarity estimation. Generally, Gloss Vector uses angles between entities' gloss vectors for relatedness calculation. After employing Pearson's chi-squared test for statistical elimination of insignificant features to optimize entities' gloss vectors, by considering concepts' taxonomy, we enrich them for better similarity measurement. Discussed measures get evaluated in the biomedical domain using MeSH, MEDLINE and dataset of 301 concept pairs. We conclude Adapted Gloss Vector similarity results are more correlated with human judgment of similarity compared to other measures.

Keywords: Semantic web · Similarity measure · Relatedness measure · UMLS · MEDLINE · MeSH · Bioinformatics · Pearson's Chi-squared test · Text mining · Ontology alignment · Computational linguistics

1 Introduction

Considering human ability for understanding different levels of similarity and relat-edness among concepts or documents, similarity and relatedness measures attempt to estimate these levels computationally. Developing intelligent algorithms, which automatically imitate this human acquisition of cognitive knowledge, is the chal-lenging task for many computational linguistics studies. Generally, the output of a similarity or relatedness measure is a value indicating how much two given concepts (or documents) are semantically similar (or related). Ontology matching and ontology alignment systems primarily employ this quantification of similarity and relatedness [1] to establish interoperability among different ontologies. These measures have also wide usage in machine translation [2], information retrieval from the Web [3].

W. Kim et al. (Eds.): JIST 2013, LNCS 8388, pp. 129–145, 2014.
DOI: 10.1007/978-3-319-06826-8_11, © Springer International Publishing Switzerland 2014

Ontologies, as representations of shared conceptualization for variety of domains [4], by aiming at organizing data for the purposes of 'understanding', are the vital part of the Semantic Web. Considering importance of ontologies for web services, to improve and facilitate interoperability across underlying ontologies, we need an automatic mechanism which aligns or mediates between vocabularies of them. Ontology alignment or ontology matching is the process for driving correspondences among entities (concepts) within two different aligning ontologies.

Formally, a correspondence with unique identifier of *id* is a five-tuple (1) by which relation R (denoting specialization or subsumption, $\leq$; exclusion or disjointness, $\perp$; instantiation or membership, $\in$; and assignment or equivalence, $=$) with the confidence level of n between entities (concepts) e_1 and e_2 of two different ontologies holds. To find this level of confidence, majority of ontology alignment systems rely on similarity and relatedness measures for the following decision of acknowledgment regarding to the examined relation.

$$\langle id, e_1, e_2, n, R \rangle \tag{1}$$

Statistically, among a variety of proposed graph-based, corpus-based, statistical, and string-based similarity (or distance) and relatedness measures, those measures which tend to consider the semantic aspects of the concepts, are proven to be more reliable [1]. For this main reason, these measures are separately known as semantic similarity or relatedness measures which are the topics of interest in this paper.

The remainder of the paper is organized as follows. We enumerate existing measures of semantic similarity and semantic relatedness in Sect. 2 with explanation of concomitant problems for each one. In Sect. 3, we list the dataset and resources employed for our experiments in the biomedical domain. In Sect. 4, we represent our measure as a solution to overcome the shortcomings of semantic similarity measures. In brief, our proposed similarity measure is optimized version of Gloss Vector semantic relatedness measure adapted for an improved semantic similarity measurement. Regarding to the experiments in the given domain, in Sect. 5, we evaluate and discuss the calculated results of similarity for different semantic similarity measures by employing a reference standard of 301 concept pairs. The reference standard is already scored by medical residents based on concept pairs' similarity. Finally, we state the conclusions and the future studies in Sect. 6.

In the following section we will discuss about two main computational classes which semantic similarity and relatedness methods get mostly divided into.

2 Distributional vs. Taxonomy-Based Similarity Measures

Different proposed approaches for measuring semantic similarity and semantic relatedness get largely categorized under two classes of computational techniques. These two classes are known for their distributional versus taxonomy-based attitudes for calculation of their estimations. By applying a technique belonging to any of these classes, it would be possible to construct a collection of similar or related terms automatically. This would be achievable by either using information extracted from a large corpus in the distributional methods or considering concepts' positional information exploited from an ontology (or taxonomy) in the taxonomy-based approaches. In the following subsections, we will explain each of these classes and functions of their kind by details along with the challenges coming in any case. In Sect. 4, we will present our method for overcoming these difficulties.

2.1   Distributional Model

The notion behind methods of this kind comes from Firth idea (1957) [5] indicating "a word is characterized by the company it keeps". In practical terms, such methods learn the meanings of words by examining the contexts of their co-occurrences in a corpus. While these co-occurred features (of words or documents) get represented in a vector space the distributional model is also recognized as the vector space model (VSM). The values of a vector belonging to this vector space get calculated from applied function in the model. The final intention of this function is to measure the relatedness of term-term, document-document or term-document based on their corresponding vectors. The result would be achieved through computing the cosine of the angle between two input concepts' vectors.

Definition-based Relatedness Measures - A number of studies try to augment the concept's vectors by taking the definition of the input term into consideration. In this cases, a concept will be indicative of one sense for the input term and its definition can be accessible from an external resource (or resources) such as a dictionary or thesaurus. Be aware that in these methods the full positional and relational information for a concept from the thesaurus is still unexploited and thesaurus here performs just as an advanced dictionary.

WordNet[1] is the known thesaurus which often gets used in non-specific contexts for this type of relatedness measures. In WordNet, terms are characterized by a synonym sets called synsets while each has its own associate definition or gloss. Synsets are connected to each other through semantic relations such as hypernym, hyponym, meronym and holonym. The Unified Medical Language System[2] (UMLS) is a mega-thesaurus specifically designed for medical and biomedical purposes. The scientific definitions of the medical concepts can be derived accordingly from the resources (thesauruses) included in the UMLS.

Lesk [6] calculate the strength of relatedness between a pair of concepts as a function of the overlap between their definitions by considering their constituent words. Banerjee and Pedersen [7] for an improved result proposed the Extended Gloss Overlap measure (also known as Adapted Lesk) which augment concepts definitions with the definitions of senses that are directly related to it in WordNet. The main drawback in these measures is their strict reliance on concepts definitions and negligence of other knowledge source.

---

[1]   http://wordnet.princeton.edu

[2]   http://www.nlm.nih.gov/research/umls

To address forging limitation in the Lesk-type measures, Patwardhan and Pedersen [8] introduced the Gloss Vector measure by joining together both ideas of concepts' definitions from a thesaurus and co-occurrence data from a corpus. In their approach, every word in the definition of the concept from WordNet is replaced by its context vector from the co-occurrence data from a corpus and the relatedness get calculated as the cosine of the angle between the two input concepts associated vectors (gloss vectors). This Gloss Vector measure is highly valuable as it: (1) employs empirical knowledge implicit in a corpus of data, (2) avoids the direct matching problem, and (3) has no need for an underlying structure. Therefore, in another study, Liu et al. [9] extended the Gloss Vector measure and applied it as Second Order Context Vector measure to the bio-medical domain. The UMLS was used for driving concepts definitions and biomedical corpuses were employed for co-occurrence data extraction. In brief, this method gets completed by five steps which sequentially are, (1) constructing co-occurrence matrix from a biomedical corpus, (2) removing insignificant words by low and high frequency cut-off, (3) developing concepts extended definitions using UMLS, (4) constructing definition matrix using the results of step 2 and 3, and (5) estimating semantic relatedness for a concept pair. For a list of concept pairs already scored by biomedical experts, they evaluated Second Order Context Vector measure through Spearman's rank correlation coefficient. Figure 1 illustrates the entire procedure.

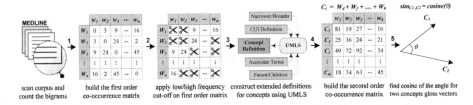

Fig. 1. 5 steps of Gloss Vector relatedness measure

Even though Gloss Vector semantic relatedness measure has proven its application in wide variety of intelligent tasks, and so various studies try to improve it [10, 11], these measures' low/high frequency cut-off phase is one of their week points as they do not take frequency of individual terms into account and merely consider frequency of bigrams for the cut-off. This notion is problematic because, first, we believe when a concept is specific (or concrete) in its meaning it needs more specific and informative terms in its definition describing that concept. Second, it is believed when one term is more specific in the meaning has less frequency in the corpus compared to those terms which are general (or abstract). Therefore, removing terms (features) only based on the frequency of bigrams can lead to elimination of many descriptive terms of fea-tures. In Sect. 4, in the first stage of our proposed measure for an enhanced measure of semantic similarity built on the Gloss Vector relatedness measure, we introduce Pearson's chi-squared test (X^2) as a measure of association to eliminate insignificant features (on which one concept is not dependent) in a statistical fashion.

Distributional Model Challenges - At times relatedness and similarity measures can be used interchangeably; however, the results may show inconsistency for several involved pairs. This issue is more considerable in ontology alignment and database schema matching as finding the relation of equivalences (or assignment) for examined concepts is at highest priority. As a realistic example, result of semantic correspondence for entities "*blood*" and "*hemorrhage*" represented high degree of relatedness when low level of similarity between these terms was really intended. This drawback is more intense for negations comparison ("*biodegradable*" vs. "*non-biodegradable*"). For this reason, we need to look for a semantic measure which specifically takes similarity of two compared entities (concepts) into account.

Semantic similarity measures typically are dependent on a hierarchical structure of concepts derived from taxonomies or thesauruses. A concept, which is equal to entity in ontology language, is a specific sense of a word (polysemy). For example, "*ventricle*" is representative word for two different concepts of "*a chamber of the heart that pumps blood to the arteries*" and "*one of the four cavities of the brain*". One concept can also have different representative words (synonymy). For instance, words "*egg*" and "*ovum*" both can imply "*the reproductive cell of the female, which contains genetic information and participates in the act of fertilization*" concept. Normally, all representative terms of a concept in a taxonomy (or an ontology) comprise a set with an assigned unique identifier for that concept (sense). In the following subsection we categorize existing semantic similarity measures into three groups, those which are taxonomy-based, generic and domain-independent for a variety of applications.

2.2   Taxonomy-Based Model

The main factor showing the unique characteristic of measures belonging to this class is their capability for estimating semantic similarity (rather than relatedness). The functions of this kind, as the model's name implies, rely on positional information of concepts derived from a taxonomy or ontology. Depending on different degrees of information extraction from taxonomy and option of employing a corpus for enriching it, the methods for similarity measurement get categorized into three main groups.

Path-based Similarity Measures - For these methods distance between concepts in the thesaurus is generally important. It means the only issue for similarity measurement of input concept pair is the shortest number of jumps from one concept to the other one. The semantic similarity measures based on this approach are:

- Rada et al. 1989 [12]

$$\text{sim}_{\text{path}}(c_1, c_2) = \frac{1}{\text{shortest is} - \text{a path}(c_1, c_2)}$$ (2)

In path-based measures we count nodes (not paths)

- Caviedes and Cimino 2004 [13]

It is sound to distinguish higher concepts in the thesaurus hierarchy, which are abstract in their meaning and so less likely to be similar, from those lower concepts, which are more concrete in their meanings (e.g. *physical_entity/matter* two high level concepts vs. *skull/cranium* two low level concepts). This is the main challenge in path-based methods as they do not consider this growth in specificity when we go down from the root towards leaves.

Path-based and Depth-based Similarity Measures - These methods are proposed to overcome the abovementioned drawback of path-based methods. They are generally based on both path and depth of concepts. These measures are:

- Wu and Palmer 1994 [14]

$$\text{sim}_{\text{wup}}(c_1, c_2) = \frac{2 \times \text{depth}(\text{LCS}(c_1, c_2))}{\text{depth}(c_1) + \text{depth}(c_2)} \tag{3}$$

LCS is the least common subsumer (lowest common ancestor) of the two concepts

- Leacock and Chodorow 1998 [15]

$$\text{sim}_{\text{lch}}(c_1, c_2) = -\log\left(\frac{\text{min path}(c_1, c_2)}{2D}\right) \tag{4}$$

minpath is shortest path, and D is the total depth of taxonomy,

- Zhong et al. 2002 [16]

$$\text{sim}_{\text{zhong}}(c_1, c_2) = \frac{2 \times \text{m}(\text{LCS}(c_1, c_2))}{\text{m}(c_1) + \text{m}(c_2)} \tag{5}$$

$m(c)$ is computable by $m(c) = 1/k^{(\text{depth}(c)+1)}$,

- Nguyen and Al-Mubaid 2006 [17]

$$\text{sim}_{\text{nam}}(c_1, c_2) = \log(2 + (\text{min path}(c_1, c_2) - 1) \times (D - d)) \tag{6}$$

D is the total depth of the taxonomy, and $d = \text{depth}(LCS(c_1, c_2))$,

From the information retrieval (IR) viewpoint when one concept is less frequent than the other concepts (especially its siblings) we should consider it more informative as well as concrete in meaning than the other ones. It is a notable challenge in path-based and depth-based similarity functions. For example when two siblings get compared with another concept in these measures, they end up with the same similarity estimations while it is highly possible one of the siblings (more frequent or the general one) has greater similarity to the compared concept.

Path-based and Information Content Based Similarity Measures - As mentioned, depth shows specificity but not frequency, meaning low frequent concepts often are much more informative so concrete than high frequent ones. This feature for one concept is known as information content (IC) [18]. IC value for a concept in the taxonomy is defined as the negative log of the probability of that concept in a corpus.

$$IC(c) = -\log(P(c)) \tag{7}$$

P(c) is probability of the concept c in the corpus

$$P(c) = \frac{tf + if}{N} \tag{8}$$

tf is term frequency of concept c, *if* is inherited frequency or frequency of concept's descendants summed together, and N is sum of all augmented concepts' (concept plus its descendants) frequencies in the ontology.

Different methods based on information contents of concepts are proposed:

• Resnik 1995 [18]

$$sim_{res}(c_1, c_2) = IC(LCS(c_1, c_2)) \tag{9}$$

• Jiang and Conrath 1997 [19]

$$sim_{jcn}(c_1, c_2) = \frac{1}{IC(c_1) + IC(c_2) - 2 \times IC(LCS(c_1, c_2))} \tag{10}$$

• Lin 1998 [20]

$$sim_{lin}(c_1, c_2) = \frac{2 \times IC(LCS(c_1, c_2))}{IC(c_1) + IC(c_2)} \tag{11}$$

Typically path-based and IC-based Similarity measures must yield better result comparing with path-based and path and depth-based methods. Nonetheless, in order to produce accurate IC measures to achieve a reliable result from IC-based similarities many challenges are involved.

The first challenge is that IC-based similarities are highly reliable on an external corpus, so any weakness or imperfection in the corpus can affect the result of these similarity measures. This issue is more intense when we work on a specific domain of knowledge (e.g. healthcare and medicine or disaster management response) when accordingly a corpus related to the domain of interest is needed. Even when the corpus is meticulously chosen, absence of corresponding terms in the corpus for the concepts in the thesaurus is highly likely. This would hinder to compute IC for many concepts (when the term frequencies of their descendants are also unknown) and consequently this will avoid semantic similarity calculation for those concepts.

The second challenge is associated with the ambiguity of terms (terms with multiple senses) in their context during the parsing phase. As we really need concept frequency rather than term frequency for IC computation, this type of ambiguity for numerous encountered ambiguous terms in the corpus will lead to miscounting and consequently noisiness and impurity of produces ICs for concepts.

The third problem arises for concepts having equivalent terms that are multi-word rather than just one word (e.g. *semantic web*). In this case the parser should be so smart to distinguish these terms (multi-word terms) in corpus from times when they appear alone; otherwise the frequency count for this type of concepts will be also unavailable. As a result, there will be no chance for IC measurement and similarity estimation for the concept.

The last setback regarding to the usage of IC-based measure is extremely associated with those specific domains of knowledge which are in the state of rapid evolution and new concepts and ideas get added to their vocabularies and ontologies every day. Unfortunately, the speed of growth for documentation regarding to these new topics is not the same as their evolution speed. Therefore, it will be another hindrance for calculating required ICs for these new concepts.

Pesaranghader and Muthaiyah in their similarity measure [21] try to deal with these drawbacks to some extent, however, these issues in semantic similarity measures cause measures of semantic relatedness like Gloss Vector outperform them in the application of similarity estimation, even though as we previously stated relatedness measures are not perfectly tuned for this kind of task. In this research, by considering high reliability, performance and applicability of Gloss Vector semantic relatedness measure in the task under study, after optimizing it using Pearson's Chi-squared test (X^2) in the first stage of the proposed methods, in the second stage, by taking a hierarchical taxonomy of concepts into account, we attempt to adapt this measure for even more accurate estimation of semantic similarity (instead of relatedness).

Our approach for computation of semantic similarity using Gloss Vector idea is fully discussed in Sect. 4. Through experiments of our study on the specific domain of biomedical, in Sect. 5, the result of our similarity estimation gets compared with other similarity measures. For this purpose, we have applied these measures on the Medical Subject Headings (MeSH) from the Unified Medical Language System (UMLS) by using MEDLINE as the biomedical corpus for computation of optimized and enriched gloss vectors. The produced similarity results get evaluated against a subset of concept pairs as a reference standard which is previously judged by medical residents. Following section will explain each of abovementioned resources employed in our experiments.

3 Experimental Data

Some of external resources available for the study act as taxonomy; other resources are a corpus to extract required information fed to the semantic similarity measures, and a dataset used for testing and evaluation. While MEDLINE Abstract is employed as the corpus, the UMLS are utilized for construction of concepts definitions used in our proposed Adapted Optimized Gloss Vector semantic similarity measure.

3.1   Unified Medical Language System (UMLS)

The Unified Medical Language System[3] (UMLS) is a knowledge representation framework designed to support biomedical and clinical research. Its fundamental usage is provision of a database of biomedical terminologies for encoding information contained in electronic medical records and medical decision support. It comprises over 160 terminologies and classification systems. The UMLS contains more than 2.8 million concepts and 11.2 million unique concept names. The Metathesaurus, Semantic Network and SPECIALIST Lexicon are three components of the UMLS.

Basically this research focuses on the Metathesaurus and Semantic Network since for calculation of all semantic similarity measures examined in the study we need to have access to the biomedical concepts resided in the UMLS Metathesaurus and know relations among these concepts through Semantic Network. In the UMLS there are 133 semantic types and 54 semantic relationships. The typical hierarchical relations are parent/child (PAR/CHD) and broader/narrower (RB/RN). Medical Subject Headings (MeSH) is one of terminologies (sources) contained in the UMLS. While all the concept pairs tested in this study are from MeSH, the accessibility to MeSH on the UMLS is requisite for our experiments. Basically MeSH is a comprehensive controlled vocabulary which aims at indexing journal articles and books in the life sciences; additionally, it can serve as a thesaurus that facilitates searching. In this research we have limited the scope to 2012AB release of the UMLS.

3.2   MEDLINE Abstract

MEDLINE[4] contains over 20 million biomedical articles from 1966 to the present. The database covers journal articles from almost every field of biomedicine including medicine, nursing, pharmacy, dentistry, veterinary medicine, and healthcare. For the current study we used MEDLINE article abstracts as the corpus to build a first order term-term co-occurrence matrix for later computation of second order co-occurrence matrix. We used the 2013 MEDLINE abstract.

3.3   Reference Standard

The reference standard[5] used in our experiments was based upon a set of medical pairs of terms created specifically for testing automated measures of semantic similarity, freely provided by University of Minnesota Medical School as an experimental study [22]. In their study the pairs of terms were compiled by first selecting all concepts from the UMLS with one of three semantic types: disorders, symptoms and drugs. Subsequently, only concepts with entry terms containing at least one single-word term were further selected for potential differences in similarity and relatedness responses.

---

[3] http://www.nlm.nih.gov/research/umls

[4] http://mbr.nlm.nih.gov/index.shtml

[5] http://rxinformatics.umn.edu/data/UMNSRS_similarity.csv

Eight medical residents (2 women and 6 men) at the University of Minnesota Medical School were invited to participate in this study for a modest monetary compensation. They were presented with 724 medical pairs of terms on a touch sensitive computer screen and were asked to indicate the degree of similarity between terms on a continuous scale by touching a touch sensitive bar at the bottom of the screen. The overall inter-rater agreement on this dataset was moderate (Intraclass Correlation Coefficient - 0.47); however, in order to reach a good agreement, after removing some concept pairs, a set of 566 UMLS concept pairs manually rated for semantic similarity using a continuous response scale was provided.

A number of concepts from original reference standard are not included into the MeSH ontology. More importantly, some of them (and their descendants) are not found in the MEDLINE, which means no chance for information content calculation used in IC-based measures. Therefore, after removing these concepts from this dataset, a subset of 301 concept pairs for testing on different semantic similarity functions including new measure in this study was available.

4   Methods

As already mentioned, the main attempts of this paper are, first, to avoid concomitant difficulties from implementing the low and high frequency cut-off of bigrams in Gloss Vector relatedness measure, and second, to improve this effective relatedness measure in a way which is more appropriate for semantic similarity measurement. According to these two main objectives of the study, the method implemented to address forgoing statements consists of two very important stages which provide us in the final point with the Adapted Gloss Vector semantic similarity measure.

4.1   Stage 1 - Optimizing Gloss Vector Semantic Relatedness Measure

We use Pearson's chi-squared test (X^2) for test of independence. By independence we mean in a bigram two given words are not associated with each other and merely have accrued together by chance if the result of X^2 test for that bigram is lower than a certain degree. The procedure of the test for our case includes the following steps: (1) Calculate the chi-squared test-statistic, X^2, using (12), (2) Determine the degree of freedom, d_f, which is 1 in our case, (3) Compare X^2 to the critical value from the chi-squared distribution with d degree of freedom (again 1 in our case).

$$X^2 = \sum_{i=1}^{n} \frac{(O_i - E_i)^2}{E_i} \tag{12}$$

where X^2 is Pearson's cumulative test statistic, O_i is an observed frequency, E_i is an expected frequency, and n is the number of cells in the contingency table. In order to show what contingency table for a bigram means and how X^2 gets computed using it, we represent "*respiratory system*" as a bigram sample on which X^2 has been applied:

In Table 1, by considering an external corpus, $O_1 = 100$ represents that "*system*" has accrued after "*reparatory*" 100 times, accordingly $O_2 = 400$ denoted

Table 1. 2×2 Contingency Table for "respiratory system"

	system	!system	TOTAL
respiratory	$O_1 = 100$ $E_1 = 2$	$O_2 = 400$ $E_2 = 498$	500
!respiratory	$O_3 = 300$ $E_3 = 398$	$O_4 = 99'200$ $E_4 = 99'102$	99'500
TOTAL	400	99'600	100'000

"*respiratory*" has been seen in the corpus without "*system*" 400 times, so in total "*respiratory*" has the frequency of 500 in the corpus. The total number of bigrams collected from the corpus is 100'000. To calculate expected value for each cell we just need to multiply the total values of the column and row to which that cell belong together and then divide the result with the total number of bigrams. The expected value for O_3 is:

$$E_3 = (99'500*400)/100'000 = 398$$

The other expected values are calculable using the same way accordingly. Now, for calculating chi-squared test-statistic we need only put numbers into the (12).

X^2(respiratory-system)

$$= \sum (O_i - E_i)^2 \Big/ E_i = (O_1 - E_1)^2 \Big/ E_1 + (O_2 - E_2)^2 \Big/ E_2 + (O_3 - E_3)^2 \Big/ E_3 + (O_4 - E_4)^2 \Big/ E_4$$

$$= (100 - 2)^2 \Big/ 2 + (400 - 498)^2 \Big/ 498 + (300 - 398)^2 \Big/ 398 + (99'200 - 99'102)^2 \Big/ 99'102$$

$$= 4845.5$$

The calculated number of 4845.5 represents level of dependence (association) between "*respiratory*" and "*system*" which is rather high. This number is calculable for any bigram extracted from the corpus. Comparing these calculated chi-squared test statistic values to the critical value from the chi-squared distribution, by considering 1 as the degree of freedom, the appropriate level of cut-off for removing those bigrams which are really unwanted could be found.

We take advantage of Pearson's chi-squared test represented above for our proposed approach of insignificant features (terms) removal. In order to integrate this statistical association measure into the Gloss Vector measure procedure, in our approach for optimizing gloss vectors computed for each concept in Gloss Vector relatedness measure, we (1) ignore the low and high frequency cut-off step in Gloss Vector measure (2) construct normalized second order co-occurrence matrix using first order co-occurrence matrix, (3) build X2-on-SOC matrix by enforcing Pearson's chi-squared test on the normalized second order co-occurrence matrix to find relative association of concepts (rows of matrix) and terms (columns of matrix), and finally (4) apply level of association cut-off on X2-on-SOC matrix. Figure 2 illustrates the entire procedure of our proposed semantic relatedness measure using Pearson's chi-squared test (X^2).

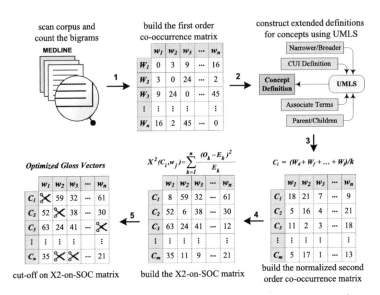

Fig. 2. 5 steps of Gloss Vector measure with optimization with X^2

There are three important points considered in this stage: (1) we count co-occurrences from corpus instead of bigrams, (2) we apply X^2 on second order co-occurrence matrix cells, in other words concept-term elements instead of bigrams (co-occurrences), (3) to find the appropriate point for cut-off of independent values (features), we gradually increase the critical point and examine correlation of calculated results of similarity using these different cut-off points to find the optimum threshold of cut-off. After this stage, we have Optimized Gloss Vector relatedness measure which still has some difficulties for similarity calculation. In the next stage, by taking a taxonomy of concepts into account, we adapt this optimized relatedness measure for the semantic similarity calculation.

4.2   Stage 2 - Adapted Gloss Vector Semantic Similarity Measure Using X^2

Here, by considering the concepts' taxonomy, we would construct enriched second order co-occurrence matrix used for final measurement of similarity between concepts. For this purpose, by using optimized X2-on-SOC matrix, we can calculate enriched gloss vectors of the concepts in a taxonomy. Our proposed formula to compute the enriched gloss vector for a concept is:

$$\text{Vector}(c_j) = \frac{ca_i(c_j) + ia_i(c_j)}{\sum_{k=1}^{m} (ca_i(c_k) + ia_i(c_k))} \quad \forall i <= n$$

$$\text{Vector} \in R^n, \text{ n is the quantity of features} \tag{13}$$

$$1 \leq j \leq m, \text{ m is the quantity of concepts}$$

Where $ca_i(c_j)$ (concept association) is the level of association between concept c_j and feature i (already computed by X^2 in the previous stage); $ia_i(c_j)$ (inherited association) is the level of association between concept c_j and feature i inherited from c_j's descendants in the taxonomy calculable by summation of all concept's descendants' levels of association with feature i; and finally, the denominator is summation of all augmented concepts (concepts plus their descendants) levels of association with feature i. All of these levels of association' values are retrievable from optimized X2-on-SOC matrix. The collection of enriched gloss vectors built enriched gloss matrix which will be used for final semantic similarity measurement.

The intuition behind this extra stage is that the higher concepts in a taxonomy (or an ontology) tend to be more general and so associated with lots of terms (features in their vectors). On the other hand, the lower concepts should be more specific and therefore associated with less number of features. Accordingly, the concepts in the middle have relative number of association with the features (terms). For this reason, by adding up the computed optimized second order vector of one concept with its descendants' optimized vectors we attempt to enforce the forgoing notion regarding to the abstractness/concreteness of the concepts; which by itself would implicitly imply the idea of path and depth for similarity measurement as well.

In order to measure a similarity between two concepts we only need enriched gloss matrix loaded on the memory. By having enriched gloss vectors of the two concepts available this way, the measurement of semantic similarity for them by calculating the cosine of the angle between these two vectors would be possible.

5   Experiments and Discussions

In the following subsections, first, in one place, we would compare the best results generated for all the semantic similarity measures discussed in this paper. Then, by considering our Adapted Gloss Vector semantic similarity measure, we focus on the Optimized X2-on-SOC matrix to illustrate the influence of the level of association cut-off using statistical approach of Pearson's chi-squared test (X^2) for more accuracy of the final similarity estimation. For the evaluation, Spearman's correlation coefficient to assess the relationship between the reference standards and the auto-generated semantic similarity results is employed.

5.1   Adapted Gloss Vector Similarity Measure vs. Other Similarity Measures

Figure 3 represents the result of the Spearman's rank correlation coefficients for 301 concept pairs against biomedical expert similarity score of them for nine similarity measures proposed in other studies and our Adapted Gloss Vector semantic similarity in this paper. To see the reliability of Gloss Vector semantic relatedness the different form of it are represented. The test is conducted on MeSH ontology from the UMLS. In both IC-based and Gloss Vector-based measures MEDLINE's single-words and bigrams for calculation of the ICs and vectors are extracted. In the Gloss Vector-based measures window size of 2 for construction of first order co-occurrence matrix after removal of stop words and porter stemming implementation is considered. In these cases we have used the finest cut-off points in removal of insignificant features.

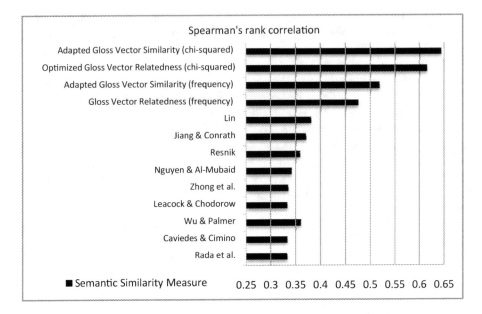

Fig. 3. Spearman's rank correlation of similarity measures

Because of these considerations, which help to produce the best estimated results of similarity, for other measures the highest produced correlation results are presented.

Table 2 shows precisely the Spearman's correlation results for 5 highest semantic similarity measures. Adapted Gloss Vector Similarity (frequency cut-off) means we applied the idea of taxonomy (stage 2) on Gloss Vector measure while instead of X^2 the bigram frequency cut-off (in the original Gloss Vector measure) is implemented. This is done to show the effectiveness of X^2 on insignificant features elimination.

Table 2. 5 highest Spearman's correlation results of similarity measures

Semantic similarity measure	Spearman's rank correlation
Lin	0.3805
Gloss Vector relatedness (frequency cut-off)	0.4755
Adapted Gloss Vector similarity (frequency cut-off)	0.5181
Optimized Gloss Vector relatedness (chi-squared cut-off)	0.6152
Adapted Gloss Vector similarity (chi-squared cut-off)	**0.6438**

With comparing Adapted Gloss Vector similarity measure in optimized the form with other measures, we can see that this approach has achieved higher Spearman's rank correlation. On the other hand, due to some drawbacks in semantic similarity measures, Gloss Vector relatedness yields more reliable estimation of similarity. We also discussed setting absolute thresholds of low and high cut-off without considering the level of dependency of two terms (words) involved in a bigram can waste some beneficial as well as expressive amount of information. Therefore, before considering any strict low and high cut-off points to remove insignificant features, we see there is a need to measure the relative level of association between terms in a bigram. The results of X^2 implementation prove efficiency of this consideration.

5.2   Pearson's Chi-squared Test and Level of Association Cut-off

If we consider words w_1, w_2, w_3, ... and w_n as an initial set of features which construct $n \times n$ first order co-occurrence matrix, in practice, only a subset of these words is significant and descriptive for a concept (to which the concept is really dependent or associated). To find these levels of association, after applying X^2 on second order co-occurrence matrix, the critical value from the chi-squared distribution with 1 as the degree of freedom can be considered to remove those features to which concept is not dependent. In order to find this critical value, as we mentioned, we gradually increase this value, by going upward from 0 until we find the best point where the best result of correlation between computed similarity and human judgment of them is achieved.

Figure 4 demonstrates the correlation changes in different cut-off points for the X^2 level of association. For example, after removing X^2 results with values less than 15 from X2-on-SOC matrix, we achieved a correlation score of 0.6425. In these represented ranges, all 26'958 concepts in MeSH have their associated vectors. Further cutoffs beyond the last point may cause loss of concepts adapted and optimized gloss vectors which will hinder future measurement of similarity for these concepts (depending on whether that concept has any descendant with an available optimized gloss vector).

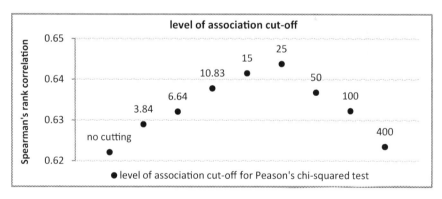

Fig. 4. Adapted Gloss Vector similarity measure and different levels of association cut-off point

Considering the results of experiments we found insignificant features elimination in the frequency cut-off approach in Gloss Vector relatedness measure is problematic as it does not consider frequency of individual terms and merely take frequency of bigrams into account. This can cause loss of many valuable features deciding for more a reliable estimation. Through a slight change in this measure's procedure using Pearson's chi-squared test we find insignificant terms statistically and more properly.

6 Conclusion

In this study we attempted to adapt the Gloss Vector relatedness measure for an improved similarity estimation calling it the result Adapted Gloss Vector semantic similarity measure. Generally, Gloss Vector uses angles between entities' gloss vectors for calculation of relatedness. In order to adapt this measure for similarity measurement, first, through altering Gloss Vector's procedure by using Pearson's chi-squared test (X^2) for determining the level of association between concepts and terms (features), we could eliminate insignificant features which lead us to the optimized gloss vectors; second, by taking a taxonomy of concepts into account and also having optimized gloss vectors of concepts, by augmenting concepts' optimized gloss vector with their descendants' optimized gloss vectors we produced enriched gloss vectors for all concepts in the taxonomy suitable for semantic similarity measurement. Our evaluation of Adapted Gloss Vector similarity measure in the biomedical domain proved this measure outperforms other similarity and Gloss Vector based measures.

For the future works, in order to implement insignificant features removal, other statistical measure of association such as Pointwise Mutual Information (PMI) and Log Likelihood Statistic (G^2) can be examined. Additionally, our proposed approach can be tested on extrinsic tasks such as ontology alignment and data base schema matching, or get evaluated in other domains of knowledge.

References

1. Muthaiyah, S., Kerschberg, L.: A hybrid ontology mediation approach for the semantic web. Int. J. E-Bus. Res. **4**, 79–91 (2008)
2. Chen, B., Foster, G., Kuhn, R.: Bilingual sense similarity for statistical machine translation. In: Proceedings of the ACL, pp. 834–843 (2010)
3. Pesaranghader, A., Mustapha, N., Pesaranghader, A.: Applying semantic similarity measures to enhance topic-specific web crawling. In: Proceedings of the 13th International Conference on Intelligent Systems Design and Applications (ISDA'13), pp. 205–212 (2013)
4. Gruber, T.R.: Toward principles for the design of ontologies used for knowledge sharing. Int. J. Hum Comput Stud. **43**, 907–928 (1995)
5. Firth, J.R.: A synopsis of linguistic theory 1930–1955. In: Firth, J.R. (ed.) Studies in Linguistic Analysis, pp. 1–32. Blackwell, Oxford (1957)
6. Lesk, M.: Automatic sense disambiguation using machine readable dictionaries: how to tell a pine cone from an ice-cream cone. In: Proceedings of the 5th Annual International Conference on Systems Documentation, New York, USA, pp. 24–26 (1986)
7. Banerjee, S., Pedersen, T.: An adapted Lesk algorithm for word sense disambiguation using WordNet. In: Gelbukh, A. (ed.) CICLing 2002. LNCS, vol. 2276, pp. 136–145. Springer, Heidelberg (2002)
8. Patwardhan, S., Pedersen, T: Using WordNet-based context vectors to estimate the semantic relatedness of concepts. In: Proceedings of the EACL 2006 Workshop (2006)
9. Liu, Y., McInnes, B.T., Pedersen, T., Melton-Meaux, G., Pakhomov. S.: Semantic relatedness study using second order co-occurrence vectors computed from biomedical corpora, UMLS and WordNet. In: Proceedings of the 2nd ACM SIGHIT IHI (2012)

10. Pesaranghader, A., Pesaranghader, A., Rezaei, A.: Applying latent semantic analysis to optimize second-order co-occurrence vectors for semantic relatedness measurement. In: Proceedings of the 1st International Conference on Mining Intelligence and Knowledge Exploration (MIKE'13), pp. 588–599 (2013)
11. Pesaranghader, A., Pesaranghader, A., Rezaei, A.: Augmenting concept definition in gloss vector semantic relatedness measure using Wikipedia articles. In: Proceedings of the 1st International Conference on Data Engineering (DeEng-2013), pp. 623–630 (2014)
12. Rada, R., Mili, H., Bicknell, E., Blettner, M.: Development and application of a metric on semantic nets. IEEE Trans. Syst. Man Cybern. **19**, 17–30 (1989)
13. Caviedes, J., Cimino, J.: Towards the development of a conceptual distance metric for the UMLS. J. Biomed. Inf. **372**, 77–85 (2004)
14. Wu, Z., Palmer, M.: Verb semantics and lexical selections. In: Proceedings of the 32nd Annual Meeting of the Association for Computational Linguistics, (1994)
15. Leacock, C., Chodorow, M.: Combining local context and WordNet similarity for word sense identification. In: Fellbaum, C. (ed.) WordNet: An Electronic Lexical Database, pp. 265–283. MIT press, Cambridge (1998)
16. Zhong, J., Zhu, H., Li, J., Yu, Y.: Conceptual graph matching for semantic search. In: Priss, U., Corbett, D.R., Angelova, G. (eds.) ICCS 2002. LNCS (LNAI), vol. 2393, pp. 92–106. Springer, Heidelberg (2002)
17. Nguyen, H.A., Al-Mubaid, H.: New ontology-based semantic similarity measure for the biomedical domain. In: Proceedings of IEEE International Conference on Granular Computing GrC'06, pp. 623–628 (2006)
18. Resnik, P.: Using information content to evaluate semantic similarity in a taxonomy. In: Proceedings of the 14th International Joint Conference on Artificial Intelligence (1995)
19. Jiang, J.J., Conrath, D.W.: Semantic similarity based on corpus statistics and lexical taxonomy. In: International Conference on Research in Computational Linguistics (1997)
20. Lin, D.: An Information-theoretic definition of similarity. In: 15th International Conference on Machine Learning, Madison, USA, (1998)
21. Pesaranghader, A., Muthaiyah, S.: Definition-based information content vectors for semantic similarity measurement. In: Noah, S.A., Abdullah, A., Arshad, H., Abu Bakar, A., Othman, Z.A., Sahran, S., Omar, N., Othman, Z. (eds.) M-CAIT 2013. CCIS, vol. 378, pp. 268–282. Springer, Heidelberg (2013)
22. Pakhomov, S., McInnes, B., Adam, T., Liu, Y., Pedersen, T., Melton, G.: Semantic similarity and relatedness between clinical terms: an experimental study. In: Proceedings of AMIA, pp. 572–576 (2010)

Advanced Semantic Web Services for Diet Therapy with Linked Data and Mediation Rules

Yusuke Tagawa[1(✉)], Arata Tanaka[1(✉)], Yuya Minami[2],
Daichi Namikawa[2], Michio Simomura[2], and Takahira Yamaguchi[1]

[1] Keio University, 3-14-1 Hiyoshi, Kohoku-ku, Yokohama-shi,
Kanagawa 223-8522, Japan
`yamaguti@ae.keio.ac.jp, k_a_y_h_j_s@z6.keio.jp`
[2] NTT Service Evolution Laboratories, NTT Service Innovation Laboratory
Group, 1-1 Hikarinooka, Yokosuka-shi, Kanagawa 239-0847, Japan

Abstract. Instead of conventional Semantic Web Services, such as OWL-S and WSMO, LOSs (Linked Open Services) recently come up with more LOD (Linked Open Data) activities. Here is discussed how to construct advanced LOS architecture with LOD and mediation rules that bridge the gaps between different LOD endpoints and then how well the service goes well with the questionnaire survey of a diabetic. Furthermore, the LOS architecture should be evaluated by function extendibility and modifiability to another task.

Keywords: Linked Open Data · Mediator · Linked Open Service · Diabetic

1 Introduction

LOD activities, such as publishing and consuming, started about five years ago in USA and Europe. However, we have just started them since 2011 here in Japan, organizing LOD Challenge Japan 2011 and 2012 [1] and Open Data Promotion Consortium [2]. Thus we have more interest in Linked Open Data (LOD) here in Japan now.

On the other hand, Semantic Web Services, such as OWL-S [3] and WSMO [4], came up with us about ten years ago, but they are not yet popular with us. That is why they did not go well for automating high-level tasks such as discovery, composition and mediation. Instead of them, much concern goes to LOS (Linked Open Services) that take LOD endpoints and other technique, such as RESTful services.

Looking at health care domain, the number of patients suffering from life style is increasing among the aged or middle-aged people. They take several therapies and diet therapy works well there. However, we need much knowledge to manage it because recipes have no linkage with calculated scores for diet therapy in most cases. In particular, on eating out, most recipes have no information even on the recipe's nutrition. Therefore, it is difficult for them to do diet therapy well. From the background, this paper discusses how to construct advanced semantic web services with LOD and mediation rules that bridge the gaps between different LOD end points and

W. Kim et al. (Eds.): JIST 2013, LNCS 8388, pp. 146–156, 2014.
DOI: 10.1007/978-3-319-06826-8_12, © Springer International Publishing Switzerland 2014

then how well the service goes well with the questionnaire survey of a diabetic. Furthermore, the LOS architecture should be evaluated by function extendibility and modifiability to another task.

This paper comes as follows: We introduce related work about A System for Supporting Dietary Habits by Analyzing Nutritional Intake Balance in Sect. 2. We explain the structure of our systems as a Linked Open Service in Sect. 3. In Sect. 4 we show the results of questionnaire survey. In Sect. 5 we show discussion about the case studies. Finally we present conclusion of this paper and future work in Sect. 6.

2   Related Work

Traditional Semantic Web Service aims to automate the following high-level tasks: Web Service discovery, combination, and execution based on the technology of the Semantic Web. This is expected for offering the foundation of automated Web Service combination and execution. In particular, WSMO (Web Service Modeling Ontology) [4] has been proposed to achieve the goal of semantic Web service. However, Traditional Semantic Web Service did not go well, focusing on automating the high-level tasks too much. Instead of automating them, we should take alternative technology to link fine-grain size of services into coarse grain size of services or applications. Thus we discuss how to exploit mediation rules that bridge the gaps between different LOD endpoints.

Looking at the cooking recipe search system for ingestion balance in [5], such as a nutrient, when a certain recipe and material are submitted, the menu in consideration of nutritional balance is searched by the search system [6], those meal logs are taken, and the user's meal support is enabled by "recording diet". However, the search system is concentrated on the balance of a meal. Thus it is difficult to apply the system to other types of users and system functions.

In the Building Sustainable Healthcare Knowledge Systems by Harnessing Efficiencies from Biomedical Linked Open Data [3], the Knowledge work Support System (KwSS) in the case of the examination by a doctor is explained. Reference [1] is the data set (Diseasome, Diseases Database, Dailymed, DrugBank, Slider) of a specific domain and the general data set (DBPedia, OpenCyc and WordNet) which exist on Web are used and analysis of information required for the system (KwSS) which supports the knowledge which is needed at the time of prescription of medical examination and medicine is conducted at the time of diagnosis. Reference [1] claims that even if they use a data set independently alone, KwSS does not function well, but using combining those data sets it is very useful to realize KwSS.

3   System Architecture

The service is based on the structure of service-oriented architecture. The data part of the service is Linked Data. Since the collision for every Linked Data takes place, this is solved by inserting the mediator. Figure 1 is the system architecture of the general service which united service-oriented architecture and some mediator.

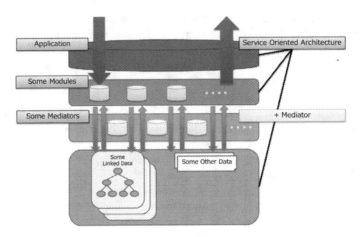

Fig. 1. System architecture

The mediator began to be used by WSMO. Although the module discovered from the UDDI registry was combined and web service was constituted at WSMO, the mediator was playing the role which bridges the gap of the input for every modules of those, and output. On the other hand, the mediator proposed this time bridges the difference in the structure, some kinds of Linked Data.

In this paper, we implemented the service-oriented architecture application with mediator, and verified mediators' usefulness. We took up the Dietary therapy support as a case study this time. The following figures are these system general drawing.

4   System Overview

We explain the system's architectures here. Figure 2 shows the overall system architecture. The system architecture of the whole proposal system is shown in Fig 2. There are some kinds of mediators. We created, firstly, one of the mediators. It performs unit conversion of the quantity of the material. A system is large and is divided into "recipe presumption module" - "nutrition calculation module" - "score calculation module." By a recipe presumption module, the material included in a recipe based on Recipe Linked Data from the recipe name sent by the user is guessed. By a nutrition calculation module, the nutrition of the material county presumed by the recipe presumption module is calculated based on the food composition table Linked Data. Finally, by a score calculation module, the rule according to illness is used, and when a certain sick, the index which should take care in the case of a meal is calculated.

4.1   Data Set

Here, we explain the Data Set used at each modules.

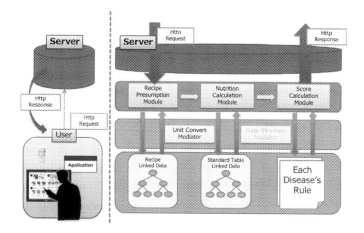

Fig. 2. System overview

Recipe Linked Data. We retrieve data on the web page such as the followings. The data have recipe name, materials, how to cook, for how many people, category, picture. Table 1 shows the number of recipes and where it's from. We made the data on the web into Linked Data here. Figure 3 shows the architecture of the Linked Data.

We connect Food Class by rdf:type property based on the recipe categories the web page has. For example, Fig. 3 shows that a Coconut Chicken Carry is at Chicken Carry category. Besides, Materials the recipe has is also connected to Material Class. These let us know Coconut Chicken Carry is a kind of Chicken Carry and it has coconut as a material.

Standard Tables Linked Data. We made the data in Standard Tables of Food Composition in Japan into Linked Data. A food composition table is Standard Tables of Food Composition in Japan[1] which the Ministry of Education, Culture, Sports, Science and Technology exhibits uses. Since the data which the Ministry of Education, Culture, Sports, Science and Technology releases is PDF form, it is unsuitable for treating by a system. Therefore, standard tables of food composition 2010 changed

Table 1. Where recipe's from and number of recipes

Web page	Number of recipes
Cookpad[a]	2,644
Nusluck kitchen[b]	2,740
Recipe encyclopedia[c]	813

[a] Cookpad http://cookpad.com/

[b] Nusluck Kitchen http://www.nasluck-kitchen.jp/

[c] Recipe Encyclopedia http://park.ajinomoto.co.jp/

[1] Ministry of Education, Culture, Sports, Science and Technology. (2013, Jan.) [Online]. http://www.mext.go.jp/b_menu/shingi/gijyutu/gijyutu3/houkoku/1298713.htm.

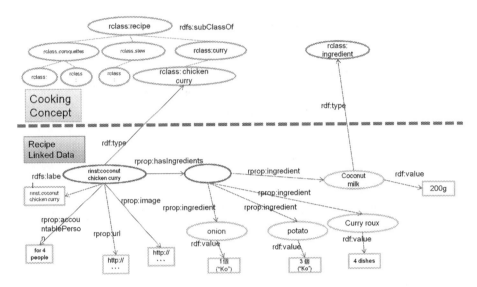

Fig. 3. Example of Recipe Linked Data

Table 2. Number of triples

Linked data	Number of triples
Recipe Linked Data	79423 triples
Food Composition Linked Data	107412 triples

into the CSV file[2] was read and the Standard Tables Linked Data was created. Standard Tables Linked Data has food ingredients and its nutrition. There are lots of kinds of the nutrition, such as protein, calcium etc. The number of Linked Data's triples finally created was as follows (Table 2).

Each Disease's Rule. Each Disease's Rule is the rule which summarized what kind of nutrient is thought as important by referring to books etc. in the case of a certain illness. In this paper, the rule to diabetes was created and the diet therapy service to a diabetic patient was proposed.

For example, if patients are diabetes, the rule will be something like that six scores of the List 1, List 2, ... List 6 should be taken into consideration, and List 1 contains grain, potatoes, vegetables and seeds with many cereals, and non-fibrous carbohydrates. Moreover, this system can respond to other illnesses easily by changing this rule. The contents of the concrete score are like written on Table 3. This scores is created consulting Tonyobyo shokuji ryoho no tameno shokuhin kokanhyo[3].

---

[2] Eisan WEB page. (2013, Jan.) [Online]. http://www.geocities.jp/eisan001588/labo/FoodComposition Table_2010.zip.

[3] Japan Diabetes Society, *Tonyobyo shokuji ryoho no tameno shokuhin kokanhyo* (The list of food exchange for a diabetic diet therapy), Bunkodo, 2007.

Table 3. Scores for diabetic

List 1	List 2	List 3
Grain	Fruit	Fish
Tubers and roots		Meat
Vegetables and seed (high-carbohydrate)		Egg and cheese
Bean (except soybean)		Soybean

List 4	List 5	List 6
Milk and its products	Oils and fats The Greasy	Vegetables Seaweeds Mushrooms Konjak

4.2 Mediator – Unit Convert Mediator

We created the mediator for performing unit conversion of the quantity of the material for nutrition calculation. In the Recipe Linked Data created this time, expression of the unit of material is scattering. A certain thing becomes one ball and a certain thing has become one etc. expression. However, all the units used in the food composition table Linked Data are gram units. Therefore, the output by a recipe presumption module is not applicable to nutrition calculation as it is after that in a module. Then, the mediator who changes the unit of Recipe Linked Data was created this time.

For example, it is as follows (Table 4).

We created a unit conversion mediator by retrieving of the web page writing information about the weight of the foods this time, a convertible number is 1080.

4.3 Modules

Recipe Presumption Module. At this module, a similar recipe is guessed from the inputted menu name of the restaurant. The menu name of the restaurant is given as a menu text which carried out OCR recognition from the picture. Therefore, it pretreats first by a Recipe Estimate Module. Here, the text of the OCR recognition result of a restaurant menu is called a raw text. Two or more menu names, a price, a noise, etc. are contained in the raw text. Generally, because the menu name in a restaurant menu is divided and described, signs, such as a "line feed mark", a "blank space", "-", divide a character string, and change the divided character string into a recipe name.

Table 4. Examples of mediation

Materials	Before conversion	After conversion (g)
Onion	1 "Ko" (item)	200
Salt	1 teaspoon	5
Rice	1 bowl	150

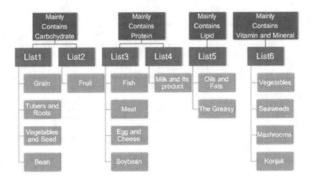

Fig. 4. Nutrition and Each Disease's Rule

Then, a recipe instance is acquired from Recipe Linked Data based on a menu name. It is the basis quoted as an example of a recipe instance by Fig. 4. The material which the recipe has with a *rprop:hasIngredients* property and a *rprop:ingredients* property is expressed. Since the unit of the material which a recipe instance has various forms, it uses the Mediator here. All the units change to a gram unit by using the Mediator. Besides, the quantity of material passes the recipe instance used as a gram unit to a nutrition calculation module.

Nutrition Calculation Module. At the nutrition calculation module, we perform nutrition calculation of the recipe instance passed by the recipe presumption module. Since all the materials of the recipe are changed into the gram unit by the unit conversion Mediator, We can acquire easily how much nutrition is contained in the material of this recipe instance from the Standard Tables Linked Data. Nutrition is acquired from Standard Tables Linked Data, and it adds to a recipe instance, and passes to the Score Calculation Module.

Score Calculation Module. At this module, the recipe instance passed from the Nutrition Calculation Module and the Each Disease's Rule are used, and score calculation according to illness is performed. The score of List 1 to List 6 is calculated this time based on nutrition for the rule for diabetes. There are six scores consulted by the diabetic. We call them List 1, 2,...6 this time. Figure 4 shows that what kinds of ingredient each List has. When a recipe has an ingredient which type is grain, the ingredient will increase the score "List 1" based on its amount.

4.4   System Flow

At this phase, we explain the process when user input the recipes. The process has the following steps.

1. User Application
2. Recipe Presumption Module
3. Nutrition Calculation Module
4. Score Calculation Module
5. Web Server

Table 5. Each module and used data set

Modules	Using rule
Recipe Presumption Module	Unit Mediator
	Recipe Linked Data
Nutrition Calculation Module	Standard Tables Linked Data
Score Calculation Module	Each Disease's Rule

Each modules use some data sets and the following table, Table 5, shows the corresponding.

1. *User Application.*
   The system needs user data such as age, height, sex, weights and exercise level. Based on these data, the system set the desired values. So, the user sends these information by our application.
2. *Recipe Presumption Module.*
   Here, a similar recipe is guessed from the menu name of the inputted restaurant, and a recipe with a material county is outputted.
3. *Nutrition Calculation Module.*
   Here, a recipe instance with the material county outputted from Recipe Presumption Module is considered as an input, the nutrition of a material county is calculated, and nutrition calculation of a recipe is performed.
4. *Score Calculation Module.*
   Here, the recipe instance which was outputted from Nutrition Calculation Module and which has been nutrition calculated is considered as an input, and the score which a diabetic should carry out for the rule according to illness to reference at reference is calculated.
5. *Web Server.*
   The recipe instance which the above module calculated is returned to a user's application, and a result is displayed.

5   Case Study

We conducted the questionnaire survey which has our system tried on a web. Universe person is like followings.

(a) Started and Suffer from Diabetes.
(b) Netsurf 1 30 Minutes or More per Day with Smart Phone or Mobile Phone.
(c) Those who are performing dietary therapy, who have not carried out although dietary therapy is got to know and it enters or who stopped although diet therapy was performed.

Under these conditions, classified by the conditions of 3, it performed every 100 sample recoveries each.

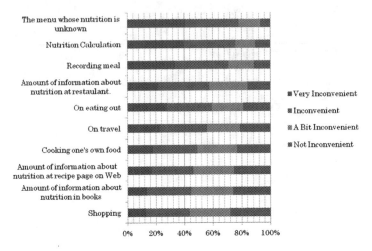

Fig. 5. User's need

This time, evaluation by a web questionnaire was performed. First, we confirmed user's needs, and had our system actually used. The following evaluations were obtained in the inquiry of needs (Fig. 5).

Those who think that it is inconvenient to do diet therapy especially by each item are about 80 %. In particular, in the item of "The menu whose nutrition is unknown", and "nutrition calculation", there were those who have answered "It is very inconvenience" 40 %, and the person who answered "Not inconvenience" brought a result of 10 % or less. From these things, although the information on the nutrient of a menu or a daily dish wants, it turns out that nutrition calculation is inconvenient.

Next, it is the evaluation after having our system actually used. The following results were obtained when we asked which they would like to use in comparison with the sources of information currently usually used from the questionnaire candidate for every pattern (Fig. 6).

About those "who stopped although dietary therapy was performed", about 70 % of people wants to use the proposal service and about those "who are not performing dietary therapy", about 80 % wants to use the proposal service. However, about those "who are performing dietary therapy", only about half of them would like to use the way of proposal service. The opinion that the usual site to which it is used is more intelligible was also seen. The difference of evaluation was born among questionnaire candidates. Although many of causes of those who have already stopped dietary therapy were the time and effort of nutrition calculation, since the time and effort was saved by proposal service, high evaluation was able to be obtained. However, since practice had arisen in the nutrition calculation for dietary therapy for those who have already performed dietary therapy, those who estimated that new proposal service was good became approximately the whole half.

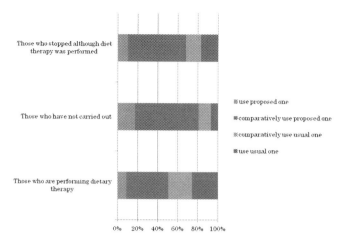

Fig. 6. Results of questionnaire

6   Discussion

In this paper, we developed Recipe Linked Data and Standard Tables Linked Data, we canceled the mismatching during the output and input for every Module by a unit conversion mediator and, by applying the Each Disease's Rule, proposed the service which can present a suitable menu to a user as one example of Linked Open Service. In Sect. 3, we described that there are the easiness of extension and functional change as a feature of Linked Open Service, we discuss the usefulness of these and a mediator, and the usefulness as a service from a questionnaire result.

First, we discuss the easiness of extension. Since what is necessary is just to only add the new module which uses the output from Score Calculation Module as input, it can be said that it also has extended easiness.

Second, we discuss the easiness of functional change. Using Each Disease's Rule allows us to change the type of outputs from the module and all we have to do is just to change the rule. So, it can be said the service has the easiness of functional change.

Third, we discuss the usefulness of mediator. In Linked Open Service, we have to make the input and output even and it is needed to change into the state where there are no semantic gaps. However, the mediator allows us solve this problem. This result allows us to step forward to making the linkage between some Linked Data automatically without minding fine grain size of differences. Therefore, it can be said that a mediator's existence was useful.

Finally, we discuss the usefulness as a service, when we saw the web questionnaire results; the difference of evaluation was born between those who are continuing diabetic dietary therapy, those who stopped and those who are not doing. Many of causes of those who have already stopped dietary therapy were the time and effort of nutrition calculation, and since the time and effort was saved by a proposed service, they were able to obtain high evaluation. As mentioned above, it was shown that Linked Open Service serves as very useful service now, while Linked Open Data is increasing. Besides, it was shown that existence of the mediator who connects the output and input of each module for making Linked Open Service useful is also indispensable. It will be expected by uniting these from now on that better service will be produced.

7   Future Work

In order to show the extended easiness which is the feature of Linked Open Service, we aim at changing Each Disease's Rule for a diabetic to for a hypertensive patient, and show a concrete example about the easiness. Moreover, in order to show functional change easiness, we discover or create Allergy Linked Data, and aim at extension of a function. Figure 7 shows the architecture of the future work.

We'll change the calculation method in Score Calculation Module by changing Each Disease's Rule, and change into service for a hypertensive patient.

Moreover, we'll create Remove Allergy Recipe Module using Allergy Linked Data, and change into the service which removes the recipe containing a certain allergy-causing substance. Besides, we'll discuss the cost of these extensions.

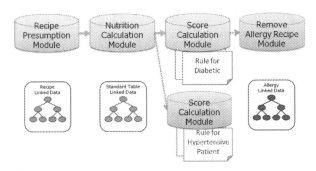

Fig. 7. Future work architecture

References

1. Linked Open Data Challenge Japan 2012, Linked Open Data Challenge Japan 2012, 5 April 2013. http://lod.sfc.keio.ac.jp/challenge2012/ (2012)
2. Open Data ryutu suishin consortium (A consortium promoting the distribution of open data), Open Data Promotion Consortium, 5 May 2013. http://www.opendata.gr.jp/ (2013)
3. OWL-S: Semantic Markup for Web Services, France Telecom et al, 19 May 2013. http://www.w3.org/Submission/OWL-S/ (2004)
4. Web Service Modeling Ontology (WSMO), Innsbruck at the Leopold-Franzens-Universität et al, 19 May 2013. http://www.w3.org/Submission/WSMO/ (2005)
5. Karikome, S., Fujii, A.: A system for supporting dietary habits by analyzing nutritional intake balance. DBSJ J. **8**(4), 1–6 (2010)
6. Karikome, S., Fujii, A.: A retrieval system for cooking recipes considering nutritional intake balance. Trans. Inst. Electron. Inf. Commun. Eng. **J92-D**(7), 975–983 (2009)

Ontology Construction

Ontology Construction Support
for Specialized Books

Yuki Eguchi[1(⊠)], Yuri Iwakata[1], Minami Kawasaki[1],
Masami Takata[2], and Kazuki Joe[2]

[1] Graduate School of Humanities and Sciences, Nara Women's University,
Nara, Nara, Japan
eguchi-yuki0910@ics.nara-wu.ac.jp
[2] Academic Group of Information and Computer Sciences,
Nara Women's University, Nara, Nara, Japan

Abstract. In this paper, we present a support system for ontology construction just based on a given specialized book with lowest possible cost. The system tries to combine minimum hand and mostly automatic constructions. It extracts required information for the ontology design from the specialized book with presenting yes-no selections to an expert of the specialized book. The ontology construction is performed automatically just using the answers of the expert. The constructed ontology is reliably and highly technical since it is constructed on the basis of the specialized book. In addition, since user operations are restricted only to yes-no selection, any expert can make use of our system without any special knowledge about ontology.

Keywords: Ontology · Ontology construction support · Knowledge engineering

1 Introduction

Information on the Internet has been increasing rapidly according to the quick spread of data communication infrastructures. While the Internet enables us to get widely diverse information, it is true that there is unignorable amount of unuseful and/or unreliable information on the Internet. The useful and/or reliable information we want to get might be overloaded with other huge amount of meaningless information. Such inconsistent information acquisition prevents us from obtaining correct and required information on the Internet. To exploit such information from huge amount of chaotic information, knowledge about the target information has to be well defined and given.

Ontology has been known as a model to systematize and express explicitly how we understand objects. Information of an object is saved in a form where computers understand the object. There are two development methods for the construction of ontology: manual by hand or automatic by computers. In the case of manual construction, experts about the target field design the ontology. Though high cost is required, a useful and reliable ontology is obtainable. In the case of automatic construction, computers generate a huge ontology automatically with low cost. However, when the automatically generated ontology is to be practically used, the usability and the reliability should be checked by human, hopefully, an expert.

W. Kim et al. (Eds.): JIST 2013, LNCS 8388, pp. 159–174, 2014.
DOI: 10.1007/978-3-319 06826 8_13, © Springer International Publishing Switzerland 2014

The primary requirement of practical ontology is its reliability even if it requires high cost. Actually, ontology is actively constructed in research areas with sufficient budget, such as bioinformatics, pharmaceutical sciences, and clinical medicine. On the other hand, the construction of practical ontology is very difficult in research areas without a certain amount of budget. Although the existence of ontology is not essential for the target research area, the use of ontology may increase research possibility such as collaboration with other research area and/or analysis of large amount of data processed by computers. Therefore, if it is possible to construct reliable ontology with reasonably low cost, various research areas without enough research funds get benefits with utilizing their own ontology.

In this paper, we present a support system for ontology construction just based on a given specialized book with lowest possible cost. The system tries to combine minimum hand and mostly automatic constructions of an ontology to extract required information for the ontology design from the specialized book with presenting yes-no selections to an expert of the specialized book. The ontology construction is performed automatically just using the resultant answers of the expert. Specialized books are reliable since they are written by experts. Therefore, the constructed ontology is considered reliable. In addition, since user operation during the ontology construction is restricted only to answer yes or no, the constructor can use the system without any special knowledge about ontology construction, but with appropriate knowledge about the target specialized area as an expert.

2   Automatic Construction of Ontology

In this section, we introduce existing research about automatic ontology constructions. As an automatic ontology construction system, DODDLE-OWL (a Domain Ontology rapiD Development Environment - OWL extension) [1] has been reported. It consists of a pre-processing part and a quality improvement part. The pre-processing part generates prototype ontology from Japanese or English document semi-automatically. The quality improvement part helps the user to append new knowledge to the ontology and improve specialty of the prototype ontology by hand interactively. Note that the interactions suppose that the user does not have special knowledge about the documents to be processed but enough knowledge about ontology design. Therefore, it is difficult for users without enough knowledge of ontology to use the system.

OntoLT [2] constructs ontology automatically from sentences with literal annotations attached by XML. The user has to define the rules that generate the framework between literal annotations and an ontology language. Therefore the user needs technical knowledge about ontology.

OntoLearn [3] extracts terms as words from English documents through semantic interpretation of the words. The system constructs ontology automatically with existing concept-dictionaries and a corresponding corpus. Since the ontology is not checked up on after the automatic construction, the reliability of the resultant ontology is low in potential.

Text2Onto [4] constructs ontology automatically from diverse corpuses which are not only plain text but also structured text such as HTML, DTD, and XML, machine readable dictionary, and database. A user selects a corpus and a learning method while confirming the automatic-constructed ontology. Since the ontology construction depends on the user selection basis, the user acquires desired ontology. However, the user cannot modify the ontology even a bit, so an ontology expert should modify it by hand in such case.

OntoGen [5] constructs ontology using two methods: system supporting automatic construction and hand operations for concept learning and visualization. The system integrates text mining and machine learning to suggest concept relation, and the user can select the system suggestion interactively. Since the system is developed with the assumption that the user is not an ontology expert, it is difficult for the user to select the main operations with too many functions. User operations need to be simpler in order to make easy use of the system.

Ontosophie [6] extracts ontology population from existing documents semi-automatically. The system learns extraction rules to extract ontology population from plain text or HTML annotated with XML tags as supervised learning. A user determines the classification as threshold value of supervised learning before the extraction, and accepts, rejects, or modifies the extraction rules after that. While the supervised learning improves the ontology precision, it depends on the threshold value.

3   Definition of Specialized Book Ontology

In this section, we define the specialized book ontology. Figure 1 shows the outline of the definition. This ontology is defined by conceptualizing nouns and verbs. These words are found in a specialized book. The concept of this ontology is semantic contents by systematizing nouns and verbs. In this definition, concepts of a noun and a verb are defined as a class and as a property, respectively. Each class and property is expressed in a hierarchical relationship. The relationship is defined as a broader concept and a narrower concept. The property is used for the assign of a class to its domain and range. The domain and range is a link of their property to a class. Domain represents that the subjects of the property must belong to the indicated class. Range asserts the values of the property must belong to the indicated class. If a property is regarded as a predication of a sentence, its domain and its range are equivalent to a subject and an object, respectively. In the case of intransitive verb, its range is assigned to a class that represents an object against the verb. Domain and range establish a relationship between classes except hierarchical ones.

URIs (Uniform Resource Identifier) given to each concept are used for understanding the concepts. The relationship between concepts is defined by assigning the relationship between the corresponding URIs. A natural language label, which is a metadata that accompanies a concept, is attached to the URI of the concept as multi names. A metadata gives information about the target data. A comment of metadata also gives the explanation for the concept.

Figure 2 shows an example of the specialized book ontology. This example is defined from a sentence, "The volcanic ash accumulates." In this figure, a rectangle, an ellipse, and an arrow represent a literal, a URI, and URI's metadata, respectively. A noun "Volcanic ash" is defined as a class, and a verb "pile" is defined as a property. "Pile" is an intransitive verb, so it is assigned to "volcanic ash" as its range.

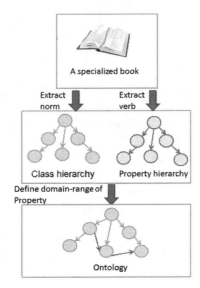

Fig. 1. Define ontology from a specialized book

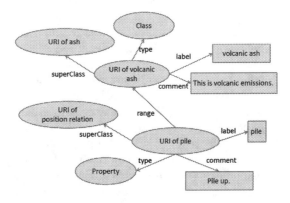

Fig. 2. Example of ontology

4   An Ontology Construction Support System for Specialized Books

4.1   System Requirements

There are three requirements to design our ontology construction support system. First, the ontology must be constructed from the source of reliable information. The reliable information may be a media generated by experts for some area, such as specialized books. User can also select the area of ontology to be constructed by choosing a specialized book. Second, user without any knowledge about ontology can use this system to construct the suitable ontology. It restricts user's action just to answer yes or no that offers a proposal about the definition of the constructing ontology.

Finally, the system must construct ontology with lower cost than hand construction. It performs semi-automatic construction to reduce user's load with asking users for simple yes-no questions.

To satisfy the above three requirements, we develop a system that constructs ontology semi-automatically from a specialized book with asking users for yes-no questions.

4.2   Ontology Construction Process

Figure 3 shows the overview of our system. At the preprocessing stage, the system extracts required information for the ontology construction. Then it defines classes, properties, and domains/ranges. At each definition, it offers proposals where user can select yes-no answers. Given the user's answers, it constructs ontology automatically with. In this point, a general ontology is needed to construct more reliable ontology.

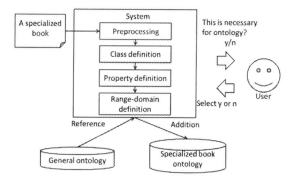

Fig. 3. Overview of the system

At first, the system performs preprocesses for the preparation of ontology construction. It extracts text data from a given specialized book. The data is applied with language morphological analysis. Morphological analysis is a well-known natural language processing method. It splits up a given sentence written in a natural language to several rows of morpheme. The word class of each row is parsed with information source such as the rules of grammar and dictionaries. A row of morpheme means the smallest unit when the language has meaning. A lot of unnecessary words for ontology are included in the results of morpheme analysis. Hence, the system needs to extract only necessary nouns and verbs. At this time, some word often appears in the specialized book, so the system avoids the duplication of the word. Moreover, some compound word combined with two or more morphemes tends to have many nouns with high specialty in the specialized book. Therefore, a natural language processing which combines a morpheme and the extracted compound word is required in advance. After the preprocessing, the general ontology is loaded in the system.

At the class definition stage, the system treats nouns which are extracted at the preprocessing stage. First, it presents a noun to ask the user whether the noun is necessary for the ontology with the form of yes or no. In the case of no, it presents the next noun and asks the user. In the case of yes, it searches the concept that has the label with the same name as the presented noun. Then, it refers the noun to the general ontology. When there is a coincided concept, it presents the information about the concept, and asks the user whether the concept should be added to the class in the specialized book ontology by yes or no. In the case of yes, it extracts the terms from the coincided concept to its broadest concept in the general ontology, and merges them into an appropriate class hierarchy of the constructing ontology. In the case that user's answer is no, the system searches the other coincided concepts to ask the user by yes or no in the same way. If this repetition is continued to find all answers are no, the system defines the noun as the narrower concept of an unknown word into a class hierarchy. The unknown word is defined as a brother concept of the broadest concept. A brother concept has the same broader concept. In the case of a compound word, although the coincided concept does not exist in the general ontology because the compound of morphemes is complicated, a broader concept might exist. In this case, the system asks the user whether the compound word should be added as a narrow concept of the noun which is the last of morphemes.

At the property definition stage, the system treats verbs which are extracted at the preprocessing stage. First, it presents a verb to ask the user whether the verb is necessary for the ontology by yes or no. In the case of no, it presents the next verb and asks the user. In the case of yes, it searches the concept that has the label with the same name as the presented verb. Then, it refers the verb to the general ontology. When there is a coincided concept, it presents the information about the concept, and asks the user whether the concept should be added to the property in the specialized book ontology by yes or no. In the case of yes, it extracts the terms from the coincided concept to its broadest concept in the general ontology, and merges them into an appropriate class hierarchy of the constructing ontology. In the case that user's answer is no, the system searches the other coincided concept to ask the user by yes or no in the same way. If this repetition is continued to find all answers are no, the system defines the verb as the narrower concept of an unknown word into a property hierarchy.

At the domain-range definition stage, the system defines domain and range of defined properties. First, the system selects a property from defined properties. Next, it searches a class which can be the domain or the range for the property. Then, it refers defined classes to the domain/range of the concept which is the same as the property in the general ontology. If there is a class for the domain/range, it shows the information about the class and asks whether the domain/range should be added by yes or no. In the case of yes, the domain/range of the property is added. In the case of no, it searches other domain/range. If there is no other domain/range, it proceeds to the next property.

5   Implementation and Execution Example

5.1   Implementation

In this section, we explain the implementation issues of the proposed system. Our system is implemented in JAVA to be configured with the UTF-8 character code for output and input in order to avoid character corruption.

For the preprocessing stage, we use Perl to execute xdoc2txt[1], MeCab[2] [7], TermExtract[3]. Xdoc2txt and MeCab as external programs. Xdoc2txt is used for extracting text data from a given specialized book. It is a free tool to extract text data from binary document such as PDF and Office documents. For the morphological analysis, we use MeCab, which is a language morphological analysis engine for Japanese. Since the grammar differs by natural language, a morphological analysis engine suitable for the language is needed. Figures 4 and 5 show some MeCab outputs in Japanese and in English, respectively. TermExtract, which is a perl module for extracting technical terms, is used for extracting nouns. It generates compound words from the morphological analysis results of MeCab. For verb extraction, we use class words, namely the parse-fine-sort-1 of MeCab outputs. It extracts just verbs whose word class is not verbal auxiliary and whose parse-fine-sort-1 is neither a non-independent word nor a suffix word.

input wordparse, class word, parse-fine-sort-1,parse-fine-sort-2, parse-fine-sort-3,conjugate form, conjugate tyep, basic form,reading,pronunciation

血液が通る
血液名詞, 一般, *, *, *, *, 血液, ケツエキ, ケツエキ
が助詞, 格助詞, 一般, *, *, *, が, ガ, ガ
通る動詞, 自立, 五段, ラ行, 基本形, 通る, トオル, トール

Fig. 4. Output of MeCab

After the preprocessing stage, Jena[4] is used for ontology related processing, which is a JAVA framework for Semantic Web. As the general ontology, we use three kinds of ontologies: Japanese word ontology, hierarchy relation ontology, and domain-range ontology, which are converted from the EDR general vocabulary dictionary[5,6] by a converter. The EDR general vocabulary dictionary includes a Japanese word dictionary and a concept dictionary. The converter is provided by DODDLE project[7]. The keyboard input is used for user's yes-no answers. User can answer yes or no by pushing y or n key. At the class and property definition stage, when there is already an offered concept or its broader concept in the constructing ontology, our system does not add it in order to reduce the execution time. After the domain-range definition stage, our system writes the constructed specialized book ontology to a file.

---

[1] http://www31.ocn.ne.jp/~h_ishida/xdoc2txt.html

[2] http://mecab.googlecode.com/svn/trunk/

[3] http://gensen.dl.itc.u-tokyo.ac.jp/termextract.html

[4] http://jena.apache.org/index.html

[5] http://www2.nict.go.jp/out-promotion/techtransfer/EDR/J_index.html

[6] http://www2.nict.go.jp/outpromotion/techtransfer/EDR/JPN/TG/TG.html

[7] http://doddleowl.sourceforge.net

血液 Blood	Noun	general	*	*	*	*	血液 (basic form)	ketueki (reading)	ketueki (pronunciation)
が	Particle	case particle	general	*	*	*	が	Ga	ga
通る flow	Verb	Independent	*	*	conjugation of 5dan-ra verbs	Basic	通る	tooru	toru

Fig. 5. Output of MeCab in English

5.2 Execution Examples

Figures 6, 7, 8, 9, 10, 11, and 12 show some execution examples of our system. Although our system generates Japanese notations, we show them in English in the figures for easy understanding. We explain the figures along with the procedure of the book ontology definition.

Figure 6 shows an example of execution of a class definition. In the class definition, the system shows the name of a noun and asks the user whether the noun is required by Yes/No. The user judges that "water" is a necessary noun and pushes 'y'. Since the concept named "water" exists in the general ontology, the system presents its direction words with explanations. Then, it asks the user whether the presented concept should be added to the class by Yes/No. In Fig. 6, since the user judges the

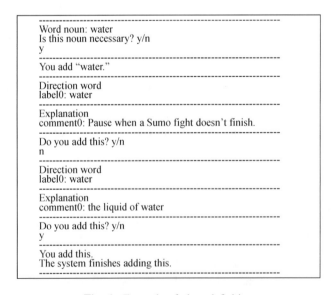

```
---------------------------------------------
Word noun: water
Is this noun necessary? y/n
y
---------------------------------------------
You add "water."
---------------------------------------------
Direction word
label0: water
---------------------------------------------
Explanation
comment0: Pause when a Sumo fight doesn't finish.
---------------------------------------------
Do you add this? y/n
n
---------------------------------------------
Direction word
label0: water
---------------------------------------------
Explanation
comment0: the liquid of water
---------------------------------------------
Do you add this? y/n
y
---------------------------------------------
You add this.
The system finishes adding this.
---------------------------------------------
```

Fig. 6. Example of class definition

Fig. 7. Visualization of water and its broader concepts

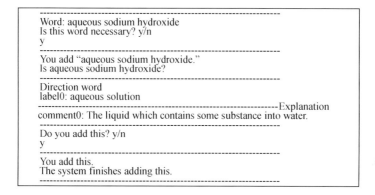

```
----------------------------------------------------------------------
Word: aqueous sodium hydroxide
Is this word necessary? y/n
y
----------------------------------------------------------------------
You add "aqueous sodium hydroxide."
Is aqueous sodium hydroxide?
----------------------------------------------------------------------
Direction word
label0: aqueous solution
---------------------------------------------------------Explanation
comment0: The liquid which contains some substance into water.
----------------------------------------------------------------------
Do you add this? y/n
y
----------------------------------------------------------------------
You add this.
The system finishes adding this.
----------------------------------------------------------------------
```

Fig. 8. The case of a compound word

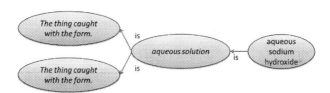

Fig. 9. Visualization of aqueous sodium hydroxide and its broader concepts

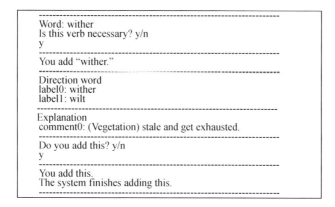

```
----------------------------------------------------------------------
Word: wither
Is this verb necessary? y/n
y
----------------------------------------------------------------------
You add "wither."
----------------------------------------------------------------------
Direction word
label0: wither
label1: wilt
----------------------------------------------------------------------
Explanation
comment0: (Vegetation) stale and get exhausted.
----------------------------------------------------------------------
Do you add this? y/n
y
----------------------------------------------------------------------
You add this.
The system finishes adding this.
----------------------------------------------------------------------
```

Fig. 10. Example of property definition

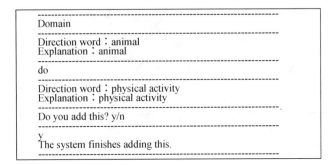

Fig. 11. Example of domain-range definition

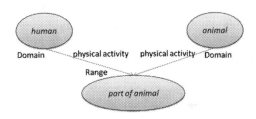

Fig. 12. Visualization of the domain-range among physical activity

second concept is more suitable than the first one, the user pushes 'y' for the second concept. By these user interactions, "water" and its broader concepts are defined as a class of the book ontology. Figure 7 visualizes some of those classes.

Figure 8 shows an execution in the case of compound word. When a concept which has the same name as a compound word does not exist in the general ontology, as shown in Fig. 9, the system shows a candidate of a broader concept and asks the user whether the compound word should be added to a narrower concept of the candidate by Yes/No. Figure 9 visualizes aqueous sodium hydroxide is added as a narrower concept of aqueous solution.

Figure 10 expresses an execution example for a property definition. The same procedure as the class definition is performed, and the system defines several properties. Figure 11 illustrates a domain-range definition example. In the domain-range definition, the system asks the user whether the combination of a subject, a predicate, and an object is appropriate by Yes/No. In the case of Fig. 11, the system asks whether animal should be added to the domain of physical activity. Since the user judges that animal is physical activity, the user pushes 'y'. Thereby, animal is added to the domain of physical activity. Figure 12 expresses the visualization of the added relations.

6 Experiments

6.1 Overview

We have two objectives validated by some experiments. First, we measure required time for user's yes-no answer with the following three methods.

(a) Just references to concepts
(b) Automatic construction just by the system
(c) Automatic construction with user's interaction

In the method (a), in order to measure the time required for a reference to a concept, two ways are provided. First, the system refers to just the first candidate for the concept, and second, it refers to all candidates for the concept. In the method (b), it assumes that all of user's answers are yes, and constructs ontology automatically.

In the method (c), the system constructs an ontology automatically with user's yes-no answers.

Second, we discuss whether the system is adequate to define and construct ontology of a given specialized book. Table 1 shows the experiment environment. As the experiment data set, we use a book of "Exciting Science 6" published by Keirindo[8], which is a science textbook for the sixth grade of elementary schools. The textbook is a rudimentary specialized book, and science is appropriate to systemize concepts.

Table 1. Experiment environment

Data	A science textbook of the sixth grader in an elementary school
OS	Windows7 Enterprise
CPU	Intel(R)Core(TM)i5-2500 K
Mounted memory	8.00 GB

6.2 Results of Experiment

Tables 2 and 3 are experimental results of the class definition and the property definition in the method (a), respectively. In the class and property definition in the method (c), the system might refer to two or more concepts per word. Therefore, the time required for a reference to a concept is measured. The sixth row expresses the time required for a reference which is calculated from the difference of execution times.

Tables 4 and 5 are the experimental results in the method (b). Table 4 is the results of the class and the property definitions, and Table 5 is the results of the domain-range definition. In Table 4, the time for a reference to a concept is taken from the execution time, and the time required to define a concept is calculated.

Table 6 shows the items for nouns and verbs. From Table 6, we find about 50 % of nouns and about 59 % of verbs which are included in the textbook are unnecessary.

---

[8] http://www.shinkokeirin.co.jp/

Table 2. Experiment results of class definition by the method (a)

Reference method	All	Only first
Referred concept	3307	966
Runtime(second)	9728.55	3848.05
Time of Noun presentation(second)	1421.50	
Time of one referred concept(second)	2.51	

Table 3. Experiment result of property definition by the method (a)

Reference method	All	Only first
Referred concept	2182	424
Runtime(second)	6053.20	1616.59
Time of verb presentation(second)	546.55	
Time of one referred concept(second)	2.52	

Table 4. Experiment result of class and property definition by the method (b)

Definition	Class	Property
Referred concepts	931	405
Defined concepts	2086	947
Runtime(second)	5519.26	2492.24
Time of referred concepts(second)	3760.13	1568.64
Time of defined concepts(second)	1759.13	923.60
Time of one defined concept(second)	0.84	0.98

Table 5. Experiment result of domain-range definition by the method (b)

Domain	497
Range	4747
Runtime(second)	553.70
Time of one definition(second)	0.11

Table 6. Number of word

Word class	Noun	Verb
Word included in the textbook	966	424
Compound word	58	
Word which is not contained in the general ontology	35	19
Number of Yes about word by method c)	484	173

They are particular words in the textbook. In the case of nouns, such words include "impression", "message", "hint", etc. In the case of verbs, we have "talk", "secure", "realize", etc. These unnecessary words are included in the sentences which call students' attention or encourage them to study with own motivation. Therefore, these words are not appropriate as contents of science.

Tables 7 and 8 are the experimental results in the method (c). Table 7 is the results of the class and the property definitions, and Table 8 is the result of the domain-range

Table 7. Experiment result of class and property definition by the method (c)

Definition	Class	Property
Defined concepts	1010	463
Runtime(second)	5915.51	2946.61
Time of referred concepts(second)	3079.39	1616.59
Time of defined concepts(second)	851.74	451.56
Number of Yes/No answers	1626	848
Time of all Yes/No answer(second)	1984.38	878.46
Time of one Yes/No answer(second)	1.22	1.04

Table 8. Experiment result of domain-range definition by the method (c)

Domain	61
Range	1265
Runtime(second)	2680.05
Time of all definition	140.01
Time of all Yes/No answer(second)	2540.04
Number of all Yes/No answer	1692
Time of one Yes/No answer(second)	1.50

definition. In order to measure the time for user's choice of Yes/No, the time required for a reference to a concept and the time required for defining the concept are taken from the total execution time. As the result, we find selecting Yes/No requires about one second.

Figures 13, 14, 15, and 16 visualize some part of the constructing ontology. The relative classes to "plant" in Fig. 13 and to "part of plant identified by organs" in Fig. 14 are visualized, respectively. In Fig. 15, "saliva", "food", and relative classes to them are visualized. Figure 16 shows relative concepts to the property of "mix", and "Sub Property of" means a broader concept.

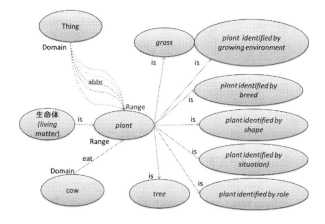

Fig. 13. Class "plant" and relative class

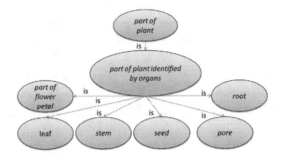

Fig. 14. Class "part of plant identified by organs" and relative class

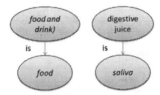

Fig. 15. Class "food" and "saliva" and relative class

Fig. 16. Relative concepts to property "mix"

7 Discussions

In this subsection, we discuss whether the proposed system is adequate to define and construct specialized book ontology with the method (c). From Figs. 13 and 14, we find that there are no arrows between "plant" and "part of plant identified by organs,"

```
食べ物は，口の中で歯にかみくだれ，唾液と混じる。
(Food is masticated and crushed in the month by teeth and mix saliva.)
            食べ物は(Food)，  -----------D
                  口の(month)-D  |
                      中で(in)---D  |
                  歯に(by teeth)-D  |
    かみくだかれ(is masticated and crushed) ---D
                      唾液と(saliva)-D
                      混じる。(mix.)
      EOS
```

Fig. 17. CaboCha output

so we know that these classes are not relative. However, "plant" consists of "part of plant identified by organs" such as leaf, seed, and petal in the real world. It is called part-of relationship. At the present ontology definition, we have not implemented the part-of relationship yet. Therefore, our system needs the function to define the part-of relationship. It is possible to use semantic interpretation of the nominal "B of A" which is included in sentences of the specialized book and labels of concepts.

Figure 17 is the results of dependency parsing in a sentence included in the textbook by using CaboCha[9] [8]. From Fig. 15, "food" is a subject, "saliva" is an object, and "mix" is a predicate in this sentence. In the case of the definitions described in Sect. 2, "food" and "saliva" are classes while "mix" is a property. "Food" and "saliva" are a domain and a range of the property, respectively. However, from Fig. 16, we find there is not such a domain-range definition in the constructed ontology. In order to construct more suitable ontology for the content of the specialized book, we should take account of sentence structures. Therefore, our system needs the function to append domain-range using dependency parsing.

8 Conclusions

In this paper, we proposed a support system for ontology construction just based on a given specialized book with lowest possible cost. The proposed system defines ontology by conceptualizing nouns and verbs included in the specialized book. It extracts information for ontology and offers a selection in the form where user can answer by yes or no. After user's answer, it constructs the ontology automatically with the answers. From the experiments, we found the time for a yes-no answer at class, property, and domain-range definitions are about one second. We analyzed the constructed ontology to find our system needs three more functions to define part-of relationship, and to append domain-range using dependency parsing. Future work includes that we apply the improved system to a real specialized book such as healthcare or medicine.

---

[9] http://code.google.com/p/cabocha/

References

1. Morita, T., Fukuta, N., Izumi, N., Yamaguchi, T.: DODDLE-OWL: interactive domain ontology development with open source software in Java. IEICE Trans. Inf. Syst.: Special Section on Knowledge-Based Software Engineering **E91-D**(4), 945–958 (2008)
2. Buitelaar, P., Olejnik, D., Sintek, M.: A protégé plug-in for ontology extraction from text based on linguistic analysis. In: Bussler, C.J., Davies, J., Fensel, D., Studer, R. (eds.) ESWS 2004. LNCS, vol. 3053, pp. 31–44. Springer, Heidelberg (2004)
3. Velardi, P., Navigli, R., Cucchiarelli, A., Neri, F.: Evaluation of OntoLearn, A Methodology for Automatic Population of Domain Ontologies. Ontology Learning from Text: Methods Applications and Evaluation, pp. 92–106. IOS Press, Amsterdam (2006)
4. Montoyo, A., Munoz, R., Metais, E.: Text2Onto - a framework for ontology learning and data-driven change discovery. In: Proceedings of the 10th International Conference on Applications of Natural Language to Information Systems, pp. 227–238 (2005)
5. Fortuna, B., Grobelnik, M., Mladenic, D.: OntoGen: semi-automatic ontology editor. In: Smith, M.J., Salvendy, G. (eds.) HCII 2007. LNCS, vol. 4558, pp. 309–318. Springer, Heidelberg (2007)
6. Celjuska, D., Vargas-Vera, M.: Ontosophie: a semi-automatic system for ontology population from text. In: International Conference on Natural Language Processing ICON (2004)
7. Kudo, T., Yamamoto, K., Matsumoto, Y.: Applying conditional random fields to Japanese morphological analysis. In: Proceedings of the 2004 Conference on Empirical Methods in Natural Language Processing (EMNLP-2004), pp. 230–237 (2004)
8. Kudo, T., Matsumoto, Y.: Fast methods for kernel-based text analysis. In: 41st Annual Meeting of the Association for Computational Linguistics (ACL2003), pp. 24–31 (2003)

Belief Base Revision for Datalog+/- Ontologies

Songxin Wang[1,2], Jeff Z. Pan[2(✉)], Yuting Zhao[2],
Wei Li[3], Songqiao Han[1], and Dongmei Han[1]

[1] Department of Computer Science and Technology,
Shanghai University of Finance and Economics, Shanghai, China
[2] Department of Computer Science, University of Aberdeen, Aberdeen, UK
jeff.z.pan@abdn.ac.uk
[3] School of Computer Science, Fudan University, Shanghai, China

Abstract. Datalog+/- is a family of emerging ontology languages that can be used for representing and reasoning over lightweight ontologies in Semantic Web. In this paper, we propose an approach to performing belief base revision for Datalog+/- ontologies. We define a kernel based belief revision operator for Datalog+/- and study its properties using extended postulates, as well as an algorithm to revise Datalog+/- ontologies. Finally, we give the complexity results by showing that query answering for a revised linear Datalog+/- ontology is tractable.

Keywords: Datalog+/- · Ontology · Belief revision · Kernel

1 Introduction

Datalog+/- [3] has its origin from database technologies, encompasses and generalizes the tractable description logics EL and DL-Lite, which can be used to represent lightweight ontologies in Semantic Web [6]. Datalog+/- enables a modular rule-based style of knowledge representation. Its properties of decidability of query answering and good query answering complexity in the data complexity allows to realistically assume that the database D is the only really large object in the input. These properties together with its expressive power make Datalog+/- a useful tool in modelling real applications such as ontology querying, web data extraction, data exchange, ontology-based data access and data integration.

Belief revision deals with the problem of adding new information to a knowledge base in a consistent way. Ontologies are not static, e.g., they may evolve over time, so it is important to study belief revision for Datalog+/- ontologies. For both classic logic and description logic, belief revision has been studied intensively and many literature exist, such as [1,7]. There are some extensions of Datalog+/- to deal with incomplete or inconsistency information, including inconsistent handling method [4], probability extension [5,8], and well-founded semantics extension [9], however, to the best of our knowledge, there is no belief revision method for Datalog+/- before.

W. Kim et al. (Eds.): JIST 2013, LNCS 8388, pp. 175–186, 2014.
DOI: 10.1007/978-3-319-06826-8_14, © Springer International Publishing Switzerland 2014

In this paper, we address the problem of belief revision for Datalog+/- ontologies and propose a *kernel-incision* based belief revision operator. Kernel consolidation was originally introduced by Hansson [10] based on the notion of kernels and incision function. The idea is that, given a knowledge base KB that needs to be consolidated (i.e., KB is inconsistent), the set of kernels is defined as the set of all minimal inconsistent subsets of KB. For each kernel, a set of sentences is removed (i.e., an incision is made) such that the remaining formulas in the kernel are consistent. Note that it is enough to remove any single formula from the kernel because they are minimal inconsistent sets. The result of consolidating KB is then the set of all formulas in KB that are not removed by the incision function.

We adopt the kernel-based consolidation idea into Datalog+/- ontologies, and give an approach to deal with belief revision in the face of new information which contains two parts of message: (i) the *new facts A*, and (ii) the *unwanted set Ω*. With our belief revision operator, a Datalog+/- ontology KB is able to be undated by adding new facts A in its database, and removing some database element to prevent any result in the unwanted set $Ω$, such as inconsistency $\perp$. We study the properties of proposed approach using extended postulates, and then give algorithms to revise the Datalog+/- ontologies. We finally give the complexity results by showing that query answering for a revised Datalog+/- ontology is tractable if a linear KB is considered.

The paper is organized as follows. In Sect. 2 we introduce some preliminary knowledge about Datalog+/-. In Sect. 3 we give our revision operator on Datalog+/- ontologies. The properties of the operator are investigated in Sect. 4. In Sect. 5 we give the algorithm and provide complexities results. Related works and the conclusion are given in Sect. 6 and 7 respectively.

2   Preliminaries on Datalog+/-

In this section, we briefly recall some necessary background knowledge on Datalog+/-.

Databases and Queries. We assume (i) an infinite universe of *(data)constants* constants Δ (which constitute the normal domain of a database), (ii) an infinite set of *(labelled) nulls* Δ_N (used as fresh Skolem terms, which are placeholders for unknown values, and can thus be seen as variables), and (iii) an infinite set of variables $\mathcal{V}$ (used in queries, dependencies, and constraints). Different constants represent different values (unique name assumption), while different nulls may represent the same value. We also assume that there is a lexicographic order on $\Delta \cup \Delta_N$, with every symbol in Δ_N following all symbols in Δ. We denote by $\mathbf{X}$ sequences of variables $X_1, ..., X_n$ with $k \geq$ o.

We assume a relational schema $\mathcal{R}$, which is a finite set of *predicate symbols* (or simply *predicates*). A *term* t is a constant, null, or variable. An *atomic formula*(or *atom*) $\mathbf{a}$ has the form $P(t_1, ..., t_n)$, where P is an n-arry predicate, and $t_1, ..., t_n$ are terms.

We assume a *Database (instance)* D for $\mathcal{R}$ is a (possibly infinite) set of atoms with predicates from $\mathcal{R}$ and arguments from Δ. A *conjunctive query* (CQ) over R has the form $Q(\mathbf{X}) = \exists \mathbf{Y} \Phi(\mathbf{X}, \mathbf{Y})$, where $\Phi(\mathbf{X}, \mathbf{Y})$ is a conjunction of atoms (possibly equalities, but not inequalities) with the variables $\mathbf{X}$ and $\mathbf{Y}$, and possibly constants, but without nulls. A *Boolean CQ (BCQ)* over $\mathcal{R}$ is a CQ of the form $Q()$, often written as the set of all its atoms, without quantifiers. Answers to CQs and BCQs are defined via *homomorphisms*, which are mappings $\mu : \Delta \cup \Delta_N \cup \mathcal{V} \to \Delta \cup \Delta_N \cup \mathcal{V}$ such that (i) $c \in \Delta$ implies $\mu(c) = c$, (ii) $c \in \Delta_N$ implies $\mu(c) \in \Delta \cup \Delta_N$ and (iii) μ is naturally extended to atoms, sets of atoms, and conjunctions of atoms.

The set of all *answers* to a CQ $Q(\mathbf{X}) = \exists \mathbf{Y} \, \Phi(\mathbf{X}, \mathbf{Y})$ over a database D, denoted $Q(D)$, is the set of all tuples t over Δ for which there exists a homomorphism $\mu : \mathbf{X} \cup \mathbf{Y} \to \Delta \cup \Delta_N$ such that $\mu : \Phi(X, Y) \subseteq D$ and $\mu(X) = t$. The answers to a BCQ $Q()$ over a database D is *Yes*, denoted $D \models Q$, iff $Q(D) \neq \emptyset$.

Given a relational schema $\mathcal{R}$, *a tuple-generating dependency* (TGD) σ is a first-order formula of the form $\forall \mathbf{X} \forall \mathbf{Y} \, \Phi(\mathbf{X}, \mathbf{Y}) \to \exists \Psi(\mathbf{X}, \mathbf{Z})$, where $\Phi(\mathbf{X}, \mathbf{Y})$ and $\Psi(\mathbf{X}, \mathbf{Z})$ are conjunctions of atoms over $\mathcal{R}$ (without nulls), called the *body* and *head* the head of σ, denoted $body(\sigma)$ and $head(\sigma)$, respectively. Such σ is satisfied in a database D for $\mathcal{R}$ iff, whenever there exists a homomorphism h that maps the atoms of $\Phi(\mathbf{X}, \mathbf{Y})$ to atoms of D , there exists an extension h' of h that maps the atoms of $\Psi(\mathbf{X}, \mathbf{Z})$ to atoms of D. All sets of TGDs are finite here. A TGD σ is *guarded* iff it contains an atom in its body that contains all universally quantified variables of σ. A TGD σ is *linear* iff it contains only a single atom in its body. *Query answering* under TGDs, i.e., the evaluation of CQs and BCQs on databases under a set of TGDs is defined as follows. For database D for $\mathcal{R}$, and a set of TGDs Σ on $\mathcal{R}$, the set of models of D and Σ, denoted $mods(D, \Sigma)$, is the set of all (possibly infinite) databases B such that (i) $D \subseteq B$ and (ii) every $\sigma \in \Sigma$ is satisfied in B. The set of *answers* for a CQ Q to D and Σ, denoted $ans(Q, D, \Sigma)$, is the set of all tuples $\mathbf{a}$ such that $\mathbf{a} \in Q(B)$ for all $B \in mods(D, \Sigma)$. The *answer* for a BCQ Q to D and Σ is *Yes*, denoted $D \cup \Sigma \models Q$, iff $ans(Q, D, \Sigma) \neq \emptyset$. It is proved that query answering under general TGDs is undecidable, even when the schema and TGDs are fixed.

For a BCQ Q we say that $(D \cup \Sigma)$ *entail* Q if answer for a BCQ Q to D and Σ is *Yes*, $D \cup \Sigma$ *entail* a set of BCQs if it entail some element of it.

Negative constraints (NC): A negative constraints γ is a first-order formula $\forall \mathbf{X} \, \Phi(\mathbf{X}) \to \perp$, where $\Phi(\mathbf{X})$ (called the body of γ) is a conjunction of atoms (without nulls and not necessarily guarded).

Equality-generating dependencies (EGD): A equality-generating dependency σ is a first-order formula of the form $\forall \mathbf{X} \, \Phi(\mathbf{X}) \to X_i = X_j$, where $\Phi(\mathbf{X})$, called the *body* of Σ, denoted $body(\sigma)$, is a (without nulls and not necessarily guarded) conjunction of atoms, and X_i and X_j are variables from X . Such σ is satisfied in a database D for $\mathcal{R}$ iff, whenever there exists a homomorphism h such that $h(\Phi(\mathbf{X}, \mathbf{Y}) \subseteq D$, it holds that $h(X_i) = h(X_j)$.

The Chase. The *chase* was first introduced to enable checking implication of dependencies, and later also for checking query containment. It is a procedure for repairing a database relative to a set of dependencies, so that the result of the chase satisfies the dependencies. By *chase*, we refer both to the chase procedure and to its result.

⋅   The chase works on a database through TGD and EGD chase rules. Let D be a database, and Σ a TGD of the form $\Phi(\mathbf{X}, \mathbf{Y}) \rightarrow \exists \mathbf{Z}\, \Psi(\mathbf{X}, \mathbf{Z})$. Then, Σ is *applicable* to D if there exists a homomorphism h that maps the atoms of $\Phi(X, Y)$ to atoms of D. Let Σ be applicable to D, and h_1 be a homomorphism that extends h as follows: for each $X_i \in \mathbf{X}$, $h_1(X_i) = h(X_i)$; for each $Z_j \in \mathbf{Z}$, $h_1(Z_j) = z_j$ where z_j is a fresh null, i.e., $z_j \in \Delta_N$, z_j does not occur in D, and z_j lexicographically follows all other nulls already introduced. The *application of Σ on D* adds to D the atom $h_1(\Psi(X, Z))$ if not already in D.

The chase algorithm for a database D and a set of TGDs consists of essentially an exhaustive application of the TGD chase rule in a breadth-first (level-saturating) fashion, which outputs a (possibly infinite) chase for D and Σ.

The chase relative to TGDs is a *universal model*, that means, there exists a homomorphism from $chase(D, \Sigma)$ onto every $B \in mods(D, \Sigma)$. This implies that BCQs Q over D and Σ can be evaluated on the chase for D and Σ, i.e., $D \cup \Sigma \models Q$ is equivalent to $chase(D, \Sigma) \models Q$. For guarded TGDs Σ, such BCQs Q can be evaluated on an initial fragment of $chase(D, \Sigma)$ of constant depth $k|Q|$, which is possible in polynomial time in the data complexity.

Datalog+/- Ontology. A *Datalog+/- ontology* KB $= (D, \Sigma)$, where D is database, $\Sigma = \Sigma_T \cup \Sigma_E \cup \Sigma_{NC}$, consists of a database D, a set of TGDs Σ_T, a set of non-conflicting EGDs Σ_E, and a set of negative constraints Σ_{NC}. We say KB is *linear* iff Σ_T is linear. A Datalog+/- ontology KB $= (D, \Sigma)$ is *consistent*, iff $model(D, \Sigma) \neq \emptyset$, otherwise it is inconsistent.

3   Belief Base Revision in Datalog+/- Ontology

We give an approach for performing belief base revision for Datalog+/- ontology in this section. We present a framework based on kernels and incision functions to deal with the problem of belief revision for Datalog+/- ontologies.

3.1   Revision

Firstly we give the definition of kernel in Datalog+/- ontologies.

Definition 1 (Kernel). *Given a Datalog+/- ontology KB $= (D, \Sigma)$, new observation database instance A, unwanted database instance Ω. A kernel is a set $c \subseteq D \cup A$ such that (c, Σ) entail Ω, and there is no $c' \subset c$ such that (c', Σ) entail Ω. We denote $(D \cup A) \perp\!\!\!\perp \Omega$ the set of all kernels.*

Example 1. A (guarded) Datalog+/- ontology KB $= (D, \Sigma)$ is given below. Here, the formulas in Σ_T are tuple-generating dependencies (TGDs), which say that each person working for a department is an employee (σ_1), each person that directs a department is an employee (σ_2), and that each person that directs a department and works in that department is a manager (σ_3). The formulas in Σ_{NC} are negative constraints, which say that if X supervises Y, then Y cannot be a manager (ν_1), and that if Y is supervised by someone in a department, then Y cannot direct that department (ν_2). The formula Σ_E is an equality-generating dependency (EGD), saying that the same person cannot direct two different departments.

$$D = \{directs(tom, d_1), directs(tom, d_2), worksin(john, d_1), worksin(tom, d_1)\};$$
$$\Sigma_T = \{\sigma_1 : worksin(X, D) \rightarrow emp(X), \sigma_2 : directs(X, D) \rightarrow emp(X),$$
$$\sigma_3 : directs(X, D) \wedge worksin(X, D) \rightarrow Manager(X)\};$$
$$\Sigma_{NC} = \{\nu_1 : supervises(X, Y) \wedge manager(Y) \rightarrow \bot,$$
$$\nu_2 : supervises(X, Y) \wedge worksin(X, D) \wedge directs(Y, D) \rightarrow \bot\};$$
$$\Sigma_E = \{\nu_3 : directs(X, D_1) \wedge directs(X, D_2) \rightarrow D_1 = D_2\}.$$

If new observation is supervises(tom, john), unwanted sentence is emp(tom), then kernel is $c_1 = \{worksin(tom, d_1)\}$.

If new observation is supervises(tom, john), unwanted atom is $\bot$, then kernels are

$$c_1 = \{supervises(tom, john), direct(tom, d_1), worksin(john, d_1)\}$$
$$c_2 = \{supervises(tom, john), direct(tom, d_1), worksin(tom, d_1)\}$$
$$c_3 = \{direct(tom, d_1), direct(tom, d_2)\}.$$

We then give the definition of incision function in Datalog+/- ontologies.

Definition 2 (Incision Function). *Given a Datalog+/- ontology $KB = (D, \Sigma)$, new observation database instance A, unwanted database instance Ω, an incision function is a function σ that satisfies the following two properties*

1. $\sigma((D \cup A) \perp\!\!\!\perp \Omega) \subseteq \bigcup((D \cup A) \perp\!\!\!\perp \Omega)$.
2. *if $X \in ((D \cup A) \perp\!\!\!\perp \Omega)$ then $X \cap \sigma((D \cup A) \perp\!\!\!\perp \Omega) \neq \emptyset$.*

We now give the definition of revision operator. The idea is to add the new information to ontology and then cut unwanted information from it in a consistent way. Intuitively, the revision intends to add A and avoid Ω in a rational way. Note that we cut unwanted database Ω, which is a generation of just cutting $\bot$ from the ontology to avoid the contravention. In this point we are like Ribeiro *et al.* [11].

Definition 3 (Revision Operator). *Given a Datalog+/- ontology $KB = (D, \Sigma)$, new observation database instance A, unwanted database instance Ω, let $D' = (D \cup A) \backslash \sigma((D \cup A) \perp\!\!\!\perp \Omega)$, where σ is an incision function, then revised ontology $KB * A = (D', \Sigma)$.*

Example 2. We continue with example 1. If new observation is supervises(tom, john), unwanted atom is emp(tom), let incision function be $\{worksin(tom, d1)\}$, then revised database is

$D' = \{directs(tom, d_1), directs(tom, d_2), worksin(john, d_1)\}$.

If new observation is supervises(tom, john), unwanted atom is $\perp$, let incision function be $\{supervises(tom, john), direct(tom, d_2)\}$, then revised database is
$D' = \{directs(tom, d_1), worksin(john, d_1), worksin(tom, d_1)\}$.

4 General Properties

It is important to ensure the revision operator behave rationally, so we analyze some general properties of it in this section. We first propose the revision postulates for Datalog+/- ontologies, which is an adaptation of known postulates for semi-revision, then prove that these new postulates are indeed satisfied by the operator.

Definition 4 (Postulates). *Given a Datalog+/- ontology $KB = (D, \Sigma)$, a new observation database instance A, an unwanted database instance Ω, let $KB \Diamond A = (D', \Sigma)$ be the revised ontology, then the postulates are:*

1. *(Consistency) Any element of Ω is not entailed by $KB \Diamond A$.*
2. *(Inclusion)) $D' \subseteq D \cup A$.*
3. *(Core-retainment) if $\beta \in D$ and $\beta \notin D'$, then there is D'' such that $D'' \subseteq D \cup A$, $(D'' \cup \{\beta\}, \Sigma)$ entail Ω, but (D'', Σ) does not entail Ω.*
4. *(Internal change) If $A \subseteq D, B \subseteq D$ then $KB \Diamond A = KB \Diamond B$.*
5. *(Pre-expansion) $(D \cup A, \Sigma) \Diamond A = KB \Diamond A$.*

The intuitive meaning of the postulates are as follows: *consistency* means no atom in the unwanted database should be entailed; *inclusion* means that no new atoms were added; *core-retainment* means if an atom is deleted, then there must be some reason; *internal change* means that every time an ontology is revised by any of its own elements of database, then the result ontology should be same; *pre-expansion* means that if an ontology is expanded by a database and then revised by it, the result should be the same as revising the original ontology by the database.

We now prove that the revision operator given in the last section satisfy these postulates.

Theorem 1. *Given a Datalog+/- ontology $KB = (D, \Sigma)$, new observation database instance A, unwanted database instance Ω, let $KB * A = (D', \Sigma)$ be the revised ontology defined as above, then the revision satisfies the postulates in Definition 4.*

Proof. Inclusion, internal change, pre-expansion follows directly from the construction. To prove *consistency*, assume by contradiction that it is not. Then there is an element of Ω that is entailed by $KB \Diamond A$, as Datalog+/- is a fragments of first-order logic, from compacity of first-order logic it follows that there is a $Z \in D \cup A$, such that there is an element of Ω that is entailed by (Z, Σ). We can then infer by monotonicity that there is a $Z' \subseteq Z$ such that $Z' \in (D \cup A) \perp\!\!\!\perp \Omega$. Then we must have $Z' \neq \emptyset$, and by construction there must be $\epsilon \in \sigma((D \cup A)) \cap Z'$, but if this is true then $\epsilon \notin (D \cup A)$ and $\epsilon \in Z' \subseteq (D \cup A)$, which is a contradiction. To prove *core-retainment*, we have that if $\beta \in D$ and $\beta \notin D'$, then there is $\beta \in \sigma((D \cup A) \perp\!\!\!\perp \Omega)$, That is, there is a $X \in (D \cup A) \perp\!\!\!\perp \Omega$ such that $\beta \in X$. Considering $D'' = X/\beta$, then $D'' \subseteq D \cup A$, $(D'' \cup \{\beta\}, \Sigma)$ entail Ω, but (D'', Σ) does not entail Ω.

5   Algorithms

In the section, we first give an algorithm to compute all kernels when an atom is unwanted, and then give the algorithm of revision. We deal with *linear* ontology in this section.

5.1   Computing Kernels for an Atom

Given a linear Datalog+/- ontology $KB = (D, \Sigma)$, new observation database instance A, unwanted atom ω, we give in this subsection a method to calculate all kernels $(D \cup A) \perp\!\!\!\perp \omega$.

We first give the method when unwanted atom λ is a node in *Chase* of $(D \cup A, \Sigma)$. The idea is to travel chase graph bottom-up and then cut non-minimal ones from result sets of atoms. Note that this idea is similar to the one used in Lukasiewicz etc [4], which use a bottom-up travel of chase graph starting from the matching a body every rule of Σ_{NC} to compute the culprits of an inconsistent ontology.

We now give algorithm Justs(λ). Min(Justs(λ)) are exactly kernels $(D \cup A) \perp\!\!\!\perp \lambda$, where Min is used in the usual sense of subset inclusion. Note that *just* below is not a set of nodes, but a set of sets of nodes. Note also that step 2 can be done because we can collect all information needed from *Chase*.

We now have the following theorem.

Theorem 2. *Given a linear Datalog+/- ontology $KB = (D, \Sigma)$, new observation database instance A, Let λ be a node in chase graph of Datalog+/- ontology $(D \cup A, \Sigma)$, then $Min(Justs(\lambda)) = (D \cup A) \perp\!\!\!\perp \lambda$. The algorithm run in polynomial time in the data complexity.*

Proof. We first prove that $(Min(Justs(\lambda)), \Sigma)$ entail λ, that is, for every model M of $(Min(Justs(\lambda)), \Sigma)$, there is a homomorphism μ such that $\mu(\lambda) \subseteq M$.

In fact, we will show that for all nodes that were produced in the bottom-up travel process, it holds that the node is entailed by $(Min(Justs(\lambda)), \Sigma)$, Then, as a result, the λ is also entailed by $(Min(Justs(\lambda)), \Sigma)$ automatically.

Suppose the derivation level of λ is N. We prove level by level from 1 to N. If the node level is 1. let M be a model of $(Min(Justs(\lambda)), \Sigma)$, then $M \supseteq Min(Justs(\lambda))$, but from the definition of satisfaction of a rule, whenever there is a homomorphism that map the body to the database, there is a extended homomorphism that map the head to the database, so, this homomorphism map the node to M, the entailment holds.

Algorithm 1. Justs(λ)

Require: a linear Datalog+/- ontology $KB = (D, \Sigma)$, a new observation database instance A, and a node λ in *Chase* of $(D \cup A, \Sigma)$.
Ensure: Justs(λ)
 1. $just = \emptyset$
 2. **for all** $\Phi_i \subseteq$ (nodes in *Chase*) such that there is rule $r : \Phi(X, Y) \to \exists Z \ \Psi(X, Z)$
    which is applicable to Φ_i and produces λ **do**
 3. $just = just \cup \{\Phi_i\}$
 4. **end for**
 5. **for all** $\Phi_i \in just$ **do**
 6.   **for all** $\Phi_i^j \in \Phi_i$ **do**
 7.   **if** Justs(Φ_i^j) $\neq \emptyset$ **then**
 8. $just = $ Expand($just$, Φ_i^j)
 9.   **end if**
10.   **end for**
11. **end for**
12. **return** $just$

---

13. Expand($just$, a)
14. **for all** $\phi \in just$ **do**
15.   **if** $a \in \phi$ **then**
16. $just = just/\phi$
17.   **for all** $j\_a \in$ Justs(a) **do**
18. $just = just \cup \{\phi/a \cup j\_a\}$
19.   **end for**
20.   **end if**
21. **end for**
22. **return** $just$

Suppose for all node whose derivation level is n, it is right. That is, there is a homomorphism that map the node to M, now we consider the node whose derivation level is n+1. Consider the rule that applicable and can get this node, since all parent nodes of the node has a level smaller or equal to n, so there is a homomorphism that map these nodes to M, as the rule itself is satisfied by M, we can then construct a new homomorphism by extend the above homomorphism to map the node to M. So we have $(Min(Justs(\lambda)), \Sigma)$ entail the node.

We now show that there are no other subsets of $D \cup A$ that along with Σ entail λ and is smaller than Min(Justs(λ)). Otherwise, suppose it is not, that is, there is a subset of $D \cup A$ that is smaller than Min(Justs(λ)) and along with Σ entail λ. Then we have a *ChaseGraph* that end with λ, however, according to the construction of the Min(Justs(λ)), this set should be equal to some element of Min(Justs(λ)), this is a contraction.

We finally shows that computing $Min(Justs(\lambda))$ in the linear case can be done in polynomial time in the data complexity. Note that the *ChaseGraph* is constant-depth and polynomial-size for a linear ontology due to the result in [3]. Note also that Justs(λ) is a recursive procedure, it will be called $N \times M$ times at most, where N is the depth of the graph and M is the numbers of rules in the Σ, and that at each time the algorithm is running, the time complexity exclusive the recursive procedure is polynomial, so the algorithm $Min(Justs(\lambda))$ run in polynomial time.

We now give the algorithm $Kernels(\omega)$ to compute kernels for an atom ω. Note that by $match(\omega, Chase)$ we mean the procedure of finding the same node as ω in $Chase$, if it is successful, return this node, otherwise return $\emptyset$.

Algorithm 2. $Kernels(\omega)$

Require: a linear Datalog+/- ontology $KB = (D, \Sigma)$, a new observation A, and an
    atom ω
Ensure: $Kernels(\omega)$
1. compute the $Chase$ of $KB = (D \cup A, \Sigma)$
2. L=match(ω, $Chase$)
3. **if** L=$\emptyset$ **then**
4.   return $\emptyset$
5. **else**
6.   return $Min(Justs(L))$
7. **end if**

Theorem 3 (Correctness and Complexity of Kernels(ω)). *Given a linear Datalog+/- ontology $KB = (D, \Sigma)$, new observation A, an unwanted atom ω, algorithm Kernels(ω) compute $(D \cup A) \perp\!\!\!\perp \omega$ correctly in polynomial time in the data complexity.*

Proof. If L=$\emptyset$, then $KB' = (D \cup A, \Sigma)$ do not entail the atom as $ChaseGraph$ is sound and complete with respect to query answering. If L$\neq \emptyset$ then the atom is entailed by KB', in this case, $MinJust(L)$ are all minimal sets of atoms that belong to database $D \cup A$ and along Σ entail ω according to Theorem 1. So $Kernels(\omega)$ compute $(D \cup A) \perp\!\!\!\perp \omega$ correctly in both cases.

The complexity of the algorithm depends on the *match* procedure, as the $ChaseGraph$ is polynomial-size and constant-depth for a linear ontology, the travel of $ChaseGraph$ can be done in polynomial time, so the $Kernels(\omega)$ can be done in polynomial time.

5.2 Revision

We now give the algorithm to revise a linear Datalog+/- ontology named as *RevisionKB*.

Note that $((D \cup A), \Sigma)$ may be an inconsistent ontology, however, inconsistency can be removed in the revised ontology by making $\perp \subseteq \Omega$.

We now show that the algorithm can compute revision of ontology and give the complexity.

Algorithm 3. RevisionKB

Require: a linear Datalog+/- ontology $KB = (D, \Sigma)$, a new observation A, unwanted
   instance Ω
Ensure: Revised ontology $KB * A$
 1. **for** every atom ω, $\omega \in \Omega$ **do**
 2.   compute Kernels(ω)
 3. **end for**
 4. $(D \cup A) \perp\!\!\!\perp \Omega = minimal(\bigcup_{\omega \in \Omega} Kernels(\omega))$
 5. $KB * A = ((D \cup A) \backslash \sigma((D \cup A) \perp\!\!\!\perp \Omega), \Sigma)$

Theorem 4 (Correctness and Complexity of RevisionKB). *Given a linear Datalog+/- ontology $KB = (D, \Sigma)$, new observation database instance A, unwanted database instance Ω, algorithm RevisionKB can compute revision correctly in polynomial time in the data complexity.*

Proof. Note that $(D \cup A) \perp\!\!\!\perp \Omega$ can be obtained by combining $(D \cup A) \perp\!\!\!\perp \omega$ for every elements $\omega \in \Omega$ and cut from it the non-minimal ones, the algorithm's correctness then follows directly from the definition of revision operator.

The complexity of the algorithm depends basically on the task of finding Kernels(ω), as it run in polynomial time due to Theorem 3, so we have the conclusion.

6   Related Works

There is strong relationship between Datalog+/- ontology and description logic as they can be translated to each other in many cases. In the area of belief revision in description logic, Ribeiro *et al.* [11] bear much similarities to our work since they also used a kernels-based semi-revision method, they give a belief revision method to a monotonic logic, take description logic as a spacial case. However, there are difference between their work and this paper. They deal with monotonic logic, but Datalog+/- has a different syntax and semantic and cannot been cover by it. Furthermore, Ribeiro *et al.* [11] compute kernels by invoke classical reasoning, but this paper give a direct method to calculate kernels and prove that the complexity of computing revision is tractable.

In the area of Datalog+/-, Lukasiewicz *et al.* [4] give an inconsistency reasoning method for Datalog+/- ontologies, theirs work is close related to ours work since they use culprit to resolve the inconsistency of ontologies, and the culprit is equivalent with the kernel of this paper in the case of atom unwanted is $\perp$. However, there are some difference. Although the area of belief change is closely related to the management of inconsistent information, they are still quite different in both goals and constructions. Inconsistency can be handled by using kernels and clusters in [4], ours work can also deal with inconsistency, however revision operator given in ours work can do more except this, for example, it can choose a set of atoms as unwanted information, not only $\perp$, thus give more flexibility to resolve the inconsistency, and in this sense this work is more general than theirs work. Note also that in [4] the properties of the reasoning result are not clear even in the case of inconsistency handling because they did not study the properties of the operation from the viewpoint of belief revision.

There are still some works that extend Datalog+/- with the capability of dealing with uncertainty. In Lukasiewicz *et al.* [5], they developing a probabilistic extension of Datalog+/-. This extension uses Markov logic networks as the underlying probabilistic semantics and focus especially on scalable algorithms for answering threshold queries. Riguzzi *et al.* [8] apply the distribution semantics for probabilistic ontologies (named DISPONTE) to the Datalog+/- language. Lukasiewicz *et al.* [9] tackle the problem of defining a well-founded semantics for Datalog rules with existentially quantified variables in their heads and negations in their bodies, thus provide a kind of nonmonotonic reasoning capability to Datalog+/-. Our work also deal with the commonsense reasoning in the background of Datalog+/- language, however, we focus on the problem of belief revision, instead of adding quantitative or qualitative uncertainties to ontologies.

7 Summary and Outlook

In this paper, we address the problem of belief revision for Datalog+/- ontologies. In our approach, we introduce a kernel based belief revision operator, and study the properties using extended postulates, we then provide algorithms to revise Datalog+/- ontologies, and give the complexity results by showing that query answering for a revised linear Datalog+/- ontology is tractable.

In the future, we plan to study how to implement belief revision when some heuristic information, i.e., different trust [12] or reputation [2] levels of both database and rule set due to the different source of information, can be added to Datalog+/- ontologies.

Acknowledgments. This work is partially supported by the National Natural Science Foundation of China Grant No.61003022 and Grant No.41174007, as well as the FP7 K-Drive project (No. 286348) and the EPSRC WhatIf project (No. EP/J014354/1).

References

1. Flouris, G., Huang, Z., Pan, J.Z., Plexousakis, D., Wache, H.: Inconsistencies, negations and changes in ontologies. In: Proceedings of the 21st National Conference on Artificial Intelligence, AAAI-06, pp. 1295–1300 (2006)
2. Koster, A., Pan, J.Z.: Ontology, semantics and reputation. In: Agreement Technologies. Springer (2013). ISBN 978-94-007-5582-6
3. Lukasiewicz, T., Cali, A., Gottlob, G.: A general datalog-based framework for tractable query answering over ontologies. J. Web Semant. **14**, 57–83 (2012)
4. Lukasiewicz, T., Martinez, M.V., Simari, G.I.: Inconsistency handling in datalog+/- ontologies. In: The Proceedings of the 20th European Conference on Artificial Intelligence, ECAI 2012, pp. 558–563 (2012)

5. Lukasiewicz, T., Martinez, M.V., Simari, G.I.: Query answering under probabilistic uncertainty in Datalog+/- ontologies. Ann. Math. Artif. Intell. **69**(1), 195–197 (2013)

6. Pan, J.Z., Thomas, E., Ren, Y., Taylor, S.: Exploiting tractable fuzzy and crisp reasoning in ontology applications. IEEE Comput. Intell. Mag. **7**(2), 45–53 (2012)

7. Qi, G., Haase, P., Huang, Z., Ji, Q., Pan, J.Z., Völker, J.: A Kernel revision operator for terminologies — algorithms and evaluation. In: Sheth, A., Staab, S., Dean, M., Paolucci, M., Maynard, D., Finin, T., Thirunarayan, K. (eds.) ISWC 2008. LNCS, vol. 5318, pp. 419–434. Springer, Heidelberg (2008)

8. Riguzzi, F., Bellodi, E., Lamma, E.: Probabilistic Datalog+/- under the distribution semantics. In: The Proceedings of the 25th International Workshop on Description Logics (DL), Aachen, Germany, pp. 519–529 (2012)

9. Lukasiewicz, T., Martinez, M.V., Simari, G.I.: Well-founded semantics for extended datalog and ontological reasoning. In: The Proceedings of the 32nd ACM Symposium on Principles of Database System. ACM Press (2013)

10. Hansson, S.O.: Semi-revision. J. Appl. Non-Class. Logics **7**(1–2), 151–175 (1997)

11. Ribeiro, M.M., Wassermann, R.: Base revision for ontology debugging. J. Logic Comput. **19**(5), 721–743 (2009)

12. Sensoy, M., Fokoue, A., Pan, J.Z., Norman, T., Tang, Y., Oren, N., Sycara, K.: Reasoning about uncertain information and conflict resolution through trust revision. In: Proceedings of the 12th International Conference on Autonomous Agents and Multiagent Systems (AAMAS2013) (2013)

Constructing City Ontology from Expert for Smart City Management

Tong Lee Chung[1(✉)], Bin Xu[1], Peng Zhang[1], Yuanhua Tan[2],
Ping Zhu[2], and Adeli Wubulihasimu[3]

[1] Knowledge Engineering Group, Department of Computer Science
and Technology, Tsinghua University, Beijing, China
{tongleechung86,zpjumper}@gmail.com,
xubin@tsinghua.edu.cn
[2] Hongyou Software Co Ltd, Karamay, China
{tanyh66,hy-zhp}@petrochina.com.cn
[3] PetroChina Urumqi Petrochemical Company, Beijing, China
aslwsl@petrochina.com.cn

Abstract. City Ontology plays an important role in smart city management for data integration, reasoning decision support etc. With these managerial domain knowledge scattered among a large number of experts, researchers face a huge challenge constructing a complete ontology for city management. This paper presents a simple yet efficient method for non-computer science experts to construct an ontology. We use a middle part that acts as a transition layer called activity model which is later merged into the city managerial ontology. We prove the effectiveness of this method by constructing a managerial ontology for two departments in Karamay's Smart City Program.

Keywords: Ontology · City management · Smart city · Managerial activity model

1 Introduction

An ontology is a model of a particular domain, built for a particular purpose [1]. Domain ontologies capture knowledge of one particular domain and are usually constructed manually [1]. OWL[1] language is standardize by the World Wide Web Consortium (W3C) Web Ontology Working Group for constructing ontology and is widely used by researchers and developers. A city managerial ontology describes managerial activities and processes in a city that can be used for data integration from different database, reasoning to fill in missing links and decision support.

Yet information for constructing a city managerial ontology is scattered among many experts, mostly not from computer science background. City management is a huge domain with many different fields and experts are those who have many years of experience [2]. These experts would appreciate that their expertise be stored as ontology instead of being preserved in books. The biggest challenge faced in ontology

---

[1] http://www.w3.org/TR/owl-features/

W. Kim et al. (Eds.): JIST 2013, LNCS 8388, pp. 187–194, 2014.
DOI: 10.1007/978-3-319-06826-8_15, © Springer International Publishing Switzerland 2014

construction is to equipped domain experts with the right tool. It would be naive idea to have experts learn an ontology language and make them work together to construct a complete city ontology.

This paper puts forward a method for experts to construct a managerial ontology that doesn't request prerequisites of ontology language or knowledge. We first have expert to model their managerial knowledge as processes and activity which is used as a transition model. The managerial activity model is then transformed into an OWL form ontology where duplicate concepts are later merged.

This method is tested by constructing the ontology for Karamay's Urban Management department's management activities and process and oil exploration activities. The ontology is reasonable and rather complete that describes activates and procedures of their duty in city management.

The contributions of this paper are as followed:

- Provide experts a tool to construct ontology automatically from managerial activity model. Instead of having experts work in a field of ontology which they are not familiar with, let them model the process of managerial work in a way they are more comfortable with. This method will save developers and experts time and money constructing ontologies. This method will also reduce the communication barriers between domain experts and developers.
- Put forward an effective activity model for describing city managerial process. The managerial activity model is able to describe workflow of city management. This model gives experts abilities to describe their work in a form they are familiar with. And this model is also easy for developers to understand.

The rest of the paper is organized as follows: In Sect. 2, we will present the managerial activity model. In Sect. 3, we will discuss the transformation process of managerial activity model to ontology and two success case using our method. In Sect. 4, we will briefly look at related works. And finally, we will conclude our paper in Sect. 5.

2   Managerial Activity Model to Ontology

Before going into details, we will first describe city managerial procedures and city managerial activities and the importance of ontology in city management. Then we will define some rules and regulations for modeling city managerial activity which makes transformation process possible.

City Management and Ontology

By city management, we mean managing functional departments of city government like police department, fire department etc. In a city, managing all these department efficiently requires integrating different databases and optimizing activities and etc. This is a difficult task and requires an understanding of all domain knowledge, a retentive memory and an agile logic in one person which is almost impossible.

An ontology is a model of a particular domain, built for a particular purpose [1]. It is handy to construct an ontology to include all procedure and activities in city management. A complete ontology can be useful in many areas. Data integration can be done with the help of ontology. Activity arrangement can be better optimized with the help of ontology. Reasoning and inference can be done on ontology for decision support. A huge challenge is how to construct such an ontology. City managing experts are often those who have been in the field for many years and only process a small portion of knowledge in city management. Usually, these people do not have a strong computer science background. How to get all these knowledge together and construct a complete ontology is a real big challenge.

Management Activity Model

A managerial activity model is a model that describes the workflow of an activity. We were greatly influenced by workflow business models [3, 4] and workflow graph [5]. Experts in city management are more confident about their workflow than ontology, so it would be a good idea to have experts work on their comfort zone. There are a few requirement for the activity model that allows the model be transformed into an ontology directly.

The basic unit of the model is activity and cannot be divided. Activity consists of people, venue, time, objects etc. This is a reasonable and complete form to describe activities in city management and can be easily understood. A procedure is a set of activities in a sequential order. A function is a set of procedure in a sequential order. And the certain domain consists of a series of functions. This makes up the upper layer of city management in a hierarchical form. The bottom layer consists of objects and attributes. Objects are physical things that were involved in an activity. Objects are separated into two categories: persons and non-person. Person describes the people that were involved in the activity. We define two types of people in an activity: operator and participant. Operator by definitions is the person operating or the person in charge and participant is people that the activity involves. We use material to define other non-person objects. Material type includes infrastructure, device, document and etc. Attribute is used to enrich descriptions of objects and activities. Types of attributes include date, string, digit etc. The output of management activities are usually a report that concludes the activity.

The model contains four columns: Identifiers, terms, category and type. Identifier is used to indexing and showing hierarchical relation. Term is basically the name of the concept. Category is used to determine which category the item falls into. And the type defines the type of the item. We have already define category and type in the above section.

This model can be understood by experts easily and matches the logic they have in mind for managing a city in their own field. This model allows experts to focus only on their expert field and ignore ambiguity and communication problem they will have to face when working together. The tools used to construct this model is as simple as Excel. This makes expert's works simple and focused. This model doesn't require experts to have knowledge about ontology or computer science. It lets them better express their knowledge as it should be. Table 1 is a sample from experts in Urban Management Department. The underlined item represents an activity.

Table 1. Constructing managerial activity model using Excel

Identifier	Term	Category	Type
J	Urban Management	domain	
J.3	City appearance	function	
J.3.8	Street standardize inspection	process	
J.3.8.1	Beforehand meeting	activity	
J.3.8.1.1	Meeting date	Attribute	date
J.3.8.1.2	Meeting room	Material	infrastructure
J.3.8.1.2.1	Address	Attribute	string
J.3.8.1.3	Chief inspector	Person	operator
J.3.8.1.3.1	Name	Attribute	string
J.3.8.1.4	Inspector	Person	participant
J.3.8.1.4.1	Name	Attribute	string
J.3.8.1.5	*Meeting minutes for Inspection*	*Material*	*document*
J.3.8.1.5.1	Meeting date	Attribute	date
J.3.8.1.5.2	Meeting room	Material	infrastructure
J.3.8.1.5.2.1	Address	Attribute	string
J.3.8.1.5.3	Chairman	Person	operator
J.3.8.1.5.3.1	Name	Attribute	string
J.3.8.1.5.4	Inspector	Person	participant
J.3.8.1.5.4.1	Name	Attribute	string
J.3.9.1	On-street inspection	activity	
J.3.9.1.1	inspection date	Attribute	date
J.3.9.1.2	inspection duration	Attribute	digit
J.3.9.1.3	Chief inspector	Person	operator
J.3.9.1.3.1	Name	Attribute	string
J.3.9.1.4	Inspector	Person	participant
J.3.9.1.4.1	Name	Attribute	string
J.3.9.1.5	Vehicle	Material	device
J.3.9.1.5.1	Plate number	Attribute	digit
J.3.9.1.6	Two way radio	Material	device
J.3.9.1.6.1	Radio number	Attribute	digit
J.3.9.1.12	*Inspection report*	*Material*	*document*
J.3.9.1.12.1	inspection date	Attribute	date
J.3.9.1.12.1.1	inspection duration	Attribute	digit
J.3.9.1.12.1.2	Chief inspector	Person	operator
J.3.9.1.12.1.3	Name	Attribute	string
J.3.9.1.12.1.4	Inspector	Person	participant
J.3.9.1.12.1.5	Name	Attribute	string
J.3.9.1.12.1.6	Vehicle	Material	device
J.3.9.1.12.2	Plate number	Attribute	digit
J.3.9.1.12.2.1	Two way radio	Material	device
J.3.9.2.1	Radio number	Attribute	digit

3   Constructing Ontology from Activity Model

Design of the activity model is intended to construct a city managerial ontology without having experts look at the ontology. The design itself consist of an ontology that we consider the base of the city ontology. The requirements of activity model is based on this base model.

The activity model can be easily constructed into a tree using the identifiers. Each level has a concept in the base ontology that is the superClass. The tree is divided into two part: the upper level and the lower level. The upper level is the hierarchy tree of activity. We use the relation "hasFunction","hasProcess","hasActivity" to denote the corresponding relations. This part of the tree is preserved as to represent the flow of work. We use two relations to represent workflow: "first" and "next". In the upper level, the "first" relation indicates that it is the first child of all siblings and "next" relation to indicate that it is the next in the set of children. We use type column to define predicates between two concepts. Next we build sub-predicate relation for predicate linking to base ontology (Fig. 1).

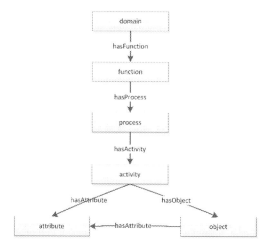

Fig. 1. Base ontology

The lower level of the tree contains objects and attributes, which is the part we have to merge same concepts. There are two strategies we use for merging the same concept. The basic strategy is using the term of the item. If it has the same term then we consider that they are the same concept. Strategy two is more complicated. We first compare the cosine similarity of the term, if it exceeds a threshold then we compare the semantic similarity based on linked node counting. Similarity is calculated as the quotient of same concept linked over the total number of linked concepts. We add a normalizing factor to smoothen the results.

But since this is a strict ontology, we won't have machine do all the job, we will enquire expert confirmation before we merge concepts. That will require experts who define the term to confirm whether the two concepts should be merged. We will lower the load for them by finding possible concepts to merge. And finally, we output this ontology in OWL format. Figure 2 show a graph ontology of the city managerial model given in Table 1. We omit the subPredicate relation from the graph to make it cleaner.

We were very fortunate to be able to test our method in Karamay. Karamay is an oil city with success in digitization. We tested our method in two departments: Oil Exploration Department and Urban Management Department. We asked experts in these field to model their management process with the management activity model. They easily understood the requirements and finished modelling rather quickly. We transformed these model into one ontology and had these experts look at the ontology. They considered it to be complete and compatible. The ontology is outputted in OWL format and will be used later in data integration and decision support engine (Table 2).

4   Related work

Most domain experts are not from computer science field, most of them do not have the understanding to computer science technologies. We wish to give them the right tools to construct an ontology. Researchers have come up with many tools and

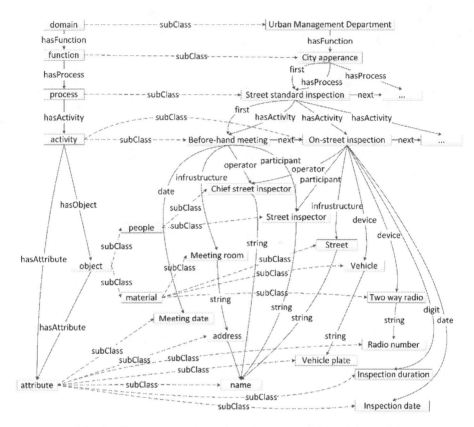

Fig. 2. City management ontology from managerial activity model

Table 2. Stats of ontology construction

Column 1	# of activity model	# of concepts	# of links
Oil exploration department	23	9890	40572
Urban management department	15	5200	24345
Total	38	15090	64917

methods to construct ontology. Jena is a java API that has full support of ontology construction and common task to perform [6]. Protégé Ontology Editor is an ontology editor and knowledge base framework that is easy to construct an ontology [7]. Bouillet propose a method where several experts can construct an ontology in a collaborative manner [8]. But these tools and methods are not the right tools for experts to construct ontology. Experts will have to construct the ontology manually, sometimes without even knowing what the ontology looks like thus increasing the burden of experts. Also, experts will have to collaborate a lot making some of them lose focus on the areas they are skilled at. Our method allows expert describe workflow and activities instead of directly constructing ontology. Our method lets expert works collaborative in a manner that they only focus on their own field.

There are many research on smart cities. Beirao construct a street description for a city ontology [9]. Jun Zhai propose a system architecture of integrated information platform for digital city based on ontology [10]. There has been a lot of breakthrough in the field of smart city with ontology, but with very little consideration of providing an efficient way to construct ontology.

5   Conclusion and Future Work

With the rapid development and digitalization of cities, there is an urgent need to construct an ontology for developing intelligent system for smart city support. But experts face the problem of learning ontology technology, computer science technology and cooperating with many other experts. In this paper, we propose a method that allows non-computer science experts to efficiently preserve their expert knowledge as ontology. We introduce a middle transition model called management activity model that is fairly easy for experts to model their expert field and can be directly converted to ontology requiring only a small process for experts.

As future work, we will try to extend the method to easily construct hierarchy in objects and attributes, for example, name is a super-class of street name and people name. We would like to find missing links within the ontology, for example, street inspector include chief street inspector. We will also have more experts to model the whole domain of city ontology of Karamay. Another important part of future is to select vocabulary so that our work can be reused properly.

Acknowledgments. This work is supported by the China National High-Tech Project (863) under grants No SS2013AA010307.

References

1. Antoniou, G., Grouth, P., van Harmelen, F., Hoekstra, R.: A Semantic Web Primer, Chap. 7, 3rd edn, pp. 193–213. MIT Press, Cambridge (2012)
2. Donahue, A.K., Selden, S.C., Ingraham, P.W.: Measuring government management capacity: a comparative analysis of city human resources management systems. J. Public Adm. Res. Theory **10**(2), 381–412 (2000)
3. Myers, K., Berry, P.: Workshop management system: an AI perspective. AIC-SRI report, pp. 1–34 (1998)
4. Morries, M., Schindehutte, M., Allen, J.: The entrepreneur's business model: towards a unified perspective. J. Bus. Res. **58**, 726–735 (2005)
5. Eder, J., Gruber, W., Pichler, H.: Transforming workflow graphs. In: INTEROP-ESA, Genf, Switzerland (2005)
6. http://jena.apache.org/documentation/ontology/
7. http://protege.stanford.edu/

8. Bouillet, E., Feblowitz, M., Liu, Z., Ranganathan, A., Riabov, A.: A knowledge engineering and planning framework based on OWL ontologies. Association for the advancement of artificial intelligence (2007)
9. Beirao, J., Monttenegro, N., Gil, J.: The city as a street system: a street description for a city ontology. In: SIGraDi 2009 sp (2009)
10. Zhai, J., Jiang, J., Yu, Y., Li, J.: Ontology-based integrated information platform for digital city. In: WiCOM 2008, 4th International Conference (2008)

Constructing Event Corpus from Inverted Index for Sentence Level Crime Event Detection and Classification

S.G. Shaila, A. Vadivel[(✉)], and P. Shanthi

Multimedia Information Retrieval Group,
Department of Computer Applications, National Institute of Technology,
Trichy, Tamil Nadu, India
{shaila,vadi}@nitt.edu, shanthicse@mamce.org

Abstract. Event detection identifies interesting events from web pages and in this paper, a new approach is proposed to identify the event instances associated with an interested event type. The terms that are related to criminal activities, its co-occurrence terms and the associated sentences are considered from web documents. These sentence patterns are processed by POS tagging. Since, there is no knowledge on the sentences for the first instances, they are clustered using decision tree. Rules are formulated using pattern clusters. Priorities are assigned to the clusters based on the importance of patterns. The importance of the patterns defines the semantic relation towards event instances. Considering the priorities, weights are assigned for the rules. Artificial Neural Network (ANN) is used to classify the sentences to detect event instances based on the gained knowledge. Here ANN is used for training the weighted sentence patterns to learn the event instances of the specific event type. It is observed that the constructed rule is effective in classifying the sentences to identify event instance. The combination of these sentence patterns of the event instances are updated into the corpus. The proposed approach is encouraging when compared with other comparative approaches.

Keywords: Event instance · Web documents · Sentence · Terms · NLP patterns · ANN · Crime

1 Introduction

In recent times, the Internet has grown rapidly in size and documents provide a wealth of information about several events in an unstructured form. There is a great need to understand the content of the information sources in the form of events such as What, When, Where, Why, etc. A system is required to quickly understand the overall relevance of documents, collections and overviews to exhibit their structure by highlighting possible subsets of interest. Cognitive scientists believe that people experiences and cognize the objective world based on events [15]. The concept of the events is used in various knowledge processing domains. In natural language, a semantic portion composed of nouns, verbs, prepositions and other forms are considered as events and text can be considered as a collection of events. The traditional

W. Kim et al. (Eds.): JIST 2013, LNCS 8388, pp. 195–208, 2014.
DOI: 10.1007/978-3-319-06826-8_16, © Springer International Publishing Switzerland 2014

view of natural language understanding refers text as a collection of concepts by combining specific forms of concepts such as nouns, verbs, prepositions, and adjectives and can be used to extract event information. The approach in [7] have used event based features to represent sentences and shown that the approach improves the quality of the final summaries compared to a baseline bag-of-words approach. The events present in the text provide greater knowledge granularity and an avenue to understand the event level based on natural language. However, it is found that classifying the sentences at a sentence level is a challenging task. For example, if the event type is "Crime" and the sentences such as "8 people killed in separate incidents of violence" and "A young couple was found dead on Monday morning" should be identified. These sentences contain terms connected with death like "kill", "die" or "murder" and is considered as interesting terms since it catches many interesting event instances. Sometimes, there are instances, where this approach may fail and this is due to the fact that it is a naive approach. Event detection in web documents is fetching attention from the researchers and many event extraction approaches have been reported for decades. The Topic Detection and Tracking (TDT) has investigated the development of technologies that could detect novel events in segmented and unsegmented news streams to track the progression of these events over time [4]. An Automatic Content Extraction (ACE) approach [2] has also been proposed for event detection, named as Event Detection and Recognition (EDR), which identifies all event instances of a pre-specified set of event types [10].

In general, events are classified as event instance and event type. An event type refers to the generic class that the event belongs to and event instance refers to a specific occurrence of a particular event. In particular event, the sentence that only describes the event is referred as event mention and the word that most clearly expresses the event's occurrence is event trigger. An approach has been proposed for automatically identifying event extent, event trigger and event argument using the bootstrapping method [14]. Nobel Prize winning domain is used to extract the events by using extraction rules from the text fragments by using binary relations as seeds. A combined approach with statistical and knowledge based technique has been proposed for extracting events by focusing on the summary report genre of a person's life incidents to extract the factual accounting of incidents [11]. The approach covers entire instance of all verbs in the corpus. The events are extracted by breaking each event into elements by understanding the syntax of each element [1]. The role played by each element in the event and the relationship between related events are identified. Events are identified by tagging the text with name and noun phrase using pattern matching techniques and resolved co-references using rule based approach [3]. However, one of the drawbacks of this approach lies in identifying semantic relatedness. The extracted events are segmented into subtasks such as anchor identification, argument identification, attribute assignment and event co-reference [6]. Each task used classifier such as memory based learners and maximum entropy learners. The language modeling approach is compared to event mention detection with an approach based on SVM classifier [9, 12]. An approach has been proposed to recognize the event in the biomedical domain as one of the conceptual recognition and analysis [5]. The method identified the concepts using named entity recognition task. An automatic event extraction system is proposed by solving a classification problem with rich features [8]. In [13], rough set theory for the terms is applied that identify a specific interested event and the terms that do not identify the specific event of interest.

It is noticed from the above discussion that most of the approaches identifies and extracts the events without considering the semantic features of the text. The drawback lies in identifying the patterns within the event instance. Thus, it is imperative that a suitable approach is required for identifying various patterns in the considered event instances. In this paper, we have addressed this issue by formulating the rules that finds semantics connected to a event instances using hand crafted list of terms collected from various crimes related web sites. The approach identifies the sentence patterns in a document that describes one or more instances of a specified event type. Since sentence meaning improves the efficiency of event detection, the terms are the basic unit of meaning and the terms in the same context have semantic relationships between them. We consider this task as a text classification problem where each sentence in a document is classified into Objective event and Subjective event. Classification And Regression Tool (CART) based decision tree is initially generated for clustering the sentences patterns at unsupervised level. Rules are generated and weights are assigned based on patterns importance. The combination of these patterns in subjective and objective sub-classes are considered for Artificial Neural Network (ANN) to train the sentence patterns by encoding the training and test instance. These sentence patterns are updated into the corpus. The remainder of this paper is organized as follows. In the next Section, we describe the proposed approach. The experimental results are presented in Sect. 4 and we conclude the paper in the last Section.

2   Proposed Work

In our proposed approach, we have identified the event instances of specific event type by building an event corpus from the inverted index. A large number of web documents from crime websites are crawled and collected. As a preprocessing step, the stop words are removed from the text and however, the terms are stored without stemming to maintain linguistic patterns. The inverted index is constructed by extracting the terms and their frequency of occurrence. For each term $<t>$, there is a posting list that contains document $id's$ and frequency of occurrence of term $<d, f>$. The information available in the inverted index such as terms, frequency of occurrences and the related document $id's$ are used for event detection.

Let D be a set of documents retrieved from WWW and T be a set of terms present in D. The presence of terms in a document may be treated as a labeling approach denoted as follows.

$$l: T \times D \rightarrow \{True, False\} \tag{1}$$

The inverted index consists of interested as well as non-interested terms. We select interested terms manually. In this context, the terms that explains the event topic are considered as interested terms. For instance, crime, kill, blood etc., relates to crime event and synonyms, hyponyms and hypernyms for the interested terms are also found. This helps to find the rest of the interested term present in the inverted index. From Eq. (1), it is assumed that an interested term $t \in T$ presents in a document $d \in D$, if $l:(t, d) = True$. In document retrieval applications, the posting list is extracted from the inverted index.

The posting list is in the form of $<t, d, f>$ where f is frequency of occurrence assigned to term t in document d. Since, a term can be physically appearing in many documents, given a interested term i_t, such that $i_t \subseteq I_T$, i_t can be defined as the relationship of $<t, d, f>$ as given in Eq. (2).

$$C^D(i_t) = \{ <i_t, d, f> \mid d \in D, i_t \in T, f \in F \text{ and } \forall i_t \subseteq I_T, (i_t = t), \ (t, d) = True\}$$
(2)

Later, the sentences associated with the interested terms are extracted from the respective documents as shown in above Fig. 1 and interested event corpus is built, which gives the first step to identify criminal activity events.

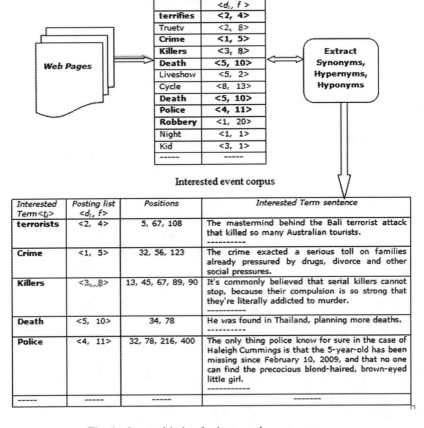

Fig. 1. Inverted index for interested event corpus

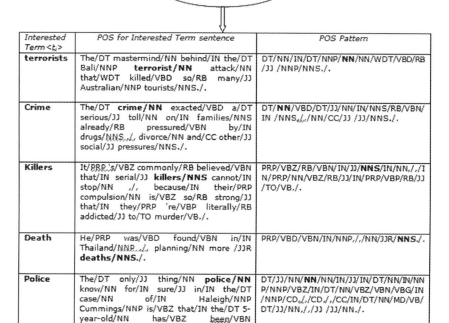

Fig. 2. Interested sentence POS tagging

The interested sentence from the event corpus is further processed for finding Part Of Speech (POS) Tagging (http://nlp.stanford.edu/software/tagger.shtml). The Stanford NLP tool is used for POS tagging to find the patterns of interested sentences. Below, Fig. 2 depicts the interested terms and their associated sentences along with their POS tags.

The sentence patterns obtained are huge and there is no knowledge. In the absences of knowledge or ground truth, it would be very difficult to classify the sentences. In this approach, initially the sentences are clustered using decision tree, which is an unsupervised approach. Upto level 3, the sentences are clustered based on POS constraints and the fourth level represents the final sentence cluster, which is depicted in Fig. 3.

Table 1 represents the sentence classes in 3 levels. The first level is displayed at INPUT1$ based on presence/absence of verb. The classes appearing on the left side of the node are considered as Objective classes, which consist of 32,263 non- verb sentences. The class appearing on the right side of the node are treated as Subjective

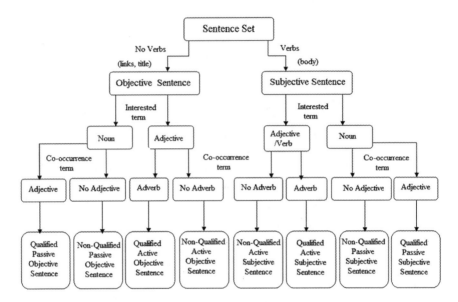

Fig. 3. Sentence clustering using descision tree

Table 1. Sentence classes using descision tree

Levels	Split	Left	Right
INPUT1\$	"Objective", "Subjective"	32263	14675
INPUT2\$	"Active", "Passive"	13420	33518
INPUT3\$	"Qualified", "Non-Qualified"	34866	12072

classes with 14,675 verb sentences. Similarly, second level is displayed at INPUT2\$ based on the presence of Noun/Adjective/verb of sentences. The left node represents 13,420 sentences of active classes having "Adjectives", "Verbs", whereas right side node represents 33,518 Passive class sentences which have "Noun". Last level is displayed at INPUT3\$ and the split based on qualified or non-qualified behavior of sentences. Left side node contains 34,866 Qualified sentences and are having Adjective/Adverbs co-occurrence words. Right side node contains 12,072 Non-Qualified sentences, which are not having Adjective/Adverbs co-occurrence words.

Now, the output of the decision tree has provided certain knowledge on the split and the decision parameters on which the values are constrained. In this work, we have used this knowledge and constructed if-then-else construct on the decision parameters. This construct is obtained for each level of the decision tree and considered as rule sets. The first level rule is based on the occurrences of verbs in sentences and is classified as Subjective and Objective sentences and lots of them have their significance. Subjective sentences are also called as dynamic sentences and usually these sentences are obtained in the *<body>* section of a HTML page and trigger the event by providing the clear information about the event or activity. On the other hand, objective sentences are called as static sentences and usually these sentences occur in *<title>*, *<anchor>* and *<h₁....h₆>* section of a HTML page. In general, they mention the event by highlighting the information in brief. The second level rule is made based on the appearance of interested keyword and it may appear as Noun/Verb/Adjective. If interested keyword is noun, the sentence is considered as passive sentence, as it gives only the objective information such as where/what/who about an event (event mention). e.g., Al-Qaeda attack. While the interested keyword is an Adjective/Verb, then the sentence is considered as active sentence. This type of keywords conveys subjective information about the event such as event happenings. e.g., he has come to kill many people. The last level is identified based on the association of the interested keyword with its co-occurrence terms.

For better understanding the concept, the rules are presented below.

For Non-Verbal Objective sentences,

Rule 1:   *if the interested term is Noun and its co-occurrence term is Adjective then the sentences is clustered as "Qualified Passive Objective Sentence" (QPOS)*
  else it is clustered as "Non-Qualified Passive Objective Sentence" (NQPOS)

Rule 2:   *if the interested term is Adjective and its co- occurrence term is Adverb, then the sentences is clustered as "Qualified Active Objective Sentence" (QAOS)*
  else it is clustered as "Non-Qualified Active Objective Sentence" (NQAOS)

For Verbal Subjective sentences,

Rule 1:   *if the interested term is Noun and its co-occurrence term is Adjective then the sentences is clustered as "Qualified Passive Subjective Sentence" (QPSS)*
  else it is classified as "Non-Qualified Passive Subjective Sentence" (NQPSS)

Rule 2:   *if the interested term is Verb/Adjective and its co-occurrence term is Adverb then the sentences is clustered as "Qualified Active Subjective Sentence" (QASS)*
  else it is classified as "Non-Qualified Active Subjective Sentence" (NQASS)

The web pages $WP = \{wp_1, wp_2,...,wp_x\}$ in the WWW consist of set of interested terms $t = \{t_1, t_2,..., t_y\}$ such that $t \subset T$ where T is the whole set of terms available in the webpage and s is a set of interested sentences where $s = \{s_1, s_2,...,s_z\}$ such that $s \subseteq S$ where S is the whole set of sentences available in the webpage. Here, x, y and z represents total number of web pages, terms and sentences respectively. The web pages are analyzed to understand the importance of each class of sentences and the information content in it. The priorities are assigned to the classes based on the generated rules. Priorities are assigned in the range of 1–4 for the subclasses of

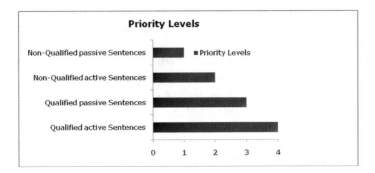

Fig. 4. Assigning priority to the various classes

Subjective and Objective classes. The highest priority 4 is given for Qualified active Sentences, and 3 for Qualified Passive Sentences, 2 for Non-Qualified Active Sentences and lowest priority 1 is assigned for Non-Qualified Passive Sentences. This is depicted in Fig. 4.

The weights are assigned for first level clusters i.e., for Objective class it is 0.25 since it provides static information and it is 0.75 for Subjective class, as it provides dynamic information, respectively. These weights and priorities are used in Eq. (3) to assign the weights for the remaining level clusters in decision tree. The key idea of assigning the weights to the classes is to understand the importance of sentences class that contributes to the event instance and to identify their semantic relation within the webpage. The weights are calculated in such a way that the sum of weights of all the classes in each level is equivalent to 1. The Eq. (3) is derived based on our knowledge on the sentences in HTML pages.

$$Wt_{(Sub-Class)} = \left(\left[\left(\frac{P_{(Sub-Class)}}{\sum_{n=1}^{4} P_n} \right) \right] * W_{(Class)} \right) \tag{3}$$

where, $Wt_{(Sub-Class)}$ is the weight assigned for sub-classes, $P_{(Sub-Class)}$ is the priority level of sub-class, P_n is summation of all priorities levels and $W_{(Class)}$ represents weight of first level classes. Highest weight is assigned to the Qualified Active Subjective Sentence since, the sentences present here is found in <body> section of a web page and it provides clear description about event. These sentences are used for emphasizing the events and these textual sentences are very important in describing the content of the web page. Likewise, the weight is distributed to other sub-classes of Subjective and Objective sentences based on the priority level of the sentence. Thus, the weights of Subjective and Objective sentences are distributed among their sub-categories for Qualified and Non-Qualified Passive Objective Sentence, Qualified and Non-Qualified Active Objective Sentence, Qualified and Non-Qualified Passive Subjective Sentence, Qualified and Non-Qualified Active Subjective Sentence. This is depicted in Fig. 5.

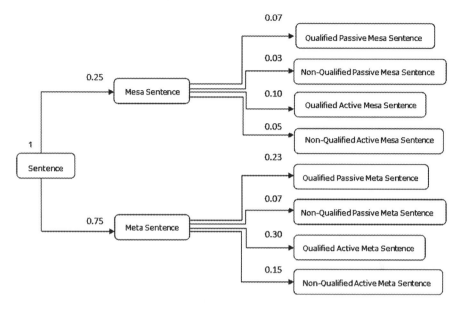

Fig. 5. Weights for various classes

We have collected 500 web documents which is a homogeneous dataset of crime event for our experimental work and 500 terms are chosen as interested term. From these terms, 46,938 interested sentences are obtained. Since we have knowledge on the decision parameter and rules are constructed, the same can be used for classifying the sentences in supervised learning mode. This is done for improving the classification at the later stage and also the decision tree based process can be dispensed. In this paper, we have used Artificial Neural Network (ANN), which is an adaptable system that can learn relationships through repeated presentation of data and is capable of generalizing a new data. Here, the network is given a set of inputs and corresponding desired outputs and the network tries to learn the input-output relationship by adapting its free parameters. During classification, the objective is to assign the input patterns to one of several categories or classes, usually represented by outputs restricted to lie in the range from 0 to 1, so that they represent the probability of class membership. Figure 6 illustrates Multi Layer Perceptron (MLP) for sentence pattern matching. The circles are the Processing Elements (PE's) arranged in layers. The left column is the input layer, the middle column is the hidden layer and right column is the output layer. The lines represent weighted connections between PE's and the PE's in the hidden layer are automatically adjusted by the neural network. The output patterns are the eight classes of sentences, namely QPOS (class A), NQPOS (class B), QAOS (class C), NQAOS(class D), QPSS(class E) and NQPSS(class F), QASS (class G), NQASS (class H).

Table 2 represent target frequency, which gives the frequency distribution of sentences among 8 classes. The table depicts the sentence count for all 8 classes. From the 46,938 interested sentences, 8155 sentences fall into a class A i.e., 17.37 % of the sentences match for the Qualified Passive Objective sentence pattern. Similarly,

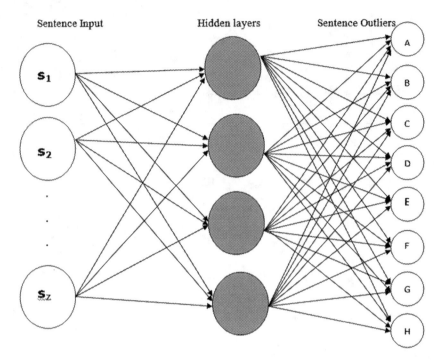

Fig. 6. MLP for sentence pattern matching

Table 2. Target frequency table

Classes	Sentence-count	Percentage (%)
A	8155	17.37
B	17846	38.02
C	997	2.12
D	5265	11.22
E	2506	5.34
F	5011	10.68
G	414	0.88
H	6744	14.37

17,846 (38.02 %) sentences match for class B (Non-Qualified Passive Objectivesentences), 997 (2.12 %) sentences match for class C (Qualified Active Objectivesentences), 5265 (11.22 %) sentences match for class D (Non-Qualified Active Objectivesentences), and so on and the classification is presented in Fig. 7.

The sentence classes patterns from the graph inferences that the combination of A, B, C, D classes in objective classes contributes more towards event instances since they store objective informations in the form of named entity recognitions. Class E, F, G, H of subjective classes contributes more towards the description of specific instances in the form of subjective information. Thus the combination of sentence patterns of objective and subjective classes are sufficient to identify the event

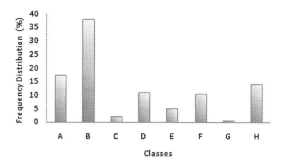

Fig. 7. Sentence classes and outliers

instances of a event type. These sentence patterns were updated into the corpus as an event instance of a specific event type.

3 Experimental Results

We have used interested event homogeneous dataset by crawling crime related web pages from various sources such as http://www.trutv.com/library/crime from WWW and developed our own corpus collection for our experiments. Table 3 shows the features of interested event corpus.

We have used the crime related events such as Die, Kill and Attack. In this paper, our results show that proposed rules for classification of the event is effective, which is dependent on the target event type. The proposed approach has focused on event detection at an interested terms level and the co-occurrence word level and then at the sentence level. From the crime library articles, all instances of the Die, Kill and Attack event type have been manually annotated at the sentence level. We have collected 500 documents, which are related to criminal activities. Among the 500 documents, we have used 50 % documents for training and remaining 50 % documents for testing in ANN. The sentences in the training documents are manually separated based on the weight of sentence and are grouped into a single document. This document is used as the training document and passed to the preprocessing stage. To evaluate the performance of the proposed approach, we extracted relevant documents based on events. The well-known performance measures such as Precision, Recall and F-measure are used and the results are in Table 4. For the event type "Die", we got precision of 90 % for Subjective sentence and 92 % for Objective sentence, Recall is 92 and 93 %

Table 3. Interested event corpus statistics

Corpus features	Homogeneous corpus
Number of documents	500
Number of sources	25
Avg. number of interested terms	500
Avg. number of interested sentences	46,938

Table 4. Performance comparison of proposed approach for various events

Event types	SubjectiveClass			ObjectiveClass		
	Precision	Recall	F1	Precision	Recall	F1
"Die"	90.19	92.09	91.09	92.22	93.73	92.49
"attack"	91.61	90.87	90.23	96.15	97.26	96.70
"Kill"	89.63	85.52	87.06	94.08	93.49	93.28

Table 5. IBC corpus statistics

Corpus features	IBC corpus
Number of documents	332
Number of sources	77
Avg. document length	25.98
Avg. events per document	4.6
Avg. events per sentence	1.14

for Subjective and Objective sentence and so on. Based on the experimental results, it is observed that the proposed classification approach is effective. While the corpus is generated using the classified events for domain specific information extraction applications, the approach provides encouraging results.

The second corpus, we have used for the evaluation purpose is a collection of articles from the Iraq Body Count (IBC) database1 [16] annotated for the "Die" event. This dataset arose from a larger humanitarian project that has focused on the collection of fatalities statistics from unstructured news data during the Iraqi War.

We use this additional corpus to augment the amount of data used for training and testing the "Die" event type and to investigate the use of extra training data on overall classification performance. The feature of the Corpus is presented in Table 5 and has got 332 documents from 77 sources. The performance of the proposed approach is compared with other similar event detection approach that used SVM classification and trigger-based classifications for the event type "Die" [12].

The performance is measured in terms of Precision, Recall and F1-measure for Subjective and Objective sentence which has given good results compared to other approaches and found to be encouraging and is presented in Fig. 8. We have got 97 % as precision and 94 % as recall, whereas SVM classification approach has achieved 96 % as precision and recall as 93 % respectively. The Trigger based classification approach has achieved 97 % as precision and recall as 92 % for Objective event. In Subjective event class, the proposed approach has gained 92 % as precision and recall as 92 %. The SVM classification approach has achieved 92 % as precision and recall as 88 % only. The Trigger based classification approach has achieved 91 % as precision and recall as 89 % for Subjective event. This is depicted in Fig. 8(a) and (b). Based on all the experimental results, it is observed that the proposed classification approach is effective and for the criminal event detection the performance is encouraging.

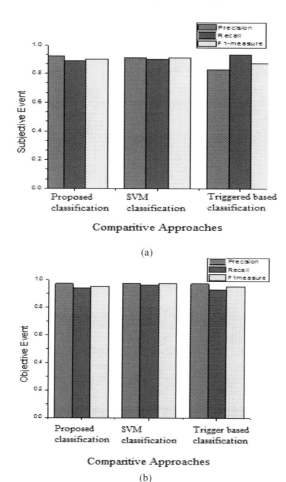

Fig. 8. Performance comparison of proposed approach with other algorithms for event type "Die". (a) Precision , recall, f-measure for subjective event instance. (b) Precision , recall, f-measure for objective event instance

4 Conclusion and Future Works

The proposed approach identifies events based on certain interested terms related to crime domain. The sentences are clustered using unsupervised learning and rules are generated. Priorities are assigned based on the importance of clusters and suitable weight is assigned for capturing sentence semantics with respect to event instance. A set of rules along with weights are used for the classification using ANN. The classified sentences are updated in a corpus and an event based corpus is generated. The performance is evaluated by detecting event instances and the approach performs well. In the future, this work will be extended to text summarization.

Acknowledgement. The work done is supported by research grant from MHRD, Govt. of India, under the Grant NITT/Dean-ID/SCSP-TSP/RP/02 dated 11-02-2014 and Indo-US 21st century knowledge initiative programme under Grant F.No/94-5/2013(IC) dated 19-08-2013.

References

1. Abuleil, S.: Using NLP techniques for tagging events in arabic text. In: 19th IEEE International Conference on Tools with AI, pp. 440–443. IEEE Press (2007)
2. ACE (Automatic Content Extraction) English Annotation Guidelines for Events Version 5.4.3 2005.07.01 Linguistic Data Consortium. http://www.ldc.upenn.edu
3. Aone, C., Ramos-Santacruz, M.: REES: a large-scale relation and event extraction system. In: 6th Conference on Applied Natural Language Processing, pp. 76–83. Morgan Kaufmann Publishers Inc, Washington (2000)
4. Allan, J., Jaime, C., George, D., Jonathon, Y., Yiming, Y.: Topic detection and tracking pilot study (final report) (1998)
5. Cohen, K.B., Verspoor, K., Johnson, H., Roeder, C., Ogren, P., Baumgartner, W., White, E., Tipney, H., Hunter, L.: High-precision biological event extraction with a concept recognizer. In: BioNLP09 Shared Task Workshop, pp. 50–58 (2009)
6. David, A.: Stages of event extraction. In: COLING/ACL 2006 Workshop on Annotating and Reasoning About Time and Events, pp. 1–8. ACL (2006)
7. Filatova, E., Hatzivassiloglou, V.: Event-based extractive summarization. In: ACL 2004 Workshop on Summarization, pp. 104–111. Barcelona, Spain (2004)
8. Makoto, M., Rune, S., Jjin-Dong, K., Junichi, T.: Event extraction with complex event classification using rich features. J. Bioinform. Comput. Biol. **8**(1), 131–146 (2010)
9. Naughton, M., Stokes, N., Carthy, J.: Investigating statistical techniques for sentence-level event classification. In: 22nd International Conference on Computational Linguistics, Association for Computational Linguistics, pp. 617–624. Stroudsburg (2008)
10. Murff, H., Vimla, P., George, H., David, B.: Detecting adverse events for patient safety research: a review of current methodologies. J. Biomed. Inform. **36**(1/2), 131–143 (2003)
11. McCracken, N., Ozgencil, N.E., Symonenko, S.: Combining techniques for event extraction in summary reports. In: AAAI 2006 Workshop Event Extraction and Synthesis, pp. 7–11 (2006)
12. Naughton, M., Stokes, M., Carthy, J.: Investigating statistical techniques for sentence level event classification. In: 22nd International Conference on Computational Linguistics, pp. 617–624 (2008)
13. Sangeetha, S., Michael, Arock., Thakur, R.S.: Event mention detection using rough set and semantic similarity. In: A2CWiC the 1st Amrita ACM-W Celebration on Women in Computing in India, no. 62. ACM, New York (2010)
14. Xu, F., Uszkoreit, H., Li, H.: Automatic event and relation detection with seeds of varying complexity. In: AAAI 2006 Workshop Event Extraction and Synthesis, pp. 491–498. Boston (2006)
15. Zhou, W., Liu, Z., Kong, Q.: A survey of event-based knowledge processing. J. Chin. J. Comput. Sci. **33**(2), 160–162 (2008). (In Chinese)
16. http://www.iraqbodycount.org/database/

Semantic Reasoning

Parallel OWL Reasoning: Merge Classification

Kejia Wu[✉] and Volker Haarslev

Department of Computer Science and Software Engineering,
Concordia University, Montreal, Canada
w_kejia@cs.concordia.ca

Abstract. Our research is motivated by the ubiquitous availability of multiprocessor computers and the observation that available *Web Ontology Language (OWL)* reasoners only make use of a single processor. This becomes rather frustrating for users working in ontology development, especially if their ontologies are complex and require long processing times using these OWL reasoners. We present a novel algorithm that uses a *divide and conquer* strategy for parallelizing OWL TBox classification, a key task in description logic reasoning. We discuss some interesting properties of our algorithm, e.g., its suitability for distributed reasoning, and present an empirical study using a set of benchmark ontologies, where a speedup of up to a factor of 4 has been observed when using 8 workers in parallel.

1 Introduction

Due to the semantic web, a multitude of ontologies are emerging. Quite a few ontologies are huge and contain hundreds of thousands of concepts. Although some of these huge ontologies fit into one of OWL's three tractable profiles, e.g., the well known Snomed ontology is in the $\mathcal{EL}$ profile, there still exist a variety of other OWL ontologies that make full use of OWL DL and require long processing times, even when highly optimized OWL reasoners are employed. Moreover, although most of the huge ontologies are currently restricted to one of the tractable profiles in order to ensure fast processing, it is foreseeable that some of them will require an expressivity that is outside of the tractable OWL profiles.

The research presented in this paper is targeted to provide better OWL reasoning scalability by making efficient use of modern hardware architectures such as multi-processor/core computers. This becomes more important in the case of ontologies that require long processing times although highly optimized OWL reasoners are already used. We consider our research an important basis for the design of next-generation OWL reasoners that can efficiently work in a parallel/concurrent or distributed context using modern hardware. One of the major obstacles that need to be addressed in the design of corresponding algorithms and architectures is the overhead introduced by concurrent computing and its impact on scalability.

Traditional divide and conquer algorithms split problems into independent sub-problems before solving them under the premise that not much

W. Kim et al. (Eds.): JIST 2013, LNCS 8388, pp. 211–227, 2014.
DOI: 10.1007/978-3-319-06826-8_17, © Springer International Publishing Switzerland 2014

communication among the divisions is needed when independently solving the sub-problems, so shared data is excluded to a great extent. Therefore, divide and conquer algorithms are in principle suitable for concurrent computing, including shared-memory parallelization and distributed systems.

Furthermore, recently research on *ontology partitioning* has been proposed and investigated for dealing with monolithic ontologies. Some research results, e.g. ontology modularization [10], can be used for decreasing the scale of an ontology-reasoning problem. The reasoning over a set of sub-ontologies can be executed in parallel. However, there is still a solution needed to reassemble sub-ontologies together. The algorithms presented in this paper can also serve as a solution for this problem.

In the remaining sections, we present our merge-classification algorithm which uses a divide and conquer strategy and a heuristic partitioning scheme. We report on our conducted experiments and their evaluation, and discuss related research.

2   A Parallelized Merge Classification Algorithm

In this section, we present an algorithm for classifying *Description Logic (DL)* ontologies. Due to lack of space we refer for preliminaries about DLs, DL reasoning, and semantic web to [3,13].

We present the merge-classification algorithm in pseudo code. Part of the algorithm is based on standard top- and bottom-search techniques to incrementally construct the classification hierarchy (e.g., see [2]). Due to the symmetry between *top-down* ($\top$_*search*) and *bottom-up* ($\bot$_*search*) search, we only present the first one. In the pseudo code, we use the following notational conventions: Δ_i, Δ_α, and Δ_β designate sub-domains that are divided from Δ; we consider a subsumption hierarchy as a partially order over Δ, denoted as $\leq$, a subsumption relationship where C is subsumed by D ($C \sqsubseteq D$) is expressed by $C \leq D$ or by $\langle C, D \rangle \in \leq$, and $\leq_i$, $\leq_\alpha$, and $\leq_\beta$ are subsumption hierarchies over Δ_i, Δ_α, and Δ_β, respectively; in a subsumption hierarchy over Δ, $C \prec D$ designates $C \sqsubseteq D$ and there does not exist a named concept E such that $C \leq E$ and $E \leq D$; $\prec_i$, $\prec_\alpha$ and $\prec_\beta$ are similar notations defined over Δ_i, Δ_α, and Δ_β, respectively.

Our merge-classification algorithm classifies a taxonomy by calculating its divided sub-domains and then by merging the classified sub-taxonomies together. The algorithm makes use of two facts: (i) If it holds that $B \leq A$, then the subsumption relationships between B's descendants and A's ancestors are determined; (ii) if it is known that $B \not\leq A$, the subsumption relationships between B's descendants and A's ancestors are undetermined. The canonical DL classification algorithms, top-search and bottom-search, have been modified and integrated into the merge-classification. The algorithm consists of two stages: divide and conquering, and combining. Algorithm 1 shows the main part of our parallelized DL classification procedure. The keyword *spawn* indicates that its following calculation must be executed in parallel, either creating a new thread in a shared-memory context or generating a new process or session in a non-shared-memory context. The keyword *sync* always follows *spawn* and suspends the current calculation procedure until all calculations invoked by *spawn* return.

Algorithm 1. $\kappa(\Delta_i)$

---

    input : The sub-domain Δ_i
    output: The subsumption hierarchy classified over Δ_i
1 **begin**
2   **if** $divided\_enough?(\Delta_i)$ **then**
3 | **return** $classify(\Delta_i)$;
4   **else**
5        $\langle \Delta_\alpha, \Delta_\beta \rangle \leftarrow divide(\Delta_i)$;
6        $\leq_\alpha \leftarrow$ **spawn** $\kappa(\Delta_\alpha)$;
7        $\leq_\beta \leftarrow \kappa(\Delta_\beta)$;
8   **sync**;
9   **return** $\mu(\leq_\alpha, \leq_\beta)$;
10   **end if**
11 **end**

Algorithm 2. $\mu(<_\alpha, <_\beta)$

---

    input : The master subsumption hierarchy $\leq_\alpha$
             The subsumption hierarchy $\leq_\beta$ to be merged into $\leq_\alpha$
    output: The subsumption hierarchy resulting from merging $\leq_\alpha$ over $\leq_\beta$
1 **begin**
2     $\top_\alpha \leftarrow select\text{-}top(\leq_\alpha)$;
3     $\top_\beta \leftarrow select\text{-}top(\leq_\beta)$;
4     $\bot_\alpha \leftarrow select\text{-}bottom(\leq_\alpha)$;
5     $\bot_\beta \leftarrow select\text{-}bottom(\leq_\beta)$;
6     $\leq_\alpha \leftarrow \top\_merge(\top_\alpha, \top_\beta)$;
7     $\leq_i \leftarrow \bot\_merge(\bot_\alpha, \bot_\beta)$;
8   **return** $\leq_i$;
9 **end**

The domain Δ is divided into smaller partitions in the first stage. Then, classification computations are executed over each sub-domain Δ_i. A classified sub-terminology $\leq_i$ is inferred over Δ_i. This divide and conquering operations can progress in parallel.

Classified sub-terminologies are to be merged in the combining stage. The told subsumption relationships are utilized in the merging process. Algorithm 2 outlines the master procedure, and the slave procedure is addressed by Algorithms 3, 4, 5, and 6.

2.1   Divide and Conquer Phase

The first task is to divide the universe, Δ, into sub-domains. Without loss of generality, Δ only focuses on *significant* concepts, i.e., concept names or atomic concepts, that are normally declared explicitly in some ontology $\mathcal{O}$, and *intermediate* concepts, i.e., non-significant ones, only play a role in subsumption

---

Algorithm 3. $\top\_search(C, D, \leq_i)$

---

    input : C: the new concept to be classified
                D: the current concept with $\langle D, \top \rangle \in \leq_i$
                $\leq_i$: the subsumption hierarchy
    output: The set of parents of C: $\{p \mid \langle C, p \rangle \in \leq_i\}$.

```
 1  begin
 2  │   mark_visited(D);
 3  │   green ← φ;
 4  │   forall the d ∈ {d | ⟨d, D⟩ ∈≺ᵢ} do /* collect all children of D that subsume C */
 5  │   │   if ≤?(C, d) then
 6  │   │   │   green ← green ∪ {d};
 7  │   │   end if
 8  │   end forall
 9  │   box ← φ;
10  │   if green = φ then
11  │   │   box ← {D};
12  │   else
13  │   │   forall the g ∈ green do
14  │   │   │   if ¬marked_visited?(g) then
15  │   │   │   │   box ← box ∪ ⊤_search(C, g, ≤ᵢ) ; /* recursively test whether C is
                        subsumed by the descendants of g */
16  │   │   │   end if
17  │   │   end forall
18  │   end if
19  │   return box; /* return the parents of C */
20  end
```

---

tests. Each sub-domain is classified independently. The *divide* operation can be naively implemented as an even partitioning over Δ, or by more sophisticated clustering techniques such as *heuristic partitioning* that may result in a better performance, as presented in Sect. 3. The conquering operation can be any standard DL classification method. We first present the most popular classification methods, top-search (Algorithm 3) and bottom-search (omitted here).

The DL classification procedure determines the most specific super- and the most general sub-concepts of each significant concept in Δ. The classified concept hierarchy is a partial order, $\leq$, over Δ. $\top\_search$ recursively calculates a concept's *intermediate* predecessors, i.e., intermediate immediate ancestors, as a relation $\prec_i$ over $\leq_i$.

2.2 Combining Phase

The independently classified sub-terminologies must be merged together in the combining phase. The original top-search (Algorithm 3) (and bottom-search) have been modified to merge two sub-terminologies $\leq_\alpha$ and $\leq_\beta$. The basic idea is to iterate over Δ_β and to use top-search (and bottom-search) to insert each element of Δ_β into $\leq_\alpha$, as shown in Algorithm 4.

Algorithm 4. $\top\_merge^-(A, B, \leq_\alpha, \leq_\beta)$

---

input : A: the current concept of the master subsumption hierarchy, i.e. $\langle A, \top \rangle \in \leq_\alpha$
B: the new concept from the merged subsumption hierarchy, i.e. $\langle B, \top \rangle \in \leq_\beta$
$\leq_\alpha$: the master subsumption hierarchy
$\leq_\beta$: the subsumption hierarchy to be merged into $\leq_\alpha$
output: The merged subsumption hierarchy $\leq_\alpha$ over $\leq_\beta$.

1 **begin**
2 $\quad$ $parents \leftarrow \top\_search(B, A, \leq_\alpha)$;
3 $\quad$ **forall the** $a \in parents$ **do**
4 $\quad\quad$ $\leq_\alpha \leftarrow \leq_\alpha \cup \langle B, a \rangle$; /* insert B into $\leq_\alpha$ */
5 $\quad\quad$ **forall the** $b \in \{b \mid \langle b, B \rangle \in \prec_\beta\}$ **do** /* insert children of B (in $\leq_\beta$) below parents of B (in $\leq_\alpha$) */
6 $\quad\quad\quad$ $\mid$ $\leq_\alpha \leftarrow \top\_merge^-(a, b, \leq_\alpha, \leq_\beta)$;
7 $\quad\quad$ **end forall**
8 $\quad$ **end forall**
9 $\quad$ **return** $\leq_\alpha$;
10 **end**

Algorithm 5. $\top\_merge(A, B, \leq_\alpha, \leq_\beta)$

---

input : A: the current concept of the master subsumption hierarchy, i.e. $\langle A, \top \rangle \in \leq_\alpha$
B: the new concept of the merged subsumption hierarchy, i.e. $\langle B, \top \rangle \in \leq_\beta$
$\leq_\alpha$: the master subsumption hierarchy
$\leq_\beta$: the subsumption hierarchy to be merged into $\leq_\alpha$
output: the merged subsumption hierarchy $\leq_\alpha$ over $\leq_\beta$

1 **begin**
2 $\quad$ $parents \leftarrow \top\_search^*(B, A, \leq_\beta, \leq_\alpha)$;
3 $\quad$ **forall the** $a \in parents$ **do**
4 $\quad\quad$ $\leq_\alpha \leftarrow \leq_\alpha \cup \langle B, a \rangle$;
5 $\quad\quad$ **forall the** $b \in \{b \mid \langle b, B \rangle \in \prec_\beta\}$ **do**
6 $\quad\quad\quad$ $\mid$ $\leq_\alpha \leftarrow \top\_merge(a, b, \leq_\alpha, \leq_\beta)$;
7 $\quad\quad$ **end forall**
8 $\quad$ **end forall**
9 $\quad$ **return** $\leq_\alpha$;
10 **end**

However, this method does not make use of so-called told subsumption (and non-subsumption) information contained in the merged sub-terminology $\leq_\beta$. For example, it is unnecessary to test $\leq?(B_2, A_1)$ when we know $B_1 \leq A_1$ and $B_2 \leq B_1$, given that A_1, A_2 occur in Δ_α and B_1, B_2 occur in Δ_β.

Therefore, we designed a novel algorithm in order to utilize the properties addressed by Proposition 1 to 6. The calculation starts top-merge (Algorithm 5), which uses a modified top-search algorithm (Algorithm 6). This pair of procedures find the most specific subsumers in the master sub-terminology $\leq_\alpha$ for every concept from the sub-terminology $\leq_\beta$ that is being merged into $\leq_\alpha$.

Proposition 1. *When merging sub-terminology $\leq_\beta$ into $\leq_\alpha$, if $\langle B, A \rangle \in \prec_i$ is found in top-search, $\langle A, \top \rangle \in \leq_\alpha$ and $\langle B, \top \rangle \in \leq_\beta$, then for $\forall b \in \{b \mid \langle b, B \rangle \in \leq_\beta\}$ and $\forall a \in \{a \mid \langle A, a \rangle \in \leq_\alpha\} \cup \{A\}$ it is unnecessary to calculate whether $b \leq a$.*

Algorithm 6. $\top\_search^*(C, D, \leq_\beta, \leq_\alpha)$

---

input : C: the new concept to be inserted into $\leq_\alpha$, and $\langle C, \top \rangle \in \leq_\beta$
        D: the current concept, and $\langle D, \top \rangle \in \leq_\alpha$
        $\leq_\beta$: the subsumption hierarchy to be merged into $\leq_\alpha$
        $\leq_\alpha$: the master subsumption hierarchy
output: The set of parents of C: $\{p \mid \langle C, p \rangle \in \leq_\alpha\}$

1 **begin**
2 | $mark\_visited(D)$;
3 | $green \leftarrow \phi$; /* subsumers of C that are from $\leq_\alpha$ */
4 | $red \leftarrow \phi$; /* non-subsumers of C that are children of D */
5 | **forall the** $d \in \{d \mid \langle d, D \rangle \in \prec_\alpha \wedge \langle d, \top \rangle \not\leq_\beta\}$ **do**
6 | | **if** $\leq?(C, d)$ **then**
7 | | | $green \leftarrow green \cup \{d\}$;
8 | | **else**
9 | | | $red \leftarrow red \cup \{d\}$;
10 | | **end if**
11 | **end forall**
12 | $box \leftarrow \phi$;
13 | **if** $green = \phi$ **then**
14 | | **if** $\leq?(C, D)$ **then**
15 | | | $box \leftarrow \{D\}$;
16 | | **else**
17 | | | $r \leftarrow \{D\}$;
18 | | **end if**
19 | **else**
20 | | **forall the** $g \in green$ **do**
21 | | | **if** $\neg marked\_visited?(g)$ **then**
22 | | | | $box \leftarrow box \cup \top\_search^*(C, g, \leq_\beta, \leq_\alpha)$;
23 | | | **end if**
24 | | **end forall**
25 | **end if**
26 | **forall the** $r \in red$ **do**
27 | | **forall the** $c \in \{c \mid \langle c, C \rangle \in \prec_i\}$ **do**
28 | | | $\leq_\alpha \leftarrow \top\_merge(r, c, \leq_\alpha, \leq_\beta)$;
29 | | **end forall**
30 | **end forall**
31 | **return** box;
32 **end**

Proposition 2. *When merging sub-terminology $\leq_\beta$ into $\leq_\alpha$, if $\langle B, A \rangle \in \prec_i$ is found in top-search, $\langle A, \top \rangle \in \leq_\alpha$ and $\langle B, \top \rangle \in \leq_\beta$, then for $\forall b \in \{b \mid \langle b, B \rangle \in \prec_\beta \wedge b \neq B\}$ and $\forall a \in \{a \mid \langle a, A \rangle \in \prec_\alpha \wedge a \neq A\}$ it is necessary to calculate whether $b \leq a$.*

Proposition 3. *When merging sub-terminology $\leq_\beta$ into $\leq_\alpha$, if $B \not\leq A$ is found in top-search, $\langle A, \top \rangle \in \leq_\alpha$ and $\langle B, \top \rangle \in \leq_\beta$, then for $\forall b \in \{b \mid \langle b, B \rangle \in \leq_\beta \wedge b \neq B\}$ and $\forall a \in \{a \mid \langle a, A \rangle \in \leq_\alpha\} \cup \{A\}$ it is necessary to calculate whether $b \leq a$.*

Proposition 4. *When merging sub-terminology $\leq_\beta$ into $\leq_\alpha$, if $\langle A, B \rangle \in \prec_i$ is found in bottom-search, $\langle \bot, A \rangle \in \leq_\alpha$ and $\langle \bot, B \rangle \in \leq_\beta$, then for $\forall b \in \{b \mid$*

$\langle B, b \rangle \in \leq_\beta \}$ *and* $\forall a \in \{a \mid \langle a, A \rangle \in \leq_\alpha\} \cup \{A\}$ *it is unnecessary to calculate whether* $a \leq b$.

Proposition 5. *When merging sub-terminology* $\leq_\beta$ *into* $\leq_\alpha$, *if* $\langle A, B \rangle \in \prec_i$ *is found in bottom-search,* $\langle \perp, A \rangle \in \leq_\alpha$ *and* $\langle \perp, B \rangle \in \leq_\beta$, *then for* $\forall b \in \{b \mid \langle B, b \rangle \in \prec_\beta \wedge b \neq B\}$ *and* $\forall a \in \{a \mid \langle A, a \rangle \in \prec_\alpha \wedge a \neq A\}$ *it is necessary to calculate whether* $a \leq b$.

Proposition 6. *When merging sub-terminology* $\leq_\beta$ *into* $\leq_\alpha$, *if* $A \not\leq B$ *is found in bottom-search,* $\langle \perp, A \rangle \in \leq_\alpha$ *and* $\langle \perp, B \rangle \in \leq_\beta$, *then for* $\forall b \in \{b \mid \langle B, b \rangle \in \leq_\beta \wedge b \neq B\}$ *and* $\forall a \in \{a \mid \langle A, a \rangle \in \leq_\alpha\} \cup \{A\}$ *it is necessary to calculate whether* $a \leq b$.

When merging a concept B, $\langle B, \top \rangle \in \leq_\beta$, the top-merge algorithm first finds for B the most specific position in the master sub-terminology $\leq_\alpha$ by means of *top-down* search. When such a most specific super-concept is found, this concept and all its super-concepts are naturally super-concepts of every sub-concept of B in the sub-terminology $\leq_\beta$, as is stated by Proposition 1. However, this newly found predecessor of B may not be necessarily a predecessor of some descendant of B in $\leq_\beta$. Therefore, the algorithm continues to find the most specific positions for all sub-concepts of B in $\leq_\beta$ according to Proposition 2. Algorithm 5 addresses this procedure.

Non-subsumption information can be told in the top-merge phase. Top-down search employed by top-merge must do subsumption tests somehow. In a canonical top-search procedure, as indicated by Algorithm 3, the branch search is stopped at this point. However, the conclusion that a merged concept B, $\langle B, \top \rangle \in \leq_\beta$, is not subsumed by a concept A, $\langle A, \top \rangle \in \leq_\alpha$, does not rule out the possibility of $b \leq A$, $b \in \{b \mid \langle b, B \rangle \in \prec_\beta\}$, which is not required in traditional top-search and may be abound in the top-merge procedure, and therefore must be followed by determining whether $b \leq A$. Otherwise, the algorithm is incomplete. Proposition 3 presents this observation. For this reason, the original top-search algorithm must be adapted to the new situation. Algorithm 6 is the updated version of the top-search procedure.

Algorithm 6 not only maintains told subsumption information by the set *green*, but also propagates told non-subsumption information by the set *red* for further inference. As addressed by Proposition 3, when the position of a merged concept is determined, the subsumption relations between its successors and the *red* set are calculated. Furthermore, the subsumption relation for the concept C and D in Algorithm 6 must be explicitly calculated even when the set *green* is empty. In the original top-search procedure (Algorithm 3), $C \prec_i D$ is implicitly derivable if the set *green* is empty, which does not hold in the modified top-search procedure (Algorithm 6) since it does not always start from $\top$ any more when searching for the most specific position of a concept.

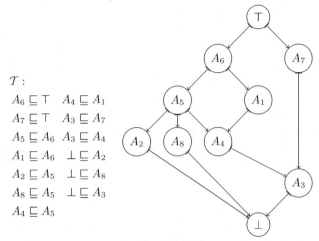

$\mathcal{T}$:

$A_6 \sqsubseteq \top$ $A_4 \sqsubseteq A_1$

$A_7 \sqsubseteq \top$ $A_3 \sqsubseteq A_7$

$A_5 \sqsubseteq A_6$ $A_3 \sqsubseteq A_4$

$A_1 \sqsubseteq A_6$ $\bot \sqsubseteq A_2$

$A_2 \sqsubseteq A_5$ $\bot \sqsubseteq A_8$

$A_8 \sqsubseteq A_5$ $\bot \sqsubseteq A_3$

$A_4 \sqsubseteq A_5$

(a) The TBox given.   (b) The classified terminology hierarchy.

Fig. 1. An example ontology.

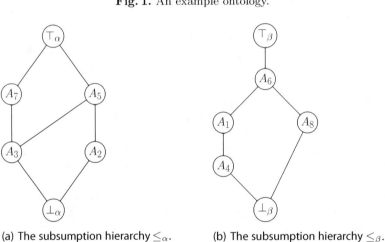

(a) The subsumption hierarchy $\leq_\alpha$.   (b) The subsumption hierarchy $\leq_\beta$.

Fig. 2. The subsumption hierarchy over divisions.

2.3   Example

We use an example to illustrate the algorithm further. Given an ontology with a TBox defined by Fig. 1(a), which only contains simple concept subsumption axioms, Fig. 1(b) shows the subsumption hierarchy.

Suppose that the ontology is clustered into two groups in the divide phase: $\Delta_\alpha = \{A_2, A_3, A_5, A_7\}$ and $\Delta_\beta = \{A_1, A_4, A_6, A_8\}$. They can be classified independently, and the corresponding subsumption hierarchies are shown in Fig. 2.

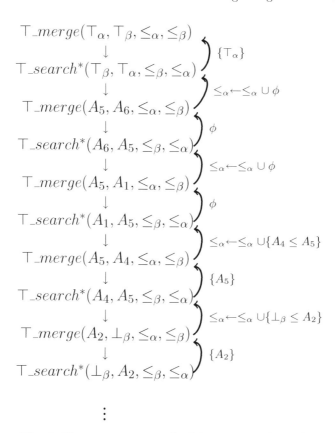

$$\top\_merge(\top_\alpha, \top_\beta, \leq_\alpha, \leq_\beta)$$
$$\downarrow \qquad \} \{\top_\alpha\}$$
$$\top\_search^*(\top_\beta, \top_\alpha, \leq_\beta, \leq_\alpha)$$
$$\downarrow \qquad \} \leq_\alpha \leftarrow \leq_\alpha \cup \phi$$
$$\top\_merge(A_5, A_6, \leq_\alpha, \leq_\beta)$$
$$\downarrow \qquad \} \phi$$
$$\top\_search^*(A_6, A_5, \leq_\beta, \leq_\alpha)$$
$$\downarrow \qquad \} \leq_\alpha \leftarrow \leq_\alpha \cup \phi$$
$$\top\_merge(A_5, A_1, \leq_\alpha, \leq_\beta)$$
$$\downarrow \qquad \} \phi$$
$$\top\_search^*(A_1, A_5, \leq_\beta, \leq_\alpha)$$
$$\downarrow \qquad \} \leq_\alpha \leftarrow \leq_\alpha \cup \{A_4 \leq A_5\}$$
$$\top\_merge(A_5, A_4, \leq_\alpha, \leq_\beta)$$
$$\downarrow \qquad \} \{A_5\}$$
$$\top\_search^*(A_4, A_5, \leq_\beta, \leq_\alpha)$$
$$\downarrow \qquad \} \leq_\alpha \leftarrow \leq_\alpha \cup \{\bot_\beta \leq A_2\}$$
$$\top\_merge(A_2, \bot_\beta, \leq_\alpha, \leq_\beta)$$
$$\downarrow \qquad \} \{A_2\}$$
$$\top\_search^*(\bot_\beta, A_2, \leq_\beta, \leq_\alpha)$$

$$\vdots$$

Fig. 3. The computation path of determining $A_4 \leq_i A_5$.

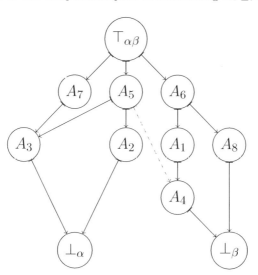

Fig. 4. The subsumption hierarchy after $A_4 \leq A_5$ has been determined.

Algorithm 7. $cluster(G)$

---

input : G: the subsumption graph
output: R: the concept names partitions

1 **begin**
2 | $R \leftarrow \phi$;
3 | $visited \leftarrow \phi$;
4 | $N \leftarrow get\_children(\top, G)$;
5 | **foreach** $n \in N$ **do**
6 | | $P \leftarrow \{n\}$;
7 | | $visited \leftarrow visited \cup \{n\}$;
8 | | $R \leftarrow R \cup \{build\_partition(n, visited, G, P)\}$;
9 | **end foreach**
10 | **return** R;
11 **end**

In the merge phase, the concepts from $\leq_\beta$ are merged into $\leq_\alpha$. For example, Fig. 3 shows a possible computation path where $A_4 \leq A_5$ is being determined.[1] If we assume a subsumption relationship between two concepts is proven when the parent is added to the set *box* (see Line 15, Algorithm 6), Fig. 4 shows the subsumption hierarchy after $A_4 \leq A_5$ has been determined.

3   Partitioning

Partitioning is an important part of this algorithm. It is the main task in the dividing phase. In contrast to simple problem domains such as sorting integers, where the merge phase of a standard merge-sort does not require another sorting, DL ontologies might entail numerous subsumption relationships among concepts. Building a terminology with respect to the entailed subsumption hierarchy is the primary function of DL classification. We therefore assumed that some heuristic partitioning schemes that make use of known subsumption relationships may improve reasoning efficiency by requiring a smaller number of subsumption tests, and this assumption has been proved by our experiments, which are described in Sect. 4.

So far, we have presented an ontology partitioning algorithm by using only told subsumption relationships that are directly derived from concept definitions and axiom declarations. Any concept that has at least one told super- and one sub-concept, can be used to construct a told subsumption hierarchy. Although such a hierarchy is usually incomplete and has many entailed subsumptions missing, it contains already known subsumptions indicating the closeness between concepts w.r.t. subsumption. Such a raw subsumption hierarchy can be represented as a directed graph with only one root, the $\top$ concept. A heuristic partitioning method can be defined by traversing the graph in a breadth-first way,

---

[1] This process does not show a full calling order of computing $A_4 \leq A_5$ for sake of brevity. For instance, $\top\_merge(A_7, A_6, \leq_\alpha, \leq_\beta)$ is not shown.

Algorithm 8. $build\_partition(n, visited, G, P)$

---

input : n: an concept name

$\qquad\quad$ $visited$: a list recording visited concept names

$\qquad\quad$ G: the syntax-based subsumption graph

$\qquad\quad$ P: a concept names partition

output: R: a concept names partition

1 **begin**

2 $\quad$ $R \leftarrow \phi$;

3 $\quad$ $N \leftarrow get\_children(n, visited, G, P)$;

4 $\quad$ **foreach** $n' \in N$ **do**

5 $\qquad$ **if** $n' \notin visited$ **then**

6 $\qquad\quad$ $P \leftarrow P \cup \{n'\}$;

7 $\qquad\quad$ $visited \leftarrow visited \cup \{n'\}$;

8 $\qquad\quad$ $build\_partition(n', visited, G, P)$;

9 $\qquad$ **end if**

10 $\quad$ **end foreach**

11 $\quad$ $R \leftarrow P$;

12 $\quad$ **return** R;

13 **end**

---

starting from $\top$, and collecting traversed concepts into partitions. Algorithms 7 and 8 address this procedure.

4   Evaluation

Our experimental results clearly show the potential of merge-classification. We could achieve speedups up to a factor of 4 by using a maximum of 8 parallel workers, depending on the particular benchmark ontology. This speedup is in the range of what we expected and comparable to other reported approaches, e.g., the experiments reported for the ELK reasoner [16,17] also show speedups up to a factor of 4 when using 8 workers, although a specialized polynomial procedure is used for $\mathcal{EL}+$ reasoning that seems to be more amenable to concurrent processing than standard tableau methods.

We have designed and implemented a concurrent version of the algorithm so far. Our program[2] is implemented on the basis of the well-known reasoner JFact,[3] which is open-source and implemented in Java. We modified JFact such that we can execute a set of JFact reasoning kernels in parallel in order to perform the merge-classification computation. We try to examine the effectiveness of the merge-classification algorithm by adapting such a mature DL reasoner.

4.1   Experiment

A multi-processor computer, which has 4 octa-core processors and 128G memory installed, was employed to test the program. The Linux OS and 64-bit OpenJDK

---

[2] http://github.com/kejia/mc

[3] http://jfact.sourceforge.net

Table 1. Metrics of the test cases.

Ontology	Expressivity	Concept count	Axiom count
adult_mouse_anatomy	$\mathcal{ALE}+$	2753	9372
amphibian_gross_anatomy	$\mathcal{ALE}+$	701	2626
c_elegans_phenotype	$\mathcal{ALEH}+$	1935	6170
cereal_plant_trait	$\mathcal{ALEH}$	1051	3349
emap	$\mathcal{ALE}$	13731	27462
environmental_entity_logical_definitions	$\mathcal{SH}$	1779	5803
envo	$\mathcal{ALEH}+$	1231	2660
fly_anatomy	$\mathcal{ALEI}+$	6222	33162
human_developmental_anatomy	$\mathcal{ALEH}$	8341	33345
medaka_anatomy_development	$\mathcal{ALE}$	4361	9081
mpath	$\mathcal{ALEH}+$	718	4315
nif-cell	$\mathcal{S}$	376	3492
sequence_types_and_features	$\mathcal{SH}$	1952	6620
teleost_anatomy	$\mathcal{ALER}+$	3036	11827
zfa	$\mathcal{ALEH}+$	2755	33024

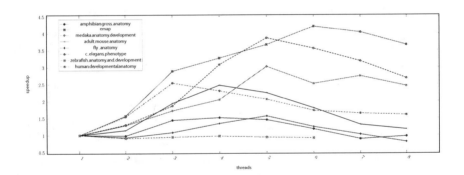

Fig. 5. The performance of parallelized merge-classification—I.

6 was employed in the tests. The JVM was allocated at least 16G memory initially, given that at most 64G physical memory was accessible. Most of the test cases were chosen from *OWL Reasoner Evaluation Workshop 2012 (ORE 2012)* data sets. Table 1 shows the test cases' metrics.

Each test case ontology was classified with the same setting except for an increased number of workers. Each worker is mapped to an OS thread, as indicated by the Java specification. Figure 5 and 6 show the test results.

In our initial implementation, we used an *even-partitioning* scheme. That is to say concept names are randomly assigned to a set of partitions. For the majority of the above-mentioned test cases we observed a small performance improvement below a speedup factor of 1.4, for a few an improvement of up to 4, and for others only a decrease in performance. Only overhead was shown in these test cases.

As mentioned in Sect. 3, we assumed that a heuristic partitioning might promote a better reasoning performance, e.g., a partitioning scheme considering subsumption axioms. This idea is addressed by Algorithms 7 and 8.

We implemented Algorithms 7 and 8 and tested the program. Our assumption has been proved by the test: Heuristic partitioning may improve reasoning performance where blind partitioning can not.

4.2   Discussion

Our experiment shows that with a heuristic divide scheme the merge-classification algorithm can increase reasoning performance. However, such performance promotion is not always tangible. In a few cases, the parallelized merge-classification merely degrades reasoning performance. The actual divide phase of our algorithm can influence the performance by creating better or worse partitions.

A heuristic divide scheme may result in a better performance than a blind one. According to our experience, when the division of the concepts from the domain is basically random, sometimes divisions contribute to promoting reasoning performance, while sometimes they do not. A promising heuristic divide scheme seems to be in grouping a family of concepts, which have potential subsumption relationships, into the same partition. Evidently, due to the presence of non-obvious subsumptions, it is hard to guess how to achieve such a good partitioning. We tried to make use of obvious subsumptions in axioms to partition closely related concepts into the same group. The tests demonstrate a clear performance improvement in a number of cases.

While in many cases merge-classification can improve reasoning performance, for some test cases its practical effectiveness is not yet convincing. We are still investigating the factors that influence the reasoning performance for these cases but cannot give a clear answer yet. The cause may be the large number of *general concept inclusion (GCI)* axioms of ontologies. Even with some more refined divide scheme, those GCI axioms can cause inter-dependencies between partitions, and may cause in the merge phase an increased number of subsumption tests. Also, the indeterminism of the merging schedule, i.e., the unpredictable

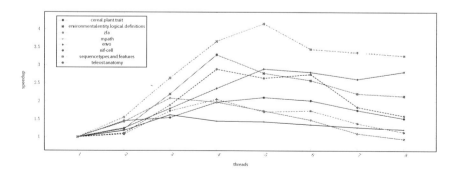

Fig. 6. The performance of parallelized merge-classification—II.

order of merging divides, needs to be effectively solved in the implementation, and racing conditions between merging workers as well as the introduced overhead may decrease the performance. In addition, the limited performance is caused by the experimental environment: Compared with a single chip architecture, the 4-chip-distribution of the 32 processors requires extra computational overhead, and the memory and thread management of JVM may decrease the performance of our program.

5   Related Work

A key functionality of a DL reasoning system is *classification*, computing all entailed subsumption relationships among named concepts. The generic top-search and bottom-search algorithms were introduced by [19] and extended by [2]. The algorithm is used as the standard technique for incrementally creating subsumption hierarchies of DL ontologies. Reference [2] also presented some basic traversal optimizations. After that, a number of optimization techniques have been explored [8,9,26]. Most of the optimizations are based on making use of the partial transitivity information in searching. However, research on how to use concurrent computing for optimizing DL reasoning has started only recently.

The merge-classification algorithm is suitable for concurrent computation implementation, including both shared-memory parallelization and (shared-memory or non-shared-memory) distributed systems. Several concurrency-oriented DL reasoning schemes have been researched recently. Reference [18] reported on experiments with a parallel $\mathcal{SHN}$ reasoner. This reasoner could process *disjunction* and *at-most cardinality restriction* rules in parallel, as well as some primary DL tableau optimization techniques. Reference [1] presented the first algorithms on parallelizing TBox classification using a shared global subsumption hierarchy, and the experimental results promise the feasibility of parallelized DL reasoning. References [16,17] reported on the ELK reasoner, which can classify $\mathcal{EL}$ ontologies *concurrently*, and its speed in reasoning about $\mathcal{EL}+$ ontologies is impressive. References [28,29] studied a parallel DL reasoning system. References [20,21] proposed the idea of applying a constraint programming solver. Besides the shared-memory concurrent reasoning research mentioned above, non-shared-memory distributed concurrent reasoning has been investigated recently by [22,25].

Merge-classification needs to divide ontologies. Ontology partitioning can be considered as a sort of *clustering* problem. These problems have been extensively investigated in networks research, such as [6,7,30]. Algorithms adopting more complicated heuristics in the area of ontology partitioning, have been presented in [5,10–12,14].

Our merge-classification approach employs the well-known *divide and conquer* strategy. There is sufficient evidence that these types of algorithms are well suited to be processed in parallel [4,15,27]. Some experimental works about parallelized merge sort are reported in [23,24].

6   Conclusion

The approach presented in this paper has been motivated by the observation that (i) multi-processor/core hardware is becoming ubiquitously available but standard OWL reasoners do not yet make use of these available resources; (ii) Although most OWL reasoners have been highly optimized and impressive speed improvements have been reported for reasoning in the three tractable OWL profiles, there exist a multitude of OWL ontologies that are outside of the three tractable profiles and require long processing times even for highly optimized OWL reasoners. Recently, concurrent computing has emerged as a possible solution for achieving a better scalability in general and especially for such difficult ontologies, and we consider the research presented in this paper as an important step in designing adequate OWL reasoning architectures that are based on concurrent computing.

One of the most important obstacles in successfully applying concurrent computing is the management of overhead caused by concurrency. An important factor is that the load introduced by using concurrent computing in DL reasoning is usually remarkable. Concurrent algorithms that cause only a small overhead seem to be the key to successfully apply concurrent computing to DL reasoning.

Our merge-classification algorithm uses a *divide and conquer* scheme, which is potentially suitable for low overhead concurrent computing since it rarely requires communication among divisions. Although the empirical tests show that the merge-classification algorithm does not always improve reasoning performance to a great extent, they let us be confident that further research is promising. For example, investigating what factors impact the effectiveness and efficiency of the merge-classification may help us improve the performance of the algorithm further.

At present our work adopts a heuristic *partitioning* scheme at the divide phase. Different divide schemes may produce different reasoning performances. We are planning to investigate better divide methods. Furthermore, our work has only researched the performance of the *concurrent* merge-classification so far. How the number of division impacts the reasoning performance in a single thread and a multiple threads setting needs be investigated in more detail.

Acknowledgements. We are grateful for Ralf Möller at Hamburg University of Technology in giving us access to their equipment that was used to conduct the presented evaluation.

References

1. Aslani, M., Haarslev, V.: Parallel TBox classification in description logics–first experimental results. In: Proceedings of the 2010 Conference on ECAI 2010: 19th European Conference on Artificial Intelligence, pp. 485–490 (2010)
2. Baader, F., Hollunder, B., Nebel, B., Profitlich, H.J., Franconi, E.: An empirical analysis of optimization techniques for terminological representation systems. Appl. Intell. **4**(2), 109–132 (1994)

3. Baader, F., et al.: The Description Logic Handbook: Theory, Implementation, and Applications. Cambridge University Press, New York (2003)
4. Cole, R.: Parallel merge sort. SIAM J. Comput. **17**(4), 770–785 (1988)
5. Doran, P., Tamma, V., Iannone, L.: Ontology module extraction for ontology reuse: an ontology engineering perspective. In: Proceedings of the Sixteenth ACM Conference on Information and Knowledge Management, pp. 61–70 (2007)
6. Ester, M., Kriegel, H.P., Sander, J., Xu, X.: A density-based algorithm for discovering clusters in large spatial databases with noise. In: Proceedings of the Second International Conference on Knowledge Discovery and Data Mining (1996)
7. Girvan, M., Newman, M.E.J.: Community structure in social and biological networks. Proc. Nat. Acad. Sci. U.S.A. **99**(12), 7821–7826 (2002)
8. Glimm, B., Horrocks, I., Motik, B., Stoilos, G.: Optimising ontology classification. In: Patel-Schneider, P.F., Pan, Y., Hitzler, P., Mika, P., Zhang, L., Pan, J.Z., Horrocks, I., Glimm, B. (eds.) ISWC 2010, Part I. LNCS, vol. 6496, pp. 225–240. Springer, Heidelberg (2010)
9. Glimm, B., Horrocks, I., Motik, B., Shearer, R., Stoilos, G.: A novel approach to ontology classification. Web Semantics: Science, Services and Agents on the World Wide Web **14**, 84–101 (2012)
10. Grau, B.C., Horrocks, I., Kazakov, Y., Sattler, U.: A logical framework for modularity of ontologies. In: Proceedings International Joint Conference on Artificial Intelligence, pp. 298–304 (2007)
11. Grau, B.C., Horrocks, I., Kazakov, Y., Sattler, U.: Modular reuse of ontologies: theory and practice. J. Artif. Intell. Res. **31**(1), 273–318 (2008)
12. Grau, B.C., Parsia, B., Sirin, E., Kalyanpur, A.: Modularizing OWL ontologies. In: K-CAP 2005 Workshop on Ontology Management (2005)
13. Hitzler, P., Krötzsch, M., Rudolph, S.: Foundations of Semantic Web Technologies. Chapman & Hall/CRC, London (2009)
14. Hu, W., Qu, Y., Cheng, G.: Matching large ontologies: a divide-and-conquer approach. Data Knowl. Eng. **67**(1), 140–160 (2008)
15. Jeon, M., Kim, D.: Parallel merge sort with load balancing. Int. J. Parallel Prog. **31**(1), 21–33 (2003)
16. Kazakov, Y., Krötzsch, M., Simančík, F.: Concurrent classification of $\mathcal{EL}$ ontologies. In: Aroyo, L., Welty, C., Alani, H., Taylor, J., Bernstein, A., Kagal, L., Noy, N., Blomqvist, E. (eds.) ISWC 2011, Part I. LNCS, vol. 7031, pp. 305–320. Springer, Heidelberg (2011)
17. Kazakov, Y., Krötzsch, M., Simančík, F.: The incredible ELK: from polynomial procedures to efficient reasoning with EL ontologies (2013) (submitted to a journal)
18. Liebig, T., Müller, F.: Parallelizing tableaux-based description logic reasoning. In: Meersman, R., Tari, Z. (eds.) OTM-WS 2007, Part II. LNCS, vol. 4806, pp. 1135–1144. Springer, Heidelberg (2007)
19. Lipkis, T.: A KL-ONE classifier. In: Proceedings of the 1981 KL-ONE Workshop, pp. 128–145 (1982)
20. Meissner, A.: A simple parallel reasoning system for the $\mathcal{ALC}$ description logic. In: Nguyen, N.T., Kowalczyk, R., Chen, S.-M. (eds.) ICCCI 2009. LNCS, vol. 5796, pp. 413–424. Springer, Heidelberg (2009)
21. Meissner, A., Brzykcy, G.: A parallel deduction for description logics with ALC language. Knowl.-Driven Comput. **102**, 149–164 (2008)
22. Mutharaju, R., Hitzler, P., Mateti, P.: DistEL: A distributed EL+ ontology classifier. In: Proceedings of The 9th International Workshop on Scalable Semantic Web Knowledge Base Systems (2013)

23. Radenski, A.: Shared memory, message passing, and hybrid merge sorts for stand-alone and clustered SMPs. In: The 2011 International Conference on Parallel and Distributed Processing Techniques and Applications, vol. 11, pp. 367–373 (2011)
24. Rolfe, T.J.: A specimen of parallel programming: parallel merge sort implementation. ACM Inroads 1(4), 72–79 (2010)
25. Schlicht, A., Stuckenschmidt, H.: Distributed resolution for ALC - first results. In: Proceedings of the Workshop on Advancing Reasoning on the Web: Scalability and Commonsense (2008)
26. Shearer, R., Horrocks, I., Motik, B.: Exploiting partial information in taxonomy construction. In: Grau, B.G., Horrocks, I., Motik, B., Sattler, U. (eds.) Proceedings of the 2009 International Workshop on Description Logics. CEUR Workshop Proceedings, Oxford, UK, vol. 477, 27–30 July 2009
27. Todd, S.: Algorithm and hardware for a merge sort using multiple processors. IBM J. Res. Dev. 22(5), 509–517 (1978)
28. Wu, K., Haarslev, V.: A parallel reasoner for the description logic ALC. In: Proceedings of the 2012 International Workshop on Description Logics (2012)
29. Wu, K., Haarslev, V.: Exploring parallelization of conjunctive branches in tableau-based description logic reasoning. In: Proceedings of the 2013 International Workshop on Description Logics (2013)
30. Xu, X., Yuruk, N., Feng, Z., Schweiger, T.A.J.: SCAN: a structural clustering algorithm for networks. In: Proceedings of the 13th ACM SIGKDD International Conference on Knowledge Discovery and Data Mining, pp. 824–833 (2007)

TLDRet: A Temporal Semantic Facilitated Linked Data Retrieval Framework

Md-Mizanur Rahoman[1(✉)] and Ryutaro Ichise[1,2]

[1] Department of Informatics, The Graduate University for Advanced Studies,
Tokyo, Japan
{mizan,ichise}@nii.ac.jp
[2] Principles of Informatics Research Division, National Institute of Informatics,
Tokyo, Japan

Abstract. Temporal features, such as date and time or time of an event, employ concise semantics for any kind of information retrieval, and therefore for linked data information retrieval. However, we have found that most linked data information retrieval techniques pay little attention on the power of temporal feature inclusion. We propose a keyword-based linked data information retrieval framework, called TLDRet, that can incorporate temporal features and give more concise results. Preliminary evaluation of our system shows promising performance.

1 Introduction

Temporal feature related information, such as information related to date and time or time of an event, is helpful in finding an appropriate result or in discovering a new relationship. Though temporal feature related information extraction of typical document-based data has quite a long history of research, the same kind of research is comparatively less investigated for linked data. The reason could be that the research community has paid little attention on this issue or the linked data has a different structure than the typical document-based data. In this study we examine these issues and propose an efficient and easy-to-use linked data information retrieval framework that can retrieve information related to temporal features.

Linked data can portray knowledge in a simple yet rich structure. Such data adapt three major technologies: uniform resource identifier (URI), which is a generic means to identify entities or concepts in the world; Hypertext Transfer Protocol (HTTP), which is a simple yet universal mechanism for retrieving resources or descriptions of resources, and; Resource Description Framework (RDF), which is a generic graph-based data model for structuring and linking data that describe facts in the world by RDF triple $<s, p, o>$ where linked data resource s, p and o represent those facts [6]. Unlike other constraint-based data such as relational data, data publishing over linked data is rather easy, because individual data publishers can publish their data with their own data schema without knowing other publishers' data schema. Data schema refers

W. Kim et al. (Eds.): JIST 2013, LNCS 8388, pp. 228–243, 2014.
DOI: 10.1007/978-3-319-06826-8_18, © Springer International Publishing Switzerland 2014

meta-information about data. Thus, either the same data publishers or another group of publishers can connect published data and construct global data. Linked data adapt this loose data publishing strategy to foster the rapid population of data. Consequently linked data presently contain a vast amount of knowledge.

Rula et al. investigated BTC 2011 dataset and showed that, though not many, contemporary linked data store temporal fact related information [10]. Evidently it can be said that along with other real world facts, linked data hold temporal fact related information. Rula et al. investigation on BTC 2011 dataset showed that contemporary linked data store temporal fact related information in two perspectives: (i) document-centric, where time points are associated to RDF triple and (ii) fact-centric, where time points or intervals are associated with facts [10]. Usually document-centric temporal information is used to validate RDF triples such as last date/time of modification of triple. On the other hand, fact-centric temporal information informs historical temporal value attachment to RDF resources such as an exemplary dataset could store a fact-centric temporal linked data for *birthday of Michael Jackson* by a RDF triple$<res:Michael\_Jackson$ $prop:birthDate29-Aug-1958>$. In this study, we investigate temporal linked data information retrieval for fact-centric perspective. We also assume that inclusion of document-centric perspective will be an interesting topic which we want to explore in future.

In linked data information retrieval, temporal facts can be mentioned in various ways such as

- data publishers could store them with various storage models (e.g., Temporal RDF [5], stRDF [1]), with various time related ontologies (e.g. Timely YAGO [17], owl-time[1], timeline[2] etc) [10]
- data consumers could seek them with various structured time-specific queries (e.g., stSPARQL [1]), or familiar keyword-based time-specific queries such as explicit time-specific queries (e.g., *on the 29th of August 1958*) and event-specific queries (e.g., *during World War I*).

So retrieval of temporal linked data information requires to handle all those issues, which is a challenge for both data publishers and data consumers. Considering the above, we adapt familiar keyword-based linked data information framework which hide all such complexities. We contribute mainly by proposing linked data information retrieval framework which can explore temporal semantics. From linked data perspective, we assume that effective exploration of temporal semantics is an important advancement towards its realistic utilization.

The remainder of this paper is divided as follows. In Sect. 2 we describe related work. In Sect. 3 we briefly introduce our adapted system and describe on its further extension to fit temporal linked data information retrieval. In Sect. 4 we describe on our proposed temporal linked data retrieval framework, TLDRet. In Sect. 5 we discuss the results of implementing our proposal in experiments. Section 6 concludes our work.

---

[1] http://www.w3.org/TR/owl-time/
[2] http://motools.sourceforge.net/timeline/timeline.html

2   Related Work

The Trans-lingual Information Detection, Extraction, and Summarization (TIDES) program in 2000, the Automatic Content Extraction (ACE) program in 2002, and the ACE Time Normalization (TERN) program in 2004 hosted several temporal feature related information extraction studies. Since then, multiple time annotation standards such as TIMEX2/3[3], and TimeML[4] have been established. Thus, temporal feature related information extraction of typical document-based content is an active research field. Though temporal feature related information extraction of typical document-based data has quite a long history of research, the same kind of research is comparatively less seen for linked data. The reason could be that the research community has overlooked this issue or the linked data has a different structure than the typical document-based data.

Usually linked data store information with graph-like data model called RDF. Though RDF describes basic structure of linked data, recently we find several other models such as Temporal RDF [5], stRDF [1] etc, which specify further specification to store temporal information. There are also established time related ontologies such owl-time, Timely YAGO [17] which are used to define temporal facts. Temporal feature specific SPARQL query languages such as τSPARQL [13], stSPARQL [1] are also defined to capture temporal fact related linked data information. So, it is understood that linked data research community recently has been paying attention on temporal feature related information over linked data. However, there is still lacking of easy-to-use such data retrieval framework.

Temporal feature related information extraction of linked data is relatively new. In 2011, a study by Vandenbussche et al. at the Detection, Representation, and Exploitation of Events in the Semantic Web (DeRiVE) workshop put forward an initiative regarding extraction of temporal feature related linked data [15]. The study by Vandenbussche et al. focused primarily on image extraction, whereas our study is not intended for any particular domain. Khrouf et al. and Troncy et al. both showed *event* related i.e., what (event), when (date/time), where (geo) and who (participant of an event), ontology integration. But these studies considered that linked data will be presented with a fixed ontology (i.e., LODE) and they ignored the possibility of accessing other temporal ontologies [7,14]. On contrary, in our study we adapt a easy-to-use keyword-based linked data information retrieval framework which is not susceptible to any specific ontology. Most importantly, for event-specific queries, our adapted framework could explore implicit temporal semantics i.e., from an event, it can explore date/time information which helps in finding more concise result.

However, from the keyword-based linked data retrieval perspective, there are several studies [9,12,16,18]. Keyword-based retrieval option is thought to be familiar to the users which motivated in such studies. In our study, we adapt

---

[3] http://www.timexportal.info/
[4] http://timeml.org/site/publications/specs.html

such a keyword-based framework which we further extend to capture temporal feature. We extend the QA system from Rahoman and Ichise [9].

3 Keyword-Based Linked Data Retrieval

For keyword-based linked data retrieval, we extend the QA system from [9]. We select this system because of its efficiency and simplicity. It takes keyword as input and generates possible information as output.

3.1 The QA System

The QA system uses template which resemble graph-like structure of linked data and tries to subsume some part of linked data to generate possible information [9]. In general, a template is pre-defined structure which holds some position holders and accomplishes tasks by setting those holders with task specific parameters. Position holders of templates that are used in the QA system are either filled-up by input keywords (or more precisely by linked data resources which represent input keywords) or they are kept by variables considering variables could be filled-up by some linked data resources. The QA system utilize these kept variables to pick possible information. For example, a query question *What is birthday of Michael Jackson?* which, suppose, is converted to input keywords *Michael Jackson, birthday.* For these input keywords, one of a template could be like $<res:Michael\_Jackson,\ prop:birthDate?uri>$ where resource *res:Michael_Jackson* represents input keyword *Michael Jackson,* resource *prop:birthDate* represents input keyword *birthday* and variable *?uri* is kept to be filled-up by some linked data resources. The QA system uses 11 such templates. Construction of template depends on input keywords and inside statistics of input keywords representing linked data resources.

For every input keyword, at first, the QA system finds input keyword representing linked resources. For such linked data resources, each of them are classified into two types: a *predicate type resource* (PR), or else a *non-predicate type resource* (NP). This classification is done by the resource appearance frequency in a dataset, i.e., a frequent *predicate* resource is considered as a PR, or else a NP. For example, linked resource *prop:birthDate* will be considered as PR, if inside a dataset most of the time *prop:birthDate* is appears as *predicate.*

For every two input keywords, by considering input keyword representing linked data resources and their classifications, templates are constructed as it shown in Table 1. Table 1 shows template category i.e., TC_1, TC_2, TC_3 and TC_4 in its first column. Template category primarily decide which templates will be used in particular case. Second column indicates a specific template number for each template. For input keyword representing linked data resources of every two input keywords, if one linked data resource appears as PR and another one appears as NP, templates are constructed by template # 1 and 2. If both linked data resources appear NP, templates are constructed by template # 3, 4, 5, 6, 7 and 8. If number of input keywords is more than two (e.g., number of input keywords is three), in some cases template needs to be constructed

Table 1. Query templates that are used in QA system [9]

Template Category	Template #	Template	SPARQL Query for template
TC_1	1	$<?uri, r_1, r_2>$	$SELECT\ ?uri\ WHERE\ \{?uri\ r_1\ r_2.\}$
	2	$<r_2, r_1, ?uri>$	$SELECT\ ?uri\ WHERE\ \{r_2\ r_1\ ?uri.\}$
TC_2	3	$<r_1, ?uri, r_2>$	$SELECT\ ?uri\ WHERE\ \{r_1\ ?uri\ r_2.\}$
	4	$<r_2, ?uri, r_1>$	$SELECT\ ?uri\ WHERE\ \{r_2\ ?uri\ r_1.\}$
	5	$<?uri, ?p_1, r_1> <?uri, ?p_2, r_2>$	$SELECT\ ?uri$ $WHERE\ \{?uri\ ?p_1\ r_1.\ ?uri\ ?p_2\ r_2.\}$
	6	$<r_1, ?p_1, ?uri> <r_2, ?p_2, ?uri>$	$SELECT\ ?uri$ $WHERE\ \{r_1\ ?p_1\ ?uri.\ r_2\ ?p_2\ ?uri.\}$
	7	$<r_1, ?p_1, ?uri> <?uri, ?p_2, r_2>$	$SELECT\ ?uri$ $WHERE\ \{r_1\ ?p_1\ ?uri.\ ?uri\ ?p_2\ r_2.\}$
	8	$<?uri, ?p_1, r_1> <r_2, ?p_2, ?uri>$	$SELECT\ ?uri$ $WHERE\ \{?uri\ ?p_1\ r_1.\ r_2\ ?p_2\ ?uri.\}$
TC_3	9	$<?uri, r_1, ?o_1>$	$SELECT\ ?uri\ WHERE\ \{?uri\ r_1\ ?o_1.\}$
TC_4	10	$<r_1, ?p_1, ?uri>$	$SELECT\ ?uri\ WHERE\ \{r_1\ ?p_1\ ?uri.\}$
	11	$<?uri, ?p_1, r_1>$	$SELECT\ ?uri\ WHERE\ \{?uri\ ?p_1\ r_1.\}$

for single input keyword, for this, template # 9, 10, and 11 are used. In such a case, if input keyword representing linked data resource is classified as PR, template is constructed by template # 9 and if input keyword representing linked data resource is classified as NP, templates are constructed by template # 10 and 11. Third column shows template where input keyword representing linked data resources are shown with r_1 and r_2 while variables are shown with question mark (?) following names $?uri$, $?p_1$, $?p_2$, $?o_1$ and $?o_2$. Forth column shows template corresponding converted SPARQL query. Selection of the most appropriate template is decided by individual template's resemblance to dataset. To measure individual template's resemblance, the QA system check how closely and how frequently a template subsume its dataset. The more close and more frequent template is picked as the most appropriate template. Then the QA system converts the most appropriate template to SPARQL query by establishing $SELECT$ query for the kept variable of template, such as, for template $<res:Michael\_Jackson, prop:birthDate\ ?uri>$, the SPARQL query would be $SELECT\ ?uri\ WHERE\ \{res:Michael\_Jackson\ prop:birthDate\ ?uri\ .\}$. This SPARQL query generates possible result for given input keywords.

The QA system works, if input keywords are more than one. This is because, QA system requires two input keywords to construct and find the most appropriate template. If input keywords are more than two which produces multiple most appropriate templates, QA system merges those most appropriate templates to construct merged template (details are described in [9]).

3.2   The Extended-QA System

To support temporal feature related linked data information retrieval, we extend the QA system, which we call the extended-QA system, in two ways

1. **In generation of information**

   In earlier study, the QA system would generate particular variable specific information. This is because, in earlier study, conversion of template to SPARQL query establishes *SELECT* query for only one kept variable of template i.e., *?uri*. So even if templates consist several kept variables such as $?p_1$, $?p_2$, $?o_1$ or $?o_2$, SPARQL query is always constructed for variable *?uri* which hide some potential information. In our observation, to adapt temporal semantics to the QA system, we require to capture all such information. So we extend template to SAPARQL query conversion for all kept variables of template. For example, if a query template is $<?uri, ?p_1, r_1> <?uri, ?p_2, r_2>$, then we construct SPARQL query like *SELECT* $?uri ?p_1 ?p_2$ *WHERE* $\{ ?uri\ ?p_1\ r_1.\ ?uri\ ?p_2\ r_2. \}$.

2. **In handling of number of input keywords**

   In earlier study, the QA system only works, if input keywords are more than one. In our observation, temporal semantics can be added with a single keyword. For example, an event information such as *World War I* can be presented with single keyword. So we extend the QA system for single keyword handling. For doing this, we find input keyword representing linked data resources, then use last three templates from Table 1 (i.e., template # 9, 10 and 11) and find appropriate template. Single input keyword handling is similar to input keyword handling of the QA system. This is because, in the QA system when number of input keywords is more than two (e.g. number of input keywords is three), the QA system would also need to find the most appropriate template for single input keyword.

   In the extended-QA system, to handle single input keyword, if one of this single input keyword representing linked data resource is classified as *PR*, we construct template by template # 9, and if input keyword representing linked data resource is classified as *NP*, we construct templates by template # 10 and 11. Then we select the most appropriate template by individual template's resemblance to dataset which, like the most appropriate template selection of the QA system, considers how closely and how frequently a template subsume its dataset. And that is how, among all such templates, we pick the more close and more frequent template as the most appropriate template which we eventually convert to its SPARQL query to generate possible result.

So the extended-QA system generates information for all kept variables of template and it can able to handle single input keyword.

4 TLDRet: Linked Data Retrieval Framework with Temporal Semantics

Temporal **L**inked **D**ata **R**etrieval framework (**TLDRet**) is our proposed system. We use the extended-QA system which takes keyword as input and generates real world facts as possible output. We adapt the extended-QA system with temporal semantics. That is, input keywords of TLDRet might hold temporal features such as date and time or time of an event.

According to [3,4,8,11], temporal feature attached query questions always hold indicator words called *signal words*, shown in the following examples.

- Example 1: *Which music artist was born on the 29th of August 1958?*
- Example 2: *Which US President born during World War I?*

Example 1 holds a temporal feature indicating the word *on* that follows the temporal feature *the 29th of August 1958*, and Example 2 holds the temporal feature indicating the word *during* that follows the event information *World War I*. So, words *on* and *during* are considered as *signal words*.

On the other hand, the extended-QA system defines input keywords that intuitively are able to present the required information need. Therefore, in this study, input keywords that hold *signal words* and their following input keywords, we define them as *temporal keywords*. Such as, in Example 2, *during World War I* is *temporal keyword* while *during* is *signal word*. Moreover, we assume that input keywords will contain *signal words* if the required information includes a temporal feature in its question. (Therefore, if input keywords do not hold *signal words*, TLDRet works like the extended-QA system works.) *Temporal keywords* could either hold explicit temporal value such as date and time or they could represent some event information related keywords that could be converted to explicit temporal values. The former type of *temporal keywords* are considered as *explicit temporal keywords*, whereas the latter type of *temporal keywords* are considered as *event temporal keywords*. If, for an exemplary linked data dataset, the Example 1 question is presented by the input keywords {*music artist, birthday, on the 29th of August 1958*} and the Example 2 question is presented by the input keywords {*US President, birthday, during World War I*}[5] then the *explicit temporal keyword* will be from Example 1, i.e., *on the 29th of August 1958* and the *event temporal keyword* will be from Example 2, i.e., *during World War I*. This is the categorization of two types of *temporal keywords*, because the question with the *event temporal keyword* needs to generate an explicit temporal value, while the question with the *explicit temporal keyword* holds an explicit temporal value. We categorize *temporal keywords* by normalizing input keywords by a parser. If normalized output of *signal words* following input keywords hold DATE/TIME then *temporal keywords* belong to *explicit temporal keywords*, otherwise they belong to *event temporal keywords*. In normalized output, annotation of DATE/TIME is done by TIMEX3 which is popular state-of-the-art time annotation standard. In TLDRet, we compile *signal words* from [3,4,11].

Figure 1 shows the overall process flow for TLDRet. It has two phases: phase 1 - query text processing, phase 2 - semantic query. In phase 1, TLDRet orders input keywords and annotates temporal value of *temporal keywords* to TIMEX3. Then in phase 2, TLDRet imposes a time filter to produce the intended result.

---

[5] the QA system [9] uses exact matching between input keywords and linked data resources currently.

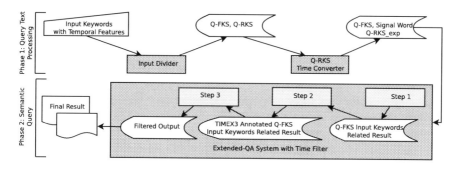

Fig. 1. Overall work flow of our proposed method

4.1   Phase 1: Query Text Processing

Here we talk about phase 1 i.e., query text processing. This phase performs pre-processing tasks before inclusion of temporal semantics. In this phase, we divide input keywords considering *signal words*. Then we annotate temporal value of *temporal keywords* by their corresponding TIMEX3 annotation. *Input divider* process and *Q-RKS time converter* process respectively accomplish these tasks.

From the perspective of answering the temporal question, Saquete et al. showed a technique using an *ordering key* [11]. The *ordering key* preserves temporal semantics of input keywords by introducing some kind of information validity constraint (detail is described in following phase). To use *ordering key*, it needs to divide input keywords into two keyword sets. Process *input divider* does this division. The two keyword sets are as follows.

1. *Q-Focus keyword set* (Q-FKS): input keywords that specify the information that user is searching. *Signal word* prior input keywords fall into Q-FKS.
2. *Q-Restriction keyword set* (Q-RKS): input keywords that specify temporal features used to restrict required information. *Signal word* and its follower input keywords fall into Q-RKS. In TLDRet, *temporal keywords* belong to Q-RKS.

If we execute the *input divider* process, it divides Example 1 into Q-FKS = {*music artist, birthday*}, Q-RKS = {*on the 29th of August 1958*}, and Example 2 into Q-FKS = {*US President, birthday*}, Q-RKS = {*during World War I*}.

As mentioned, Q-RKS input keywords hold *temporal keywords* which are further divided into *explicit temporal keywords* and *event temporal keywords*. Whatever the division, *Q-RKS time converter* finds the explicit temporal value and annotates that value to a common annotation. If Q-RKS input keywords hold *explicit temporal keyword*, i.e., keywords with a date and time, the *Q-RKS time converter* generates TIMEX3 annotated temporal value for *explicit temporal keyword*. We refer to this TIMEX3 value as Q-RKS_exp. Example 1 belongs to such case where *Q-RKS time converter* converts *explicit temporal keyword* to their corresponding TIMEX3 value as 1958-08-29. In contrast, if Q-RKS input keywords hold *event temporal keyword*, the *Q-RKS time converter* first uses our proposed the extended-QA system. Then, it parses output of the extended-QA system and captures the normalized named entity recognition (NER) value for the DATE/TIME/DURATION type, which is also able to give *event temporal keyword* corresponding TIMEX3 value. For *event temporal keyword*, this TIMEX3 value is referred to as Q-RKS_exp. Example 2 belongs to such case where, considering an exemplary dataset, *Q-RKS time converter* converts *event temporal keyword* to their corresponding TIMEX3 values as 1914-07-28, 1919-06-28, 1919-09-10, 1919-11-27, 1920-06-04 and 1920-08-10. Each of these dates are some how related to *World War I* such as date 1914-07-28 is related to armistice with Germany, date 1919-06-28 is related to signing treaty of Versailles and so on.

So execution of the *Q-RKS time converter* for Q-RKS input keywords gives explicit temporal value (Q-RKS_exp) in TIMEX3.

4.2   Phase 2: Semantic Query

This subsection describes phase 2 i.e., the semantic query. In this phase, we attache Q-RKS_exp to the output of Q-FKS input keywords in such way that temporal semantics of input keywords are preserved. *Ordering key* framework [11] supports us in preserving temporal semantics. By preserving temporal semantics, we filter out output of Q-FKS input keywords which retrieve final result. *Extended-QA System with time filter* process accomplishes the above tasks.

Before describing the *extended-QA System with time filter* process, we introduce basics of *ordering key*. *Ordering key* defines constraint of information validity that is constructed for three parameters: (i) *signal word*, (ii) temporal feature related part of Q-FKS input keywords (iii) temporal feature related part of Q-RKS input keywords. With this constraint, *ordering key* incorporates temporal semantics of input keywords which, eventually, gives option of information filtering. For every *signal word*, it introduces constraint of information validity. Such as, for Example 1 where *signal word* is *on*, temporal feature related part of Q-FKS input keywords is *birthday* and temporal feature related part of Q-RKS input keywords is *the 29th of August 1958*, so the constraint of information validity is *birthday = the 29th of August 1958*.

For simplicity, if we represent the temporal feature related part of Q-FKS input keywords to symbol F1 and the temporal feature related part of Q-RKS input keywords to symbol F2, then Table 2 shows some *signals words* and their *ordering keys*. Here the first column shows the *signal word* and the second column shows the corresponding *ordering key* between F1 and F2. To present time interval, F2 is presented with its initial (i.e., F2i) and final (F2f) points. Some of them explicitly attaches temporal feature related part of Q-RKS input keywords to two points F2 and F3. Last two rows of the table show time interval.

Now we describe the *extended-QA System with time filter* process. It takes Q-FKS, the *signal word*, and Q-RKS_exp as input and generates the final output. This process has three steps. In step 1, we use our the extended-QA system over Q-FKS input keywords to find Q-FKS input keywords related result. In step 2, we parse the step 1 result by a parser, and we replace the NER values

Table 2. Example of *signal words* and *ordering of keys*

Signal word	Ordering key
Before	F1 < F2
On, In, When, At the same time	F1 = F2
After, Since	F1 > F2
During, While, For	F2i <= F1 <= F2f
From, About	F2 <= F1 <= F3

of the DATE/TIME/DURATION type by their corresponding TIMEX3 values. In step 3, according to the *signal word* corresponding *ordering key*, we filter the TIMEX3 annotated result of step 2 that complies the temporal semantics. This filtered result is considered as our final output.

Step 1. As mentioned, Q-FKS input keywords specify the information that user is searching. Step 1 generates such information. We employ the extended-QA system over Q-FKS input keywords which generates, irrespective to temporal feature related or non-related, all such possible information.

Step 2. From temporal linked data information retrieval perspective, Q-FKS input keywords always hold temporal feature related part which is used in information filtering. As mentioned, we present this temporal feature related part by symbol F1. Step 2 finds this temporal feature related part and annotates its temporal value by its TIMEX3 annotation so that, annotated value can be used to maintain temporal semantics. For doing this, we take Step 1 result and parse, then annotate the NER values of the DATE/TIME/DURATION type by their corresponding TIMEX3 values.

Step 3. Step 3 is the last step which retains temporal semantics of input keywords and retrieves final result. According to the constraint of information validity (in *ordering key*), temporal feature related part of Q-FKS input keywords (i.e, represented by F1) and temporal feature related part of Q-RKS input keywords (i.e., represented by F2), step 3 filters the TIMEX3 annotated result of step 2 which gives final result.

As mentioned, to comply constraint of information validity, we assign

- the temporal feature related part of Q-FKS input keywords to symbol F1
- the temporal feature related part of Q-RKS input keywords to symbol F2.

Therefore, in information filtering, we require TIMEX3 value of F1 and TIMEX3 value of F2 which can implement temporal semantics. TIMEX3 value of F1 comes from TIMEX3 annotated result of step 2. Here NER value for the DATE/TIME/ DURATION type is TIMEX3 value of F1. Such as, if TIMEX3 annotated result of step 2 is *Beethoven ... 1770-12-16*, then TIMEX3 value of F1 is *1770-12-16*. Whereas, TIMEX3 value of F2 comes from Q-RKS_exps.

Table 3. *Ordering keys* and their *temporal picking points*

Ordering key type	Ordering key	Temporal Picking Point (TPP/TPPi/TPPf)
Point of time	F1 < F2	Pick lowest Q-RKS_exp value among all such values as TPP
	F1 = F2	Pick every Q-RKS_exp values as TPP
	F1 > F2	Pick highest Q-RKS_exp value among all such values as TPP
Interval	F2i <= F1 <= F2f	Pick lowest Q-RKS_exp value among all such values as TPPi
		Pick highest Q-RKS_exp value among all such values as TPPf
	F2 <= F1 <= F3	Pick Q-RKS_exp value for TIMEX3 converted value of F2 as TPPi
		Pick Q-RKS_exp value for TIMEX3 converted value of F3 as TPPf

In our observation, in some cases, we can get multiple Q-RKS_exps. Usually, multiple Q-RKS_exps either could be generated to capture interval such as: for *explicit temporal keyword, from 28th July 1914 to 10th August 1920*, or *event temporal keywords, during World War I*. Or, they could be generated when Q-RKS input keywords are not enough to define single Q-RKS_exp. In such a case, according to *ordering key*, we decide the *temporal picking point* (TPP) for the temporal feature related part of Q-RKS input keywords. This TPP helps in picking best possible Q-RKS_exp among many. For interval, the TPP is considered for two parts: initial TPP (TPPi) and final TPP (TPPf).

Table 3 shows the TPP picking strategy for various *ordering keys*. First column, the *ordering key type* divides *ordering keys* into two types: *point of time* and *interval*. Second and third column respectively show *ordering keys* and their corresponding *temporal picking points* which describe how we pick Q-RKS_exp value. *Point of time* type *ordering keys* select single Q-RKS_exp. On the other hand, *interval* type *ordering keys* select two Q-RKS_exps: initial and final. Such as, over an exemplary dataset, for Example 1 we can get TPP as 1958-08-29. This is because, the *ordering key* of Example 1 is F1 = F2 which finds only one TIMEX3 annotation for temporal feature related part of Q-RKS input keywords. On the other hand, over the same exemplary dataset, for Example 2 we can get TPPi as 1914-07-28 and TPPf as 1920-08-10. This is because, the *ordering key* of Example 2 is F2i <= F1 <= F2f where we get multiple Q-RKS_exps i.e., 1914-07-28, 1919-06-28, 1919-09-10, 1919-11-27, 1920-06-04 and 1920-08-10, among which the lowest value is 1914-07-28 and the highest value is 1920-08-10.

To filter the TIMEX3 annotated result of step 2, we find TIMEX3 value of F1, the TPP value, and construct constraint of information validity according to the *ordering key*. Then pick result of step 2 which passes constraint of information validity. That is how, we preserve temporal semantics and generate concise result.

Table 4 shows the different execution steps of the *extended-QA System with time filter* process for Example 1 over an exemplary dataset. Here the first column shows the step number, the second column shows the step description, and the third column shows the step result. Execution of step 1 selects all possible results for Q-FKS input keywords, i.e., {*music artist, birthday*}. In step 2, execution of parser over the step 1 result annotates temporal value attached parts by the corresponding TIMEX3 values. Example 1 holds *on* as the *signal word* so in step 3, by using the *signal word* corresponding to the *ordering key*, it gets

Table 4. Semantic query scenario for Example 1

Step	Step description	Result
1	Execution of proposed extended-QA system over Q-FKS input keywords i.e., {*music artist, birthday*}	Beethoven ... December, 16, 1770 Iikka Paananen ... 29th December 1960 Michael Jackson ... August 29, 1958 Ramona Maria Luengen ... December, 29, 1960 ... ...
2	Execution of parser over step 1 result for DATE/TIME/DURATION type NER values	Beethoven ... 1770-12-16 Iikka Paananen ... 1960-12-29 Michael Jackson ... 1958-08-29 Ramona Maria Luengen ... 1960-12-29 ... ...
3	Use of TIMEX3 value of F1, the TPP value, and constraint of information validity	Michael Jackson ... 1958-08-29

the *ordering key* as F1 = F2 which finds the TPP value as 1958-08-29. Finally, considering the constraint of information validity, the *extended-QA System with time filter* process filters out the possible result of Example 1.

In previous illustration, the temporal feature related part of Q-FKS input keywords (i.e, represented by F1) corresponds to a single point of time (i.e., *birthday*). But, in some cases, temporal feature related part could correspond to multiple point of times. This is because, Q-FKS input keywords could hold multiple date and time associated parts, or associated part could hold an interval. Such as, for exemplary input keywords {*US President, serving period, during World War I*}, Q-FKS input keywords are {*US President, serving period*} where date and time associated part is *serving period* which is an interval. In such a case, we consider all possible TIMEX3 value of F1, construct all possible constraint of information validity and filter possible information.

5 Experiment

We use the Question Answering over Linked Data (QALD)[6] open challenge question sets in our experiment. The QALD open challenge includes natural language question sets from DBPedia and MusicBrainz datasets, which are divided into QALD-1 and QALD-2. Moreover, both QALD-1 and QALD-2 are further divided into training question set and test question set.

Since our proposed system TLDRet depends on keywords, we intuitively retrieve keywords that simultaneously balance the datasets and the query questions. We evaluate TLDRet for efficiency achieved in temporal feature related question answering.

For temporal feature related question answering, we evaluate TLDRet for QALD questions that relate to temporal features in their answer. Such as, for MusicBrainz dataset, example of such a QALD question is *Which artists were born on the 29th of December 1960?*. We also evaluate TLDRet with four other systems.

---

[6] http://greententacle.techfak.uni-bielefeld.de/~cunger/qald/

Table 5. QALD temporal feature related question answering performance by TLDRet

Participant question set	# of questions	Performance of TLDRet		
		Precision	Recall	F1 Measure
DBPedia QALD-1	4	1.000	1.000	1.000
QALD-2	9	1.000	1.000	1.000
	Average	1.000	1.000	1.000
MusicBrainz QALD-1	18	0.722	0.722	0.722
QALD-2	20	0.750	0.750	0.750
	Average	0.737	0.737	0.737

In TLDRet, output normalization and their TIMEX3 annotation are done by Stanford parser [2].

Performance Over QALD Temporal Feature Related Questions. To show the performance of our TLDRet over QALD temporal feature related questions, we execute each set of questions and check the average precision, average recall and average F1 measure for each set of questions. Each set of questions consist both training and test questions. To calculate precision, recall and F1 measure, QALD gives each gold standard result for each participant question. So, for each participant question set, we calculate the average performance by summing up the performance of the questions and dividing them by the number of questions for each participant question set.

Table 5 shows the performance of TLDRet over QALD temporal feature related question answering. Here the first column shows the source of the participant question set, the second column shows the number of questions in each participant question set, and the third column shows the performance of TLDRet. Moreover, for each dataset, we calculate the average performance.

We find that TLDRet achieves high performance on DBPedia dataset questions. In contrast, TLDRet achieves comparatively lower performance on MusicBrainz dataset questions. Our detail investigation finds that this performance drop is not because of a lack of temporal feature attachment; rather, the extended-QA system is not able to generate Q-FKS keyword related information, which is a prerequisite for temporal feature attachment. The extended-QA system depends on some fix structured templates which, some cases, gets failed to resemble dataset. To solve this problem, the extended-QA system needs to consider other templates than those fixed structure such as, if one linked data resource appears as predicate type resource (e.g., r_1) and another linked data resource appears as non-predicate type resource (e.g., r_2), then construct template like $<?uri, r_1, ?o_1> <?uri, ?p_2, r_2>$ and so on. If this problem is solved, the TLDRet will also be able to achieve better performance on MusicBrainz dataset questions.

We adapt temporal semantics into a keyword-based QA system. According to the temporal feature related QALD questions, we conclude that our TLDRet can

Table 6. Performance comparison between TLDRet and QALD-2 open challenge participant systems over temporal feature related DBPedia QALD-2 test questions

System	Average Precision	Average Recall	Average F1 Measure*
SemSeK	0.400	0.400	0.400
Alexandria	0.000	0.000	0.000
MHE	0.400	0.400	0.400
QAKiS	0.000	0.000	0.000
TLDRet	1.000	1.000	1.000

retrieve temporal feature related information over linked data comprehensively. Moreover, unlike system described in [15], TLDRet works for any domain.

In TLDRet, successful retrieval of temporal feature related information over linked data is because of symbiotic adaptation of *temporal keyword* and *ordering key* for the given input keywords. We explore all temporal values, whether they are mentioned explicitly (by explicit date and time) or implicitly (by event-specific input), to a common format which helps to filter out concise information. Though linked data hold different structure than document-based data, by TLDRet, like temporal semantics adaptation technique in usual document-based data, we propose a technique which can capture temporal semantics over linked data.

Performance Comparison with Other Systems. For temporal feature related questions, we compare TLDRet with QALD-2 open challenge participant systems named SemSek, Alexandria, MHE, and QAKiS over DBPedia test questions[7]. Because of the availability of the challenging participant systems' results over the DBPedia QALD-2 test set[8], we use this set for comparison. We find the DBPedia QALD-2 test set holds five temporal feature related questions (i.e., Q # 7, 56, 71, 74, 92), so we collect each question's precision, recall and F1 measure for the challenge participant systems (SemSek, Alexandria, MHE, and QAKiS) and our TLDRet. We consider the precision, recall and F1 measure as 0 (zero), if any system does not participate in question answering. Table 6 shows the performance comparison (in average precision, average recall and average F1 measure*[9]) between TLDRet and the QALD-2 open challenge participant systems. It is seen that our TLDRet system decisively outperforms the challenge participant systems. Alexandria and QAKiS cannot answer any of the questions, and the other two systems can only answer 2 out of 5 questions.

Again, we conclude that our proposed method successfully adapts *temporal keyword*, *ordering key*, and explore all temporal values to a common annotation

---

[7] http://greententacle.techfak.uni-bielefeld.de/~cunger/qald/2/
dbpedia-test-questions.xml

[8] In DBPedia QALD-2 set, there are 9 temporal feature related questions, 4 are from training question set and 5 are from test question set.

[9] In average calculation, for each individual question we put F1 measure = 0 (zero), if recall = 0 (zero).

which filter out possible information efficiently. Since our proposed system TLDRet adapts temporal semantics adaptation technique like usual document-based data which is ignored by the challenge participant systems, our systems outperforms the challenge participant systems.

6   Conclusion

Linked data currently hold a very good amount of knowledge. Since linked data knowledge is connected knowledge, an efficient retrieval framework can generate more semantically enriched information. Usually temporal values such as date and time or time of an event enrich semantics and help in finding new information or new links. However, current linked data information retrieval studies pay less attention on temporal semantics for linked data information retrieval. We focus this issue and propose a linked data retrieval framework with temporal semantics. Like temporal semantics adaptation technique in usual document-based data, our proposed system TLDRet which adapts *temporal keyword, ordering key* over linked data and capture required temporal semantics. We explore all temporal values, whether they are mentioned explicitly or implicitly, to a common annotation which helps to filter out concise information. Our proposed method is not confined to a particular domain. Experiment shows the efficiency of our proposed system. Our study is influenced by Saquete et al. study in [11]. Saquete et al. advocated to adapt any general purpose QA system to find temporal feature related part of question. However finding temporal feature related part of question is not easy and not always consistent. Since linked data hold structured data which is different than usual document-based data, linked data information retrieval requires different treatment. We propose a linked data information retrieval framework which can capture temporal feature related part of question consistently. From linked data perspective, retaining of temporal semantics (both for *explicit temporal keywords* and *implicit temporal keyword*) is an important advancement towards linked data's realistic utilization. In this study, we consider retrieval of temporal linked data information for fact-centric perspective, so in the future work, we want to extend our system for document-centric perspective. We assume that inclusion of both perspectives will generate more concise output.

References

1. Bereta, K., Smeros, P., Koubarakis, M.: Representation and querying of valid time of triples in linked geospatial data. In: Cimiano, P., Corcho, O., Presutti, V., Hollink, L., Rudolph, S. (eds.) ESWC 2013. LNCS, vol. 7882, pp. 259–274. Springer, Heidelberg (2013)
2. Chang, A.X., Manning, C.: SUTime: a library for recognizing and normalizing time expressions. In: Proceedings of the 8th International Conference on Language Resources and Evaluation, pp. 23–25 (2012)
3. Derczynski, L., Gaizauskas, R.J.: A corpus-based study of temporal signals. Computing Research Repository, abs/1203.5066 (2012)

4. Fry, E.B., Kress, J.E., Fountoukidis, D.L.: The Reading Teacher's Book of Lists. Prentice Hall, Englewood Cliffs (1993)
5. Gutierrez, C., Hurtado, C., Vaisman, R.: Temporal RDF. In: Proceedings of the 2nd European Semantic Web Conference, pp. 93–107 (2005)
6. Heath, T., Bizer, C.: Linked Data: Evolving the Web into a Global Data Space. Morgan & Claypool Publishers, San Rafael (2011)
7. Khrouf, H., Milicic, V., Troncy, R.: EventMedia live: exploring events connections in real-time to enhance content. In: Proceedings of the Semantic Web Challenge, at the 11th International Conference on The Semantic Web (2012)
8. Mani, I., Wilson, D.G.: Robust temporal processing of news. In: Proceedings of the 38th Annual Meeting on Association for, Computational Linguistics, pp. 69–76 (2000)
9. Rahoman, M.-M., Ichise, R.: An automated template selection framework for keyword query over linked data. In: Takeda, H., Qu, Y., Mizoguchi, R., Kitamura, Y. (eds.) JIST 2012. LNCS, vol. 7774, pp. 175–190. Springer, Heidelberg (2013)
10. Rula, A., Palmonari, M., Harth, A., Stadtmüller, S., Maurino, A.: On the diversity and availability of temporal information in linked open data. In: Cudré-Mauroux, P., Heflin, J., Sirin, E., Tudorache, T., Euzenat, J., Hauswirth, M., Parreira, J.X., Hendler, J., Schreiber, G., Bernstein, A., Blomqvist, E. (eds.) ISWC 2012, Part I. LNCS, vol. 7649, pp. 492–507. Springer, Heidelberg (2012)
11. Saquete, E., González, J.L.V., Martínez-Barco, P., Muñoz, R., Llorens, H.: Enhancing QA systems with complex temporal question processing capabilities. J. Artif. Intell. Res. **35**, 775–811 (2009)
12. Shekarpour, S., Auer, S., Ngomo, A.-C. N., Gerber, D., Hellmann, S., Stadler, C.: Keyword-driven SPARQL query generation leveraging background knowledge. In: Proceedings of the 10th International Conference on Web, Intelligence, pp. 203–210 (2011)
13. Tappolet, J., Bernstein, A.: Applied temporal RDF: efficient temporal querying of RDF data with SPARQL. In: Aroyo, L., Traverso, P., Ciravegna, F., Cimiano, P., Heath, T., Hyvönen, E., Mizoguchi, R., Oren, E., Sabou, M., Simperl, E. (eds.) ESWC 2009. LNCS, vol. 5554, pp. 308–322. Springer, Heidelberg (2009)
14. Troncy, R., Malocha, B., Fialho, A.T.S.: Linking events with media. In: Proceedings of the 6th International Conference on Semantic Systems, pp. 1–4 (2010)
15. Vandenbussche, P.-Y., Teissédre, C.:. Events retrieval using enhanced semantic web knowledge. In: Proceedings of the Workshop on Detection, Representation, and Exploitation of Events in the Semantic Web, pp. 112–116 (2011)
16. Wang, H., Liu, Q., Penin, T., Fu, L., Zhang, L., Tran, T., Yu, Y., Pan, Y.: Semplore: a scalable IR approach to search the web of data. J. Web Semant. **7**(3), 177–188 (2009)
17. Wang, Y., Zhu, M., Qu, L., Spaniol, M., Weikum, G.: Timely yago: harvesting, querying, and visualizing temporal knowledge from wikipedia. In: Proceedings of the 13th International Conference on Extending Database Technology, pp. 697–700 (2010)
18. Zenz, G., Zhou, X., Minack, E., Siberski, W., Nejdl, W.: From keywords to semantic queries-incremental query construction on the semantic web. J. Web Semant. **7**(3), 166–176 (2009)

A Formal Model for RDF Dataset Constraints

Harold Solbrig[1]([✉]), Eric Prud'hommeaux[2],
Christopher G. Chute[1], and Jim Davies[3]

[1] Mayo Clinic, Rochester, MN 55095, USA
{solbrig.harold,chute}@mayo.edu
[2] World Wide Web Consortium, Cambridge, MA 02139, USA
eric@w3.org
[3] Department of Computer Science, University of Oxford, Oxford, UK
jim.davies@cs.ox.ac.uk

Abstract. Linked Data has forged new ground in developing easy-to-use, distributed databases. The prevalence of this data has enabled a new genre of social and scientific applications. At the same time, Semantic Web technology has failed to significantly displace SQL or XML in industrial applications, in part because it offers no equivalent schema publication and enforcement mechanisms to ensure data consistency. The RDF community has recognized the need for a formal mechanism to publish verifiable assertions about the structure and content of RDF Graphs, RDF Datasets and related resources. We propose a formal model that could serve as a foundation for describing the various types invariants, pre- and post-conditions for RDF datasets and then demonstrate how the model can be used to analyze selected example constraints.

Keywords: RDF · RDF Graph · RDF Dataset · Validation · Formal schema · Invariants · RDF validation · Z specification language

1 Introduction

RDF data sets are becoming ubiquitous. The current "Big Data" initiative is based on the conversion of huge collections of disparate data into a massive, federated collection of RDF triple stores. While these data sources have immense value as they stand, they are highly dependent on the curation and structure of the primary data sources. Today, any significant analysis effort starts with either a tacit understanding of the original source of the RDF data or of an analysis of the state of the triple store at a given point in time. Neither of these approaches is sustainable over the long term; as data stores are merged, the links to the original sources will become lost or blurred. RDF technology has reached the point that it is becoming possible to use an RDF data set as the *primary* data source rather than as a mirror. Software and tools are beginning to emerge that use RDF as a source of reliable facts rather than a source of statistics and inference. Such software needs to be able to know what it can and cannot depend on when it comes to the content of a given RDF data set. Data set federation and

W. Kim et al. (Eds.): JIST 2013, LNCS 8388, pp. 244–260, 2014.
DOI: 10.1007/978-3-319-06826-8_19, © Springer International Publishing Switzerland 2014

merging can only occur when the rules regarding the contents of the source data sets are understood and shown to be compatible. Such interface specifications can streamline the way endpoints in the Linked Data cloud get populated, get checked for consistency, and have their interfaces published.

In this paper we begin with a formal model of the foundational RDF artifacts as described in the RDF 1.1 [3] – the RDF Triple, Graph, Named Graph and Dataset. We then propose a new artifact, the *RDFDatastore* that represents a (potentially) dynamic collection of graphs that include: (1) a formal description of the invariants that a given Datastore can be guaranteed to provide; and (2) a formal description of the allowed transitions between sequential versions of the Datastore.

2 The Z Modeling Language

The Z modeling language was chosen as the formalism to be used in this paper because it provides a formal, verifiable model while remaining relatively concise. It also has a well-described, ISO standard syntax and semantics [1]. It is assumed that any implementations of the models described here will be implemented using combinations of existing and emerging tooling such as SPARQL [10], OWL [9], or OSLC Shapes [5].

2.1 Why Z?

Schmidt [7] describes three different approaches to specifying the 'meaning' of terms written in a formal language, and hence the meaning of the language itself. An *operational semantics* explains how terms may be progressively interpreted or executed; a *denotational semantics* maps each term to an expression in a mathematical language; an *axiomatic semantics* describes the meaning in terms of properties that must be satisfied: axiomatic statements and properties derived using inference rules.

These three approaches are often used in combination, with proofs of congruence presented to show that the different interpretations they place upon the language are all consistent. The axiomatic approach is the most amenable to initial, partial description; Schmidt suggests that it is suitable for "preliminary specifications for a language or to give documentation about properties that are of interest to the users."

The SPARQL query language semantics [10] is an example of the *operational* approach – it represents a set of (declarative) instructions that are interpreted by a machine. If is often difficult, however, to understand what a non-trivial SPARQL query is attempting to accomplish and more difficult yet to determine whether the query correctly implements the intent of a given specification.

Denotational semantics are often used to provide a bridge between the syntax of a given specification and its interpretation in a target mathematical space. While considerably more concise than operational semantics, it describes how a program is to be implemented, but does not directly speak to the correctness or completeness of the specification.

Our intent here is to provide an *axiomatic semantics*. The Z specification language [11] is an ideal language in which to present such a semantics. It is based upon the familiar concepts of predicate logic and set theory. It introduces the additional notion of *schemas*: a schema reference may be used as a declaration, as a constraint upon variables already in scope, or as a set – being the set of objects or 'bindings' that satisfy the declaration and constraint in question.

The ability to name and re-use patterns of declaration and constraint greatly simplifies the process of producing, maintaining, and using a mathematical specification. The fact that Z is strongly typed is another, significant consideration: many logical or conceptual errors will lead to a variable or operator being used in a way that is inconsistent with the underlying type signature of its declaration; the application of a simple type checking algorithm is enough to reveal and thus help eliminate these errors. The descriptions in this paper have been checked using the *fuzz* type checker [8].

2.2   Z Notation

The Z notation has no Boolean type. Instead, everything is a set, or an element of a set. Every expression is an element of a unique type constructed from the integers and user-defined types. A user-defined type t may be introduced by the declaration

$$[t]$$

and subsequently constrained by axiomatic definitions of the form

$$
\begin{array}{|l}
a, b : t \\
\hline
p
\end{array}
$$

This particular definition declares two elements a and b, and introduces the global constraint p on a, b, and t.

The primitive constraints $x : s$, $x \notin s$, and $s = t$ assert that x is an element of s, x is not an element of s, and that $s = t$, respectively. Constraints may be combined using conjunction ($\wedge$), disjunction ($\vee$), implication ($\Rightarrow$), and negation ($\neg$) operators, and quantified with universal and existential quantifiers ($\forall$ and $\exists$). Familiar syntax is used for set theoretic operators: for example, $\cup$ for union, $\cap$ for intersection, and $\setminus$ for set subtraction/difference.

ISO standard Z includes a language or toolkit of operators. For example: $s \rightarrow\!\!\!\!\!\!\rightarrow t$ denotes the set of all partial functions from set s to set t; and $\mathbb{P}\, s$ denotes the set of all subsets – the power set – of set s; the 'free type' declaration

$$u ::= c \langle\!\langle s \rangle\!\rangle \mid d \langle\!\langle t \rangle\!\rangle$$

introduces a set u containing a different element $c\, x$ for every element x of s, and a different $d\, y$ for every element y of t.

Functions and relations are represented as sets of pairs; for any function f, we write: 'dom f' to denote the set of all first elements of pairs in f, the 'domain'; 'ran f; to denote the set of all second elements, the 'range'; $s \lhd f$ to denote the set of pairs in f for which the first element is in s, and $s \ntriangleleft f$ to denote the set of pairs for which it is not. We write '$(\mu\, x : s \mid p)$' to denote the unique element x of s with property p: this expression is undefined if there is no such element, or if there is more than one.

Schemas are named using the $\hat{=}$ operator and a horizontal schema box, delimited by '[' and ']', or by including a name in the top line of a vertical schema box: for example,

$$
\begin{array}{|l}
\underline{e} \\
\quad a, b : t \\
\quad f : s \nrightarrow t \\
\underline{} \\
\quad q \\
\end{array}
$$

introduces a schema named e that declares three identifiers a, b, and f subject to the constraint q, together with the constraint that f is functional: associating each element of its source s with at most one element of its target t.

The schema notation can be used to describe operations or transformations: given a schema S describing a component of state, we may describe an operation using a schema that contains two copies of S, one describing the state before, the other (decorated with a prime $'$) describe the state afterwards. We write ΔS to denote the combination of decorated and undecorated versions of S.

We may decorate components with ? and ! to denote input and output, respectively. Operation schemas can be composed sequentially, using the composition operator $\S$. The primed (decorated with $'$) components of the first are identified with the unprimed components of the second – the combination represents the transformation achieved by applying one then the other. Both definitions and schemas can be generic: set parameters can be included in square brackets. Schema components can be renamed: for example, $e[n/a]$ would be a schema declaring and constraining n where e would a. Schema components can be selected using the familiar dot ('.') notation.

3   Core RDF Model

Our first step in modeling RDF constraints is to construct a formal abstract model of the RDF basics. We start by modeling the existing RDF constructs of Triple, Graph, Named graph and Dataset, as described in RDF 1.1 Concepts [3].

3.1   RDF Basics

We begin with the RDF basics, starting with the three axiomatic types:

$$[IRI, LITERAL, BNODE]$$

For our purposes is sufficient to know that these types are disjoint, and we do not need to go into the details of representation or structure at this point. There are constraints applying to the *LITERAL* type, but these are not the focus of the current paper.

We then model the basic RDF building block. We start with the (non-normative) "generalized RDF triple" [4] where the subject, predicate and object are based on the same generalized type:

$$RDF\_Term ::= iri \langle\!\langle IRI \rangle\!\rangle \mid literal \langle\!\langle LITERAL \rangle\!\rangle \mid bnode \langle\!\langle BNODE \rangle\!\rangle$$
$$GeneralizedTriple \mathrel{\widehat{=}} [\, s, p, o : RDF\_Term \,]$$

GeneralizedTriple can then constrained to model the normative *RDF_Triple* as described in [3], by restricting the subject to *IRI* or *BNODE RDF_Terms* and the predicate to only *IRI RDF_Terms*:

$$\begin{array}{l} \underline{\quad RDF\_Triple \quad\quad\quad\quad\quad\quad\quad\quad\quad\quad\quad\quad\quad} \\ \quad GeneralizedTriple \\ \underline{} \\ \quad s \in \operatorname{ran} iri \cup \operatorname{ran} bnode \\ \quad p \in \operatorname{ran} iri \end{array}$$

3.2   RDF Graph

An *RDF_Graph* as a (possibly empty) set of RDF triples:

$$RDF\_Graph == \mathbb{P}\, RDF\_Triple$$

A *NamedGraph* is the combination of a *Name* and an *RDF_Graph*. The RDF 1.1 specification states, "Despite the use of the word 'name' in 'named graph', the graph name does not formally denote the graph. It is merely syntactically paired with the graph. RDF does not place any formal restrictions on what resource the graph name may denote, nor on the relationship between that resource and the graph." [3] This is reflected in the definition of *NamedGraph* below as a cross product (vs. a partial function), meaning there could be two different graphs having the same name. The name of a graph can either be an *IRI* or a *BNODE*:

$$\begin{array}{l} \underline{\quad GraphName : \mathbb{P}\, RDF\_Term \quad\quad\quad\quad\quad\quad} \\ \quad GraphName = \operatorname{ran} iri \cup \operatorname{ran} bnode \end{array}$$

$$NamedGraph == GraphName \times RDF\_Graph$$

3.3   RDF Dataset

RDF 1.1 Concepts [4], defines *RDF_dataset* as a collection of *RDF_Graphs*, consisting of:

(a) "Exactly one default graph, being an RDF graph. The default graph does not have a name and MAY be empty."
(b) "Zero or more named graphs. Each named graph is a pair consisting of an IRI or a blank node (the graph name), and an RDF graph. Graph names are unique within an RDF dataset."

This is reflected in the type below, where *named_graphs* has been restricted to being a function.

$$\begin{array}{|l}
\hline
\_\,RDF\_Dataset \underline{\hspace{6cm}} \\
\quad default\_graph : RDF\_Graph \\
\quad named\_graphs : GraphName \nrightarrow RDF\_Graph \\
\hline
\end{array}$$

This is the point where the current RDF specification stops. It (sensibly) does not describe how *RDF_Datasets* are created and updated. From this point on, we need to introduce new material, proposing extensions to the RDF specification that would support the RDF constraint use cases.

4   RDF Extensions

The *RDF_Dataset* model could apply in a variety of different use cases, including, but not limited to:

– RDF Document – the traditional (semi-)static text document that may include a number of import statements as well as a set of assertions made directly by the document itself.
– RDF Triple Store – either a "triple store" with just a default graph or a "quad store" that can potentially represent multiple named graphs.
– Reference and metadata – named collections of statements (triples) combined with assertions about the collections themselves. See the TriG RDF Dataset Language [2] for examples.
– RDF on the internet – the default graph being the current focus and the named graphs being the graphs that are virtually available through various dereferencing mechanisms.

These use cases highlight some underspecified aspects of the current *RDF_Dataset* model, including:

1. How should multiple *RDF_Datasets* be merged?
2. How would one import one *RDF_Dataset* into a second *RDF_Dataset*?
3. Does the default graph represent *all* the triples in the *RDF_Datset* or just those that are directly asserted? How is this decision represented when a dataset is updated?
4. Can individual triples be added or removed from named graphs or should named graphs be added or removed as complete collections?

There is no obvious answer to the above questions. The behavior of a particular *RDF_Dataset* depends on the requirements of the implementing organization. It is important, however, that both the implementers *and* consumers of *RDF_Datasets* understand the decisions that have made and their ramifications. Implementers need to provide consistent (and correct!) implementations *and* publish the specific implementation characteristics using a shared syntax and semantics.

We now propose a generic model for *RDF_Dataset* initialization and update and then uses the model to investigate some of the decisions that need to be made when defining invariants, preconditions and postconditions.

4.1   RDF Dataset Initialization

The initial state of an *RDF_Dataset* is determined on creation:

```
┌─ RDF_Dataset_Init ──────────────────────────────
│ RDF_Dataset′
│ default? : RDF_Graph
│ ng? : ℙ NamedGraph
├──────────────────────────────────────────────
│ default_graph′ = default?
│ named_graphs′ = ng?
└──────────────────────────────────────────────
```

An empty *RDF_Dataset* is, by definition, one with an empty default graph and no named graphs:

$$Empty\_Dataset \mathrel{\widehat{=}} [\, RDF\_Dataset\_Init \mid default? = \emptyset \wedge ng? = \emptyset \,]$$

It is important to note that initialization does not include the update operation. Unless one starts with an empty *RDF_Dataset*, there is no guarantee that the contents of the dataset meet any preconditions for the update functions – if one wants to be certain, for instance, that all subjects of a particular `rdf:type` have certain properties, this has to be stated as an invariant on the dataset itself. As an example, the assertion that all subjects in an the default *RDF_Graph* have to be of type `foaf:Person` could be represented by:

$$\mid foaf\_firstName, foaf\_lastName, foaf\_Person : IRI$$

```
┌─ Example_Dataset_1 ─────────────────────────────
│ RDF_Dataset
├──────────────────────────────────────────────
│ ∀ g : default_graph •
│     (∃ g₂ : default_graph • g₂.s = g.s ∧
│     g₂.p = iri rdf_type ∧ g₂.o = iri foaf_Person)
└──────────────────────────────────────────────
```

4.2   RDF Dataset Update

Helper Function. The definitions in this section use the ϕ function, which takes a function from a type to a set and an instance of the type, returning the set if it is in the function, otherwise the empty set $\emptyset$

$$
\begin{array}{l}
\boxed{=[X,Y]\!=\!=\!=\!=} \\
\phi : (X \nrightarrow \mathbb{P}\ Y) \nrightarrow X \nrightarrow \mathbb{P}\ Y \\
\hline
\forall x : X;\ f : X \nrightarrow \mathbb{P}\ Y \bullet \phi\,f\,x = \textbf{if}\ x \in \text{dom}\,f\ \textbf{then}\ f\,x\ \textbf{else}\ \emptyset
\end{array}
$$

RDF Dataset Update Operations. An *RDF_Dataset* can be updated by:

1. Adding and/or removing triples from the default graph

$$
\begin{array}{l}
\underline{\textit{AddToDefault}\,} \\
\Delta RDF\_Dataset \\
addToDefault? : RDF\_Graph \\
\hline
default\_graph' = default\_graph \cup addToDefault? \\
named\_graphs' = named\_graphs
\end{array}
$$

$$
\begin{array}{l}
\underline{\textit{RemFromDefault}\,} \\
\Delta RDF\_Dataset \\
remFromDefault? : RDF\_Graph \\
\hline
default\_graph' = default\_graph \setminus remFromDefault? \\
named\_graphs' = named\_graphs
\end{array}
$$

2. Adding and/or removing triples from named graphs.
   If a graph named to *toAdd?* already exists, it is added to the existing graph. Otherwise a new graph is created.

$$
\begin{array}{l}
\underline{\textit{AddToNamedGraph}\,} \\
\Delta RDF\_Dataset \\
addToNamed? : GraphName \nrightarrow RDF\_Graph \\
\hline
\forall n : \text{dom}\ named\_graphs \cup \text{dom}\ addToNamed? \bullet \\
\quad named\_graphs'\ n = \phi\ named\_graphs\ n \cup \phi\ addToNamed?\ n \\
default\_graph' = default\_graph
\end{array}
$$

If a graph named in *toRem?* already exists, the triples are removed, otherwise they are ignored.

$\rule{0pt}{0pt}$

   RemFromNamedGraph
   $\Delta RDF\_Dataset$
   $remFromNamed? : GraphName \nrightarrow RDF\_Graph$

   $\forall\, n : \mathrm{dom}\ named\_graphs \bullet$
       $named\_graphs'\ n = named\_graphs\ n \setminus \phi\ remFromNamed?\ n$
   $default\_graph' = default\_graph$

3. Removing named graphs (which is different than removing all triples from an existing graph)

   RemNamedGraph
   $\Delta RDF\_Dataset$
   $remNamed? : \mathbb{P}\ GraphName$

   $named\_graphs' = remNamed? \lhd named\_graphs$
   $default\_graph' = default\_graph$

The above operations have been deliberately left underspecified; there are no assertions saying that members of the to remove lists must exist or those of the to add list cannot. A given *RDF_Dataset* implementation will need to decide:

(a) The order in which the operations are applied
(b) Which of the graph operations above are permitted
(c) The requirements (preconditions) for each permitted operation

For the reminder of this document, we will use a model that asserts:

(a) Triples will first be removed from the default graph, followed by adding new triples
(b) Named graphs will first be removed, followed by removing triples from the remaining graphs, followed by adding new triples
(c) All operations are permitted
(d) Triples can be added even if they already exist
(e) Triples to be removed that do not exist sill be ignored

$$RDF\_Dataset\_Update \mathrel{\widehat{=}} (RemFromDefault \mathbin{\fatsemi} AddToDefault) \land$$
$$(RemNamedGraph \mathbin{\fatsemi} RemFromNamedGraph \mathbin{\fatsemi} AddToNamedGraph)$$

Sample RDF Datatset Update Model. To provide an example of a different update model, suppose implementer publishes an *RDF_Dataset* which only allows the additions or modifications to the default graph. The implementer fist needs to decide on what "add" and "modify" mean in this context. Does it matter if the triples to be added or modified already exist? Is it necessary that the triples in the range of the modification do not exist?[1] Does the addition occur before or *after* the modification?

The specification below describes an implementation where the triples to be added or modified do not have to exist and where the order of application does not matter, because the domain of the modification is declared to be disjoint from the triples to be added.

The update operation takes two arguments – the triples to add to the *default_graph* the triples to modify. Preconditions are:

(a) The list of triples to be added and the list of triples to be modified are disjoint
(b) The triples to be removed are those in the domain of the modification list
(c) The triples to be added are the range of the triples that are *actually removed*

_Example_2_Update_____

RDF_Dataset_Update
add? : RDF_Graph
mod? : RDF_Triple $\nrightarrow$ RDF_Triple
_____

dom *mod?* $\cap$ *add?* $= \emptyset$
remFromDefault? $=$ dom *mod?*
addToDefault? $= \mathrm{ran}(($*remFromDefault?* $\cap$ *default_graph*$) \lhd$ *mod?*$)$
addToNamed? $= \emptyset \wedge$ *remFromNamed?* $= \emptyset \wedge$ *remNamed?* $= \emptyset$

Protégé OWL Editor Dataset Example. Another useful dataset description is that of a document being revised in the Protégé OWL editor. The editor starts with a document that is retrieved from an input IRI. The editor then resolves the various import statements, adding each imported document as a separate named graph. Blank nodes are mapped so that there are no blank nodes shared between named graphs or the default and the named graph. Changes may only be applied to the default graph.[2]

We first declare a couple of well-know IRIs and a URL resolution function that maps IRIs to RDF_Graphs.

owl_Ontology, owl_imports, rdf_type : IRI
URL_Resolve : IRI $\nrightarrow$ RDF_Graph

---

[1] RESTful implementations will want to minimize preconditions that prevent the operation from being idempotent. An *add* operation that states the triples cannot be present or a *remove* operation that states that the triples must exist would have to be implemented as an `http POST` operation rather than a `PUT`.

[2] While this information is undoubtedly available in the Protégé documentation, the current description is derived from the authors' experience with the editor itself and, as such, may be incomplete or inaccurate. We see this as another example where a formal specification for graph invariants would prove useful.

A Protégé dataset is constructed from an input document IRI:

$\quad$ _Protege_Dataset_Init_ _____

$\quad$ _Protege_Dataset'_
$\quad$ _RDF_Dataset_Init_
$\quad$ _iri? : IRI_

$\quad$ $default\_graph' = URL\_Resolve(iri?)$
$\quad$ $\forall\, t : default\_graph' \bullet t.p = iri\ owl\_imports\ \land$
$\qquad$ $(\exists\, t_2 : default\_graph' \bullet t.s = t_2.s \land t_2.p = iri\ rdf\_type\ \land$
$\qquad\quad$ $t_2.o = iri\ owl\_Ontology) \Rightarrow$
$\qquad\quad$ $(t.o \mapsto URL\_Resolve(\mu\, i : IRI \mid iri\ i = t.o)) \in named\_graphs'$
$\quad$ $\forall\, g : \mathrm{ran}\ named\_graphs' \bullet \forall\, t : g \bullet t.p = iri\ owl\_imports\ \land$
$\qquad$ $(\exists\, t_2 : default\_graph' \bullet t.s = t_2.s \land t_2.p = iri\ rdf\_type\ \land$
$\qquad\quad$ $t_2.o = iri\ owl\_Ontology) \Rightarrow$
$\qquad\quad$ $(t.o \mapsto URL\_Resolve(\mu\, i : IRI \mid iri\ i = t.o)) \in named\_graphs'$

The above asserts that the set of named graphs in the document is the (flattened) recursive resolution of all `owl:imports` statements.

The blank node restriction is stated as an invariant against the Protégé dataset type:

$\quad$ _Protege_Dataset_ _____

$\quad$ _RDF_Dataset_

$\quad$ $\forall\, n_1, n_2 : \mathrm{dom}\ named\_graphs \bullet n_1 \neq n_2 \Rightarrow$
$\qquad$ $\{b : BNODE \mid (\exists\, t : named\_graphs(n_1) \bullet bnode\ b = t.s\ \lor$
$\qquad\quad$ $bnode\ b = t.o)\} \cap$
$\qquad$ $\{b : BNODE \mid (\exists\, t : named\_graphs(n_2) \bullet bnode\ b = t.s\ \lor$
$\qquad\quad$ $bnode\ b = t.o)\} = \emptyset$
$\qquad$ $\{b : BNODE \mid (\exists\, t : default\_graph \bullet bnode\ b = t.s \lor bnode\ b = t.o)\} \cap$
$\qquad$ $\{b : BNODE \mid (\exists\, g : \mathrm{ran}\ named\_graphs \bullet (\exists\, t : g \bullet$
$\qquad\quad$ $bnode\ b = t.s \lor bnode\ b = t.o))\} = \emptyset$

Changes can only occur against the default graph. Imported graphs (documents) must be edited in a context where they are the primary document. Imports may be added and removed to reflect changes in import statements, but existing named graphs cannot change.

$\quad$ _Protege_Preconditions_ _____

$\quad$ $\Delta Protege\_Dataset$

$\quad$ $\forall\, ng' : named\_graphs';\ ng : named\_graphs \bullet$
$\qquad$ $first\ ng = first\ ng' \Rightarrow ng' = ng$

The above constraint could also have been expressed as a precondition on the input variables. Note that the constraint says nothing about the addition or removal of named graphs, which allows for import statements (and graphs) to be added and removed from the dataset as needed.

5   Predicates

In the context of RDF validation, a predicate is always applied in the context of a given domain of type T It must always be applicable – there can be no situations where the predicate conditions are not satisfied.

$$
\begin{array}{|l}
\_\,Predicate[T] \underline{\hspace{6cm}} \\
\hline
domain! : T \\
\hline
\end{array}
$$

Predicates can be applied to an $RDF\_Dataset$, the triples in an $RDF\_Dataset$, all of the triples in the default graph, or an arbitrary graph:

$$
\begin{array}{|l}
\_\,Dataset\_Invariant \underline{\hspace{5cm}} \\
Predicate[RDF\_Graph] \\
domain? : RDF\_Dataset \\
\hline
domain! = domain?.default\_graph \cup \\
\quad \{t : RDF\_Triple \mid (\exists\, g : \text{ran } domain?.named\_graphs \bullet t \in g)\} \\
\hline
\end{array}
$$

$$
\begin{array}{|l}
\_\,Default\_Graph\_Invariant \underline{\hspace{4cm}} \\
Predicate[RDF\_Graph] \\
domain? : RDF\_Dataset \\
\hline
domain! = domain?.default\_graph \\
\hline
\end{array}
$$

$$
\begin{array}{|l}
\_\,Graph\_Invariant \underline{\hspace{5cm}} \\
Predicate[RDF\_Graph] \\
domain? : RDF\_Graph \\
\hline
domain! = domain? \\
\hline
\end{array}
$$

5.1   Property Predicates

A *Property_Predicate* is a *Predicate* that applies to the subset of an RDF_Graph whose predicate matches a specific IRI. $d?$ is the graph to be filtered and *domain!* is the subset:

```
┌─ Property_Predicate ─────────────────────────────────────
│  d?, domain! : RDF_Graph
│  property? : IRI
│ ──────────────────────────────────────────────────────────
│  domain! = { t : d? | t.p = iri property? }
└──────────────────────────────────────────────────────────
```

A *Property_Invariant* is a *Property_Predicate* applied to an *RDF_Dataset*:

$$Property\_Invariant \mathrel{\widehat{=}} Dataset\_Invariant[d?/domain!] \wedge Property\_Predicate$$

Type Predicate. A type predicate is a *Predicate* that applies to the subset of an *RDF_Graph* whose subjects are of a specified `rdf:type`. Note that this can have two interpretations:

1. All triples whose subjects are asserted to be of a given `rdf_type`:

```
┌─ Type_Predicate ─────────────────────────────────────────
│  d?, domain! : RDF_Graph
│  type? : IRI
│ ──────────────────────────────────────────────────────────
│  domain! = { t : d? | (∃ t₂ : d? • t₂.p = iri rdf_type ∧
│                         t₂.o = iri type? ∧ t.s = t₂.s)}
└──────────────────────────────────────────────────────────
```

2. All triples whose subjects can be inferred to be of a given `rdf_type`. This leaves us with another decision – what graph(s) do we use to do the inference. In this case, we will use a complete dataset. The *subProperties* function takes an IRI and a graph and returns all of the subProperties of a given property

$$\mid subProperties : RDF\_Term \nrightarrow RDF\_Graph \nrightarrow \mathbb{P}\, RDF\_Term$$

```
┌─ Transitive_Type_Predicate ─────────────────────────────
│  Type_Predicate[directTypes/domain!]
│  Dataset_Invariant[base!/domain!]
│  domain! : RDF_Graph
│ ──────────────────────────────────────────────────────────
│  domain! = directTypes ∪ {t : d? | (∃ t₂ : d? •
│             t₂.p ∈ subProperties t₂.p base! ∧ t₂.o = iri type? ∧ t.s = t₂.s)}
└──────────────────────────────────────────────────────────
```

We can now specify a combination of invariants that asserts that all of the subjects of the default graph must be of type `rdf:Person`:

$$AllPerson\_Invariant \mathrel{\widehat{=}} Default\_Graph\_Invariant \wedge$$
$$Type\_Predicate[foaf\_Person/type?]$$

Cardinality Predicates. One of the RDF Validation patterns is of cardinality, which is applied to the number of triples in a specified graph:

$$UnlimitedNatural ::= num\langle\!\langle \mathbb{N} \rangle\!\rangle \mid none$$

Cardinality_Predicate _____
$g? : RDF\_Graph$
$min? : \mathbb{N}$
$max? : UnlimitedNatural$

$\#g? \geq min?$
$max? = none \vee (\exists_1 \, n : \mathbb{N} \bullet max? = num \, n \wedge n \geq min?)$

We can now merge the cardinality predicates with a property predicate:

$PropertyCardinality \,\hat{=}\, [\, Cardinality\_Predicate[domain!/g?]; \; Property\_Predicate \,]$
$CardinalityInvariant \,\hat{=}\, [\, PropertyCardinality; \; Dataset\_Invariant \,]$

And add cardinality to previous assertions to assert that all subjects in a given datastore are of type `foaf_Person`, and that they have exactly one `foaf_firstName`, one `foaf_lastName` and an optional fullName

---
Person_Datastore _____
AllPerson_Invariant
$CardinalityInvariant[foaf\_firstName/property?]$
$CardinalityInvariant[foaf\_lastName/property?]$
$CardinalityInvariant[foaf\_fullName/property?, fnmin!/min?]$

$min? = 1 \wedge max? = num \, 1$
$fnmin! = 0$
---

6 Preconditions

Preconditions need to be applied to the combination of the parameters of the update operation as well as the target dataset.

6.1 ReadOnly and Derived Constraints

Another interesting class of constraint that appears in the Shape model is the *ReadOnly* constraint. This could be interpreted in a number of ways, including:

(a) Once present, the predicate cannot be removed (Write Once)
(b) The predicate cannot be added to the *Datastore* if there is already an assertion in the graph with the same subject – in object oriented terms, it must be generated in the "constructor"
(c) The predicate is generated by the service and cannot be changed (Derived).

We can then model the three flavors of the write once precondition.

ReadOnly as a Write Once Constraint. The first model says that you can't delete it if it already exists.

$WriteOnce\_1$ _____
$RDF\_Dataset\_Update$
$Property\_Predicate$

$\forall\, t_1 : domain! \bullet \neg\, ((\exists\, t_2 : remFromDefault? \bullet t_1.s = t_2.s \wedge t_1.p = t_2.p) \wedge$
$\quad (\forall\, g : \mathrm{ran}\; remFromNamed? \bullet (\exists\, t_2 : g \bullet t_1.s = t_2.s \wedge t_1.p = t_2.p)))$

Note, however, that you can still *add* another predicate instance unless it is accompanied by a cardinality constraint:

$$WriteOnce\_11 \mathrel{\widehat{=}} [\; WriteOnce\_1;\; Cardinality\_Predicate\,]$$

ReadOnly as a Constructor Constraint. This second writeonly constraint says that the predicate can only be added or removed if there aren't any assertions about the subject already present in the graph unless you remove *all* assertions about the subject in the graph. At the moment, we are going to simplify this to apply only to the default graph, as including named graphs in this constraint introduces another decision tree:

$WriteOnce\_2$ _____
$Property\_Predicate$
$Default\_Graph\_Invariant$
$RDF\_Dataset\_Update$

$\exists\, t : addToDefault? \cup remFromDefault? \bullet t.p = iri\; property? \Rightarrow$
$\quad \neg\, (\exists\, t_2 : domain! \bullet t_2.s = t.s)$

6.2 Readonly as a Derived Property Constraint

This variation of Readonly is the simplest, asserting that the predicate may not be included anything to be added or deleted, unless the deletion completely removes all subjects of type *property?*

$Test\_WriteOnce\_3$ _____
$Property\_Predicate$
$RDF\_Dataset\_Update$
$fctn\_result! : RDF\_Graph$

$fctn\_result! = default\_graph' \cup$
$\quad \{t : RDF\_Triple \mid (\forall\, g : \mathrm{ran}\; named\_graphs' \bullet t \in g)\}$
$\forall\, t : addToDefault? \bullet t.p \neq iri\; property?$
$\forall\, g : \mathrm{ran}\; addToNamed? \bullet \forall\, t : g \bullet t.p \neq iri\; property?$
$\forall\, t : remFromDefault? \bullet t.p = iri\; property? \Rightarrow$
$\quad (\forall\, t_2 : fctn\_result! \bullet t\_2.s \neq t.s)$
$\forall\, g : \mathrm{ran}\; remFromNamed? \bullet \forall\, t : g \bullet$
$\quad t.p = iri\; property? \Rightarrow (\forall\, t_2 : fctn\_result! \bullet t\_2.s \neq t.s)$
---

7   Discussion and Next Steps

The formal models developed here allow us to recognize, characterize and suggest possible solutions to a number of issues that arise when trying to specify and implement RDF validation constraints. One might be forgiven for thinking that it might be more advantageous simply to capture these constraints in SPARQL, where they could be executed and tested. The advantage of a more abstract approach is that we are able to capture and reason about intentions, about the language itself, rather than working with tests of individual implementations. In some cases it is a simple matter to retrieve the abstract intention from a SPARQL query: in others, it can be difficult indeed.

In this paper, we were able to model the existing description of RDF Datasets. In the process, we uncovered several possible interpretations that would need to be resolved before the intent of any constraint model could be meaningfully discussed. We discovered a number of characteristics that were shared between predicates and invariants but, at the same time, uncovered a number of decision points regarding the scope and interpretation of said predicates. The discussions about types, read-only and cardinality predicates could serve as a baseline for the decisions that need to be taken.

The purpose of this process is not to arrive at totally precise specifications – sometimes there are real advantages to leaving things underspecified. An excellent example of this is the latest SKOS Specification [6], where the relationship between `skos:Concept` and `owl:Class` is deliberately left unspecified. The SKOS Specification *does*, however, call attention to this fact and discusses the possible choices. The formal specification process described in this paper allows developers to *recoognize* points where specifications may be imprecise so that they can consciously decide how to address them vs. potentially leaving them undiscovered for developers and users to discover.

References

1. Joint Technical Committee ISO/IEC JTC 1. Information Technology - Z formal specification notation - Syntax, type system, and semantics (2002)
2. Carothers, G., Seaborne, A.: TriG: Rdf dataset language. World Wide Web Consortium, Last Call Working Draft 19 September 2013
3. Cyganiak, R., Wood, D.: RDF 1.1 concepts and abstract syntax. World Wide Web Consortium, Working Draft WD-rdf11-concepts-20130723, August 2013
4. Cyganiak, R., Wood, D.: RDF 1.1 concepts and abstract syntax - generalized rdf triples, graphs, and datasets. World Wide Web Consortium, Working Draft WD-rdf11-concepts-20130723, August 2013
5. Johnson, D., Speicher, S.: OSLC open services for lifecycle collaboration core specification version 2.0. OSLC Specification (web page), May 2013
6. Isaac, A., Summers, E.: SKOS simple knowledge organization system primer. World Wide Web Consortium, Note NOTE-skos-primer-20090818, August 2009
7. David, A.: Schmidt. Denotational semantics - A methodology for language development, On line image (1997)
8. Spivey, J.M.: The Fuzz Manual. Computing Science Consultancy, Oxford (1988)

9. W3C OWL Working Group. OWL 2 web ontology language – document overview (second edition). World Wide Web Consortium, Recommendation REC-owl2-overview-20121211, December 2012

10. W3C SPARQL Working Group. SPARQL 1.1 overview. World Wide Web Consortium, Recommendation REC-sparql11-overview-20130321, March 2013

11. Woodcock, J., Davies, J.: Using Z. Specification, Refinement, and Proof. Prentice-Hall, Upper Saddle River (1996)

Location-Based Mobile Recommendations by Hybrid Reasoning on Social Media Streams

Tony Lee[1(✉)], Seon-Ho Kim[1], Marco Balduini[2],
Daniele Dell'Aglio[3], Irene Celino[3], Yi Huang[4], Volker Tresp[4],
and Emanuele Della Valle[2]

[1] Saltlux, 976 Deachi-dong, Kangnam-gu Seoul, Korea
{tony, shkim}@saltlux.com
http://www.saltlux.com
[2] Dip. Elettronica e Informazione, Politecnico di Milano,
via Ponzio 34/5, 20133 Milano, Italy
{marco.balduini, emanuele.dellavalle}@polimi.it
[3] CEFRIEL, ICT Institute, Politecnico di Milano,
via Fucini 2, 20133 Milano, Italy
daniele.dellaglio@polimi.it, irene.celino@cefriel.com
[4] Siemens AG, Corporate Technology, Otto-Hahn-Ring 6,
81739 München, Germany
{yihuang, volker.tresp}@siemens.com

Abstract. In this paper, we introduce BOTTARI: an augmented reality application that offers personalized and location-based recommendations of Point Of Interests based on sentiment analysis with geo-semantic query and reasoning. We present a mobile recommendation platform and application working on semantic technologies (knowledge representation and query for geo-social data, and inductive and deductive stream reasoning), and the lesson learned in deploying BOTTARI in Insadong. We have been collecting and analyzing tweets for three years to rate the few hundreds of restaurants in the district. The results of our study show the commercial feasibility of BOTTARI.

Keywords: Social media analytics · Mobile recommendation · Stream reasoning · Hybrid reasoning · Machine learning · Semantic Web · Ontology

1 Introduction

When a tourist visits new place, they would face the challenges to discover proper restaurants, shops, or other tourist attractions. Usually, if you want to have a dinner in Seoul, it's quite hard to select a preferred restaurant among a hundreds restaurants in the district they are visiting (see Fig. 1). BOTTARI is an augmented reality application for personalized and localized restaurant recommendations, experimentally deployed in the Insadong district of Seoul. At a first look, it may appear like other mobile apps that recommend restaurants, but BOTTARI is different: BOTTARI uses inductive and deductive stream reasoning [1] to continuously analyze social media streams (specifically Twitter) to understand how the social media users collectively

W. Kim et al. (Eds.): JIST 2013, LNCS 8388, pp. 261–273, 2014.
DOI: 10.1007/978-3-319-06826-8_20, © Springer International Publishing Switzerland 2014

Fig. 1. A picture of Indadong: the density of restaurants is very high

perceive the points of interest (POIs) in a given area, e.g., Insadong's restaurants. In this paper, we describe the choices we made in designing BOTTARI and the lessons we learned by experimentally deploying it in Insadong.

2   Background Work

When reasoning on massive data streams, well known artificial intelligence techniques have the right level of expressivity, but their throughput is not high enough to keep pace with the stream. The only technological solutions with the right throughput are Data Stream Management Systems (DSMS) [3] and Complex Event Processing [4], but, on the other hand, they are not expressive enough. A new type of inference engines is thus needed to reason on streams. Della Valle et al. named them *stream reasoners* [2]. A number of stream reasoning approaches have been developed. They share three main concepts: (a) they logically model the information flow as an RDF stream, i.e. a sequence of RDF triples annotated with one or more non-decreasing timestamps, (b) they process the RDF streams "on the fly", often by re-writing queries to the raw data streams, and (c) they exploit the temporal order of the streaming data to optimize the computation. BOTTARI uses both a deductive and an inductive stream reasoner. The deductive stream reasoner is based on Continuous SPARQL (C-SPARQL) [5] – an extension of SPARQL that continuously processes RDF streams observed through windows (as done in DSMS) - and exploits the Streaming Linked Data (SLD) framework [6]. The inductive stream reasoner is based on SUNS (Statistical Unit Node Set) approach [7, 8] – a scalable machine learning framework for predicting unknown but potentially true statements by exploiting the regularities in structured data. The SUNS employs a modular regularized multivariate learning approach able to deal with very high-dimensional data [9] and to integrate temporal information using Markov decomposition [10].

The LarKC platform [12] is used for orchestrating SLD, SUNS and the geo-spatial query engine and exposing their aggretated capabilities as a SPARQL endpoint. It is a pluggable Semantic Web framework that can be deployed on a high-performance computing cluster. The LarKC platform is the main result of the EU FP7 integrated project, Large Knowledge Collider [11].

3 The BOTTARI Mobile App

As shown in Fig. 2, BOTTARI is an Android application (for smart phones and tablets) in augmented reality (AR) that directs the users' attention to restaurants. In Korean language, "bottari" is a cloth bundle that carries a person's belongings while travelling. BOTTARI carries the collective perceptions of social media users about POIs in an area and uses them to recommend POIs. As shown in Fig. 2(a), BOTTARI users can search POIs in their proximity using four buttons:

1. *'For Me'* that emphasizes the personalization of POI suggestions as in local search;
2. *'Popular'* that emphasizes the presence of positive ratings of social media users;
3. *'Emerging'* that focuses on the most recent ratings posted on social media capturing seasonal effects (e.g., Insadong people seems to prefer meat restaurants in winter rather than in summer) or POIs "on fashion" for a limited period; and
4. *'Interesting'* that returns the POIs described with a category of interest for the user.

Users can see recommended POIs based on their preferences in AR, as shown in Fig. 2(a). In this view, the POIs are indicated with different icons. Thumb-up and thumb-down icons indicate social reputation as positively or negatively perceived on social media. Moreover, given the importance of the distance between the user and the recommended POIs, BOTTARI offers functionality for distance-based filtering of the recommended POIs; see the circles in the right-upper side of Fig. 2(a). The user can learn more about a POI as shown in Fig. 2(b). Figure 2(c) shows a peculiar feature of BOTTARI: the trend over time of the POI reputation as collectively perceived on social media.

A video displaying BOTTARI at work, on a mobile phone and on a tablet, is available on YouTube at http://www.youtube.com/watch?v=c1FmZUz5BOo.

4 Data Set Used in BOTTARI

BOTTARI is built on two types of data: the geo ontology for the POIs and the social media streams.

4.1 Geo Ontology for the POIs

Insadong is a 2 km^2 district with a high density of restaurants. For BOTTARI, the information about 319 restaurants and 1850 tourist attractions of Insadong were

(a)   Recommendations with AR function

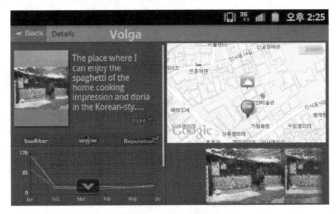

(b) Detailed information for a recommended restaurant

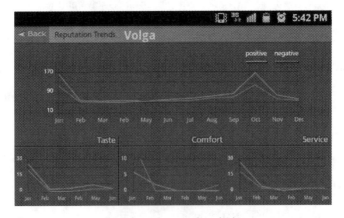

(c) Reputation trend analysis

Fig. 2. Screenshots of the BOTTARI Android application

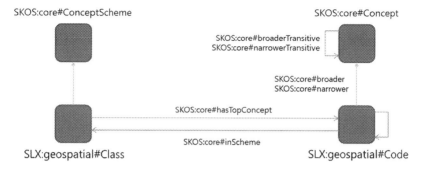

Fig. 3. Geo context model for BOTTARI

collected with a considerable manual effort. The result is a manually curated high quality geo-referenced knowledge base where each restaurant is described by 44 properties (e.g., name, images, position, address, ambiance, specialties, categories, etc.). We designed geo-context model based on RDF/OWL and SKOS. All the POIs were classified into 600 taxonomic classes described by SKOS (Figs. 3 and 7).

We also introduced Geo-SPARQL to query proper POI sets with sophisticate conditions including GPS coordinate, distance, category, ambiance, menu, price, parking, smoking and etc. For example, we can query to find "a Korean restaurant within 5 min' walk from Gallery Books which is possible parking, credit card, price is between 10,000 and 30,000, is also good for teenager or baby" like below. The query is presented hereafter, while a screenshot of the BOTTARI workbench used for testing Geo-SPARQL queries in show in Fig. 4.

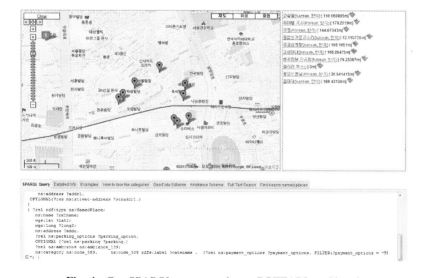

Fig. 4. Geo-SPARQL query results on BOTTARI workbench

```
PREFIX ns: <http://lod.saltlux.kr/geospatial/>
PREFIX rdf: <http://www.w3.org/1999/02/22-rdf-syntax-ns#>
PREFIX rdfs: <http://www.w3.org/2000/01/rdf-schema#>
PREFIX wgs: <http://www.w3.org/2003/01/geo/wgs84_pos#>
PREFIX f: <http://www.saltlux.com/geo/functions#>
SELECT *
WHERE {
  { ?res ns:name ?name. FILTER (?name = "갤러리 북스")
    ?res rdf:type ns:NamedPlace.
    ?res wgs:lat ?lat1;
         wgs:long ?long1;
         ns:address ?addr1.
    OPTIONAL{?res ns:street-address ?straddr1.}
  }
  { ?rel rdf:type ns:NamedPlace;
    ns:name ?relname;
    wgs:lat ?lat2;
    wgs:long ?long2;
    ns:address ?addr.
    ?rel ns:parking_options ?parking_option.
    OPTIONAL {?rel ns:parking ?parking.}
      ?rel ns:ambiance ns:ambiance_139;
      ns:category ns:code_589.        ns:code_589
      rdfs:label ?catename .  {?rel ns:payment_options
      ?payment_options. FILTER(?payment_options = "카드") }
    {?rel ns:price ?price. FILTER(f:between(?price, 10000,
      30000) )}
    {?rel ns:good_for ?goodfor. FILTER(?goodfor = "미성년"
      || ?goodfor = "유아") }
    OPTIONAL{?rel ns:street-address ?straddr.}
  }
  FILTER (f:within_distance(?lat1, ?long1, ?lat2, ?long2,
        200) )
} ORDER BY ?relname
```

4.2 Social Media Streams

The social media stream is gathered from the Web (in particular from Twitter) and converted into an RDF stream using the proprietary crawling and sentiment analysis infrastructure of Saltlux, Inc. The data used for the experiments were collected in 3 years, from February 4th, 2008 to November 23rd, 2010 (1,023 days). 200 million tweets were analyzed and, as a result, 109,390 tweets posted by more than 31,369 users were discovered to positively, neutrally or negatively talk about 245 restaurants in Insadong. More information about the technique employed for sentiment analysis is provide in Sect. 5.

This data stream is characterized by:

- *high sparsity* – defining sparsity as *1 - #Ratings / (#POIs × #Users)*. For instance, the sparsity of the positive ratings is 99.3 %;
- *incompleteness* – only 41 % of users positively rated at least one POI;
- *inconsistencies* – the same user can rate a particular POI several times expressing different opinions;
- *exponential growth over time* – data shows the exponential growth in the usage of Twitter in Korea starting from December 2009 to 2012; and
- *long-tail distribution* – ratings follows a long-tail distribution with few users that rated many POIs, and many users that rated one or two POIs.

Approaches for sentiment analysis could be divided in 2 types: machine learning based or rule (pattern matching) based. We applied hybrid model consists of syllable kernel based SVM and NLP rule engine to analyze sentiments from real-time tweet streams. We identify linguistic patterns of positive or negative sentiments in text string and encoded them into the rule set:

- adjectives (e.g. 맛있다/tasty, 재미있다/funny, 편하다/comfortable) are used to identify the polarity
- adverbs generally preceding amplify the strength of the polarity beard by the adjective.
- some noun phrase (e.g. 마음에 들어요 /like it, 문제가 많아요 /have many problems) are also used to identify polarity; and
- three different ways to make negative sentiment.

  (1)   post derivation ' '지 않다' ' (e.g. ' '마음에 들지 않아요')')
  (2)   prefix ' '안" (e.g. ' '안 불편해')')
  (3)   prefix ' '못" (e.g. ' '못 해봤어')')

More specifically, each rule consists of one linguistic pattern and two attributes like:

- Polarity (positive, neutral, negative that we encoded respectively as: +1,0,–1)
- Strength (a positive number reflecting how strong is the sentiment. We encoded the range from 0 to 5)

The grammar for defining linguistic pattern should be considered Part-Of-Speech (POS) and other linguistic features because the adjectives or verbs in Korean can have almost infinite flectional derivations (called 'eomi') and the original form of the verb has to retrieved effectively. Same canonical form of a token can correspond to several different meanings and POSs (e.g. 먹/V./V, 먹/N)/N). We designed tree based pattern matching algorithm where the match can check up to 3 consecutive tokens. For extracting linguistic patterns and sentiment words, we ran a quite big Korean corpus composed of several domains (food, restaurant, purchasing, movies, and etc.) and

Fig. 5. Workbench for sentiment analysis

applied a mutual information measure to extract the top most candidates. Figure 5 whos the workbench for sentiment analysis used in BOTTARI.

5   Ontology for BOTTARI

We designed BOTTARI following an ontology-based information access architecture [13]. BOTTARI ontology is represented in Fig. 6. It extends the SIOC vocabulary defining *TwitterUser* as a special case of *UserAccont* and the concept of Tweet as being equivalent to Post. It models the notion of POI as *NamedPlace* extending *SpatialThing* from the W3C WGS-84 vocabulary. A *NamedPlace* is enriched with a categorization (e.g., the ambience describing the atmosphere of a restaurant) and the

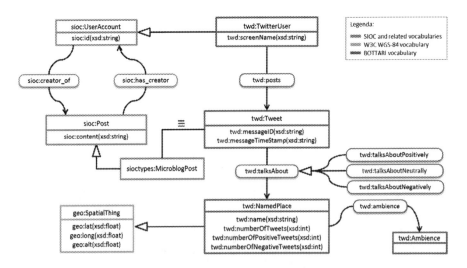

Fig. 6. Ontology model for BOTTARI

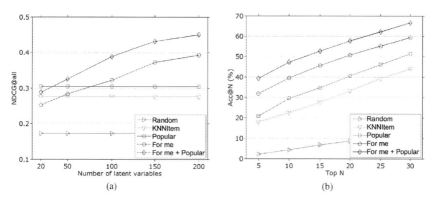

Fig. 7. BOTTARI evaluation results

count of positive/negative/neutral ratings. The most distinctive feature of BOTTARI ontology is the object property *talksAbout* – and its sub-properties for positive, negative and neutral opinions – that allows to state that a Tweet (positively, negatively or neutrally) talks about a *NamedPlace*.

6 Architecture and Components

BOTTARI architecture consists of three parts: (a) client side that interacts with the user and communicates to the back-end sending SPARQL queries, (b) a data initiated segment (PUSH) that continuously analyses the social media streams, and (c) a query initiated segment (PULL) that uses the LarKC platform to answer the SPARQL queries of the client by combining several forms of reasoning.

6.1 The PUSH Segment

The PUSH segment continuously analyses the social media streams crawled from the Web. The SEMANTIC MEDIA CRAWLER AND OPINION MINER crawl 3.4 million tweets/day related to Seoul, identifies the subset related to the Insadong restaurants (thousands per day) and extracts the users' opinions. The result is an RDF stream of positive, negative and neutral ratings of the restaurants of Insadong.

The RDF stream flows at an average rate of a hundred tweets/day and peaking at tens of tweets/minute. The RDF stream is processed in real-time by the SLD by means of the network of C-SPARQL queries where:

- a first continuous query counts the positive ratings for each POI in one day,
- a second query aggregates the result of the previous one over a week,
- a third query computes this aggregation over a month, and
- finally, a last query further aggregates the results of the queries upstream over one year.

The results of each of the four queries are published as linked data. The results of the first three queries are used in the PULL segment to answer the SPARQL queries for the Emerging recommendations and to display the trend lines illustrated in Fig. 2(c). The results of the last query are used to compute the Popular recommendations in the PULL segment.

The last component of the PUSH segment is BOTTARI inductive stream reasoner SUNS. It daily takes the results of the first query and updates the inductive materialization used for the For me recommendations.

6.2   The PULL Segment

The PULL segment is based on the LarKC platform, which acts as an ontology-based information-integration platform: BOTTARI ontology logically integrates the data models of the different plug-ins involved in computing a given type of recommendations. Whenever a user presses one of the four recommendation buttons in BOTTARI interface, the client issues a query using the BOTTARI ontology. When the LarKC platform receives the query, it decomposes it into a set of queries, one for SLD, one for SUNS and one for the Geo-SPARQL engine. The queries are executed in parallel. Each plug-in receives its re-written query and sends its partial results to the a plug-in that joins the partial results and returns the complete answer to the client, as if the query had been evaluated on a single integrated knowledge base. Caching of entire queries and intermediate results is applied in order to minimize query latency.

More specifically, the Geo-SPARQL engine, given a location, a spatial orientation and a POI category, returns a list of POIs ordered by distance from the location. It delegates the query execution to SOR, the spatial-aware RDF store by Saltlux. SUNS, given the id of the user (e.g., Alice), returns a list of POIs ordered by the estimated probability that the user will like them. SLD, given a period (i.e., a day, a week, a month, or a year), returns a list of POIs ordered by the number of tweets that talk positively about the POI in that period. As explained above, the linked data published by the PUSH segment are used.

To better clarify how we configured the LarKC platform to evaluate the BOTTARI client requests, let us consider the query that represent a mix of the queries the client sends for *Interesting* (lines 3–6), *For me* (lines 7-8) and *Emerging* (line 10) recommendations.

```
1. SELECT ?poi ?name ?lat ?long
2. WHERE {
3. { ?poi a ns:NamedPlace ; ns:name ?name ;
4.        geo:lat ?lat ; geo:long ?long ;
5.        ns:category :InterestingForForeigners . }
6.    FILTER(:within_distance(37.5,126.9,?lat,?long,200))
7. { :Alice sioc:creator_of
          [twd:talksAboutPositively ?poi]}
8.    WITH PROBABILITY ?p ENSURE PROBABILITY [0.5..1)
9. {   ?poi twd:numberOfPositiveTweetsInTheMonth ?n }
```

```
10. }
11. ORDER BY
    DESC(?n*?p*:distance(37.5,126.9,?lat,?long,200))
12. LIMIT 10
```

LarKC platform extracts lines 7-8 from the above query and rewrites them in the SPARQL with probability query below and issues it to SUNS.

```
1. CONSTRUCT {:Alice twd:talksAboutPositively
2.                [ ns:about ?poi ; ns:withProbability ?p ] }
3. WHERE {
4.   { :Alice sioc:creator_of
        [ twd:talksAboutPositively ?poi ] }
5.   WITH PROBABILITY ?p } ENSURE PROBABILITY [0.5..1)
6. }
7. ORDER BY DESC(?p)
```

The peculiarity of the query are the WITH/ENSURE PROBABILITY clauses (Line 5) and the CONSTRUCT clause (Lines 1-2). The former are part of the query language exposed by SUNS: SPARQL with probability. They allow to bind the probability that the binding exist in the inductive materialized and to ensure that such a probability is in a given range. The latter allows representing the probability values returned by SUNS without using annotations or reification.

7 Evaluation

The quality and the efficacy of BOTTARI recommendations was comparatively evaluated using the data set described in Sect. 4.

The *For me*, *Popular* and *Interesting* recommendations were compared with two baselines: random guess (Random) and k-nearest neighbour (KNNItem). The combination of *For me* and *Popular* recommendations was also considered. For all recommendations, the distance filter was not applied, because our data set does not contain the user position at twitting time (Table 1).

Table 1. Statistics of sentiment analysis results from Twitter

		Tweet	POI	User	Sparsity (%)
Ratings	Positive	19,045	213	12,863	99.30
	Negative	14,404	181	10,448	99.24
	Neutral	75,941	245	28,056	98.90
	Total	109,390	245	31,369	98.58

A key aspect of BOTTARI is the adoption of stream reasoning techniques that build on the hypothesis that a long enough window can capture all the information needed for a given task, while the rest can be forgotten. In a first set of experiments, we targeted the evaluation of the *Emerging* recommendations, which use a time

Table 2. Number of ratings with different time frames

	Nr. of ratings	%
Last day	188	0.17
Last 2 days	703	0.64
Last 7 days	5,057	4.62
Last 30 days	27,049	24.73
Last 90 days	65,600	70.01
Last 180 days	93,696	85.65
Total	109,389	100.00

window. A set of ground truths was created by withholding the newest rating for each user. Different time frames were considered: 1 day, 2 days, 7 days, 30 days, 90 days and 180 days. Table 2 shows the number of ratings in the different time frames.

As a result, we discovered that the *Emerging* recommendations with a 90 days window are as effective as the *Popular* recommendations that keep the full history (i.e., two years of data).

In a round of experiments we evaluated also the inductive stream reasoning. We followed the standard method of splitting the data into a training set and a test set was used. In this case, a ground truth contains one positive rating for each user randomly withheld from the data set. We repeated this data split five times.

We evaluated *For me* recommendations produced by SUNS with 20, 50, 100, 150 and 200 latent variables. As expected, Random was the worst. The *Emerging* recommendations on the last 90 days were slightly better than KNNItem. This might be due to the "bandwagon effect" that exists in many social communities. The *For me* recommendations significantly outperformed all the others after the number of the latent variables reached 100. The best ranking ever was produced by the combination of both *For me* and *Emerging* recommendations on the last 90 days. These results confirm the idea that a combined approach of deductive and inductive stream reasoning works best.

8   Conclusions

BOTTARI is a sophisticated application of semantic technologies that makes use of the rich and collective knowledge obtained by continuously analysing social media streams. We believe it was important to hide this complexity from the user using an intuitive and easy to use interface.

Inspired by the literature on ontology-based information access, we design BOTTARI ontology as driver of both data and service integration. It allows for combining real data sources at real scale, i.e. location-specific static information about hundreds of POIs with the results of continuous analysis of dynamic social media streams. However, we believe that the BOTTARI ontology was also crucial in handling the heterogeneous data models of the plug-ins. For instance, the inductive reasoner annotates triples in the inductive materialization with their probability to be true, but the other plug-ins cannot understand these annotations, unless they are transformed into commonly described data.

BOTTARI is engineered for scalability. Both SUNS and SLD show a scalability that goes largely beyond the actual needs of the BOTTARI deployment in Insadong. Training SUNS over two years of data takes 1.5 minutes. SLD can handle a flow of 15,000 tweets/second when the actual rate is tens of tweets/day. These results convinced Saltlux to start a large-scale deployment of BOTTARI in Korea.

Acknowledgments. This work was partially supported by the LarKC project (FP7-215535) and Mobile Cognition and Learning project in Korea.

References

1. Barbieri, D.F., Braga, D., Ceri, S., Della Valle, E., Huang, Y., Tresp, V., Rettinger, A., Wermser, H.: Deductive and inductive stream reasoning for semantic social media analytics. IEEE Intell. Syst. **25**(6), 32–41 (2010)
2. Della Valle, E., Ceri, S., van Harmelen, F., Fensel, D.: It's a streaming world! reasoning upon rapidly changing information. IEEE Intell. Syst. **24**(6), 83–89 (2009)
3. Garofalakis, M., Gehrke, J., Rastogi, R.: Data Stream Management: Processing High-Speed Data Streams. Springer-Verlag New York Inc., Secaucus (2007)
4. Luckham, D.C.: The power of events: an introduction to complex event processing in distributed enterprise systems. In: Bassiliades, N., Governatori, G., Paschke, A. (eds.) RuleML 2008. LNCS, vol. 5321, p. 3. Springer, Heidelberg (2008)
5. Barbieri, D.F., Braga, D., Ceri, S., Della Valle, E., Grossniklaus, M.: Querying rdf streams with c-sparql. SIGMOD Record **39**(1), 20–26 (2010)
6. Balduini, M., Della Valle, E., Dell'Aglio, D., Tsytsarau, M., Palpanas, T., Confalonieri, C.: Social listening of city scale events using the streaming linked data framework. In: Alani, H., et al. (eds.) ISWC 2013, Part II. LNCS, vol. 8219, pp. 1–16. Springer, Heidelberg (2013)
7. Tresp, V., Huang, Y., Bundschus, M., Rettinger, A.: Materializing and querying learned knowledge. In: Proceeings of IRMLeS 2009 (2009)
8. Huang, Y., Tresp, V., Bundschus, M., Rettinger, A., Kriegel, H.-P.: Multivariate prediction for learning on the semantic web. In: Frasconi, P., Lisi, F.A. (eds.) ILP 2010. LNCS, vol. 6489, pp. 92–104. Springer, Heidelberg (2011)
9. Huang, Y., Nickel, M., Tresp, V., Kriegel, H.-P.: A scalable kernel approach to learning in semantic graphs with applications to linked data. In: Proceedings of the 1st Workshop on Mining the Future Internet (2010)
10. Tresp, V., Huang, Y., Jiang, X., Rettinger, A.: Graphical models for relations - modeling relational context. In: International Conference on Knowledge Discovery and Information Retrieval (2011)
11. Fensel, D., van Harmelen, F., Andersson, B., Brennan, P., Cunningham, H., Della Valle, E., Fischer, F., Huang, Z., Kiryakov, A., il Lee, T.K., School, L., Tresp, V., Wesner, S., Witbrock, M., Zhong, N.: Towards LarKC: a Platform for Web-scale Reasoning. In: Proceedings of the ICSC 2008 (2008)
12. Cheptsov, A., et al.: Large knowledge collider: a service-oriented platform for large-scale semantic reasoning. In: Proceedings of the WIMS 2011 (2011)
13. Lenzerini, M.: Data integration: a theoretical perspective. In: Popa, L. (Ed.): PODS, pp. 233–246. ACM (2002)

Semantic Search and Query

Towards Exploratory Relationship Search:
A Clustering-Based Approach

Yanan Zhang, Gong Cheng$^{(\boxtimes)}$, and Yuzhong Qu

State Key Laboratory for Novel Software Technology, Nanjing University,
Nanjing 210023, People's Republic of China
ynzhang@smail.nju.edu.cn,{gcheng,yzqu}@nju.edu.cn

Abstract. Searching and browsing relationships between entities is an
important task in many domains. RDF facilitates searching by explicitly representing a relationship as a path in a graph with meaningful
labels. As the Web of RDF data grows, hundreds of relationships can be
found between a pair of entities, even under a small length constraint
and within a single data source. To support users with various information needs in interactively exploring a large set of relationships, existing
efforts mainly group the results into faceted categories. In this paper, we
practice another direction of exploratory search, namely clustering. Our
approach automatically groups relationships into a dynamically generated hierarchical clustering according to their schematic patterns, which
also meaningfully label these clusters to effectively guide exploration and
discovery. To demonstrate it, we implement our approach in the RelClus
system based on DBpedia, and conduct a preliminary user study as well
as a performance testing.

Keywords: Association discovery · Exploratory browsing · Hierarchical
clustering · Path finding · Relationship search

1 Introduction

How is Seoul related to Nara? What are the connections between Aristotle and
Confucius? Which cinematographers have cooperated with both Ang Lee and
David Lynch, and in which movies? Such information needs as well as many
others in various domains can be satisfied by an information system that supports *searching and browsing relationships* between entities. This relationship
search system can be effectively implemented based on the Semantic Web and
Linked Data, which create a Giant Global Graph where relationships between
two entities are represented as paths connecting two vertices; in particular, path
finding in RDF data has been efficiently implemented [9,14].

However, it is insufficient to simply return all the relationships found in the
data because, when the result set is large, browsing becomes a nontrivial task.
To reduce information overload, whereas efforts have been made to organize the
results by ranking [1–3,22], recent advances in *exploratory search* [15] have shown

W. Kim et al. (Eds.): JIST 2013, LNCS 8388, pp. 277–293, 2014.
DOI: 10.1007/978-3-319-06826-8_21, © Springer International Publishing Switzerland 2014

that many information needs are complex and uncertain, and search systems are expected to provide users with services beyond lookup and ranking, to facilitate cognitive processing and interpretation of the results via continuous and exploratory interaction. Among two popular methods for realizing exploratory search [10], *faceted categories* have been incorporated into practical relationship search systems like RelFinder [11,12], whereas to the best of our knowledge, *clustering* has not been explored, and will be practiced in this paper. Our main contribution is threefold.

– As the first attempt to realize exploratory relationship search by using clustering, we discuss key challenges to be met when representing and building a hierarchical clustering of relationships.
– To meet the challenges, we exploit the schematic patterns of relationships to characterize their commonalities and label a cluster, leverage class and property hierarchies to model a hierarchy of such patterns, and use information theory to measure their similarity for conducting agglomerative clustering.
– We implement our approach in RelClus,[1] a relationship search system based on DBpedia, and carry out a preliminary user study to compare it with two baseline approaches. We also test the performance of our implementation.

In the remainder of this paper, Sect. refsect:problem discusses the problem of exploratory relationship search, and Sect. 3 introduces our solution. Sections 4 and 5 report the results of user study and performance testing, respectively. Section 6 compares related work before we conclude the paper in Sect. 7.

2   Exploratory Relationship Search

In this section, we review exploratory search and its implementations: faceted categories and clustering. In particular, we discuss three challenges faced in designing a clustering algorithm for relationship search.

2.1   Exploratory Search

Exploratory search is distinguished from common lookup activities (e.g. fact retrieval, question answering) by requiring strong human participation in a more continuous and exploratory process [15]. It usually deals with complex information needs such as information aggregation and planning, which cannot be satisfied by a single lookup but require users to continuously interact with the system and explore the results. The user may be unfamiliar with the domain of her goal, but via exploration, learns concepts and progressively achieves the goal. In some cases, the user is even unsure about her goal in the beginning, but via exploration, progressively makes the goal clear and achieves it.

To implement exploratory search, information retrieval techniques need to be combined with more work on human-computer interaction. Existing efforts

---

[1] http://ws.nju.edu.cn/relclus/

mainly focus on organizing search results into *meaningful groups* labeled with concepts relevant to the domain of the goal, to help the user understand the domain and narrow the search [10]. These groups are often organized into a hierarchy, to enable iterative exploration.

So far, two methods for grouping search results have been widely adopted. One method employs a hand-crafted category system, known as *faceted categories*, to classify search results by one or many facets, each corresponding to a dimension or feature type. By comparison, *clustering* lets the search results speak for themselves; the results are grouped based on some measure of similarity. The label of each cluster is also dynamically generated from the results.

Whereas faceted categories have been practiced [11] to enhance relationship search, to the best of our knowledge, no attention has been paid to clustering-based relationship search, which is the focus of this paper. In the following we will highlight several challenges faced in this direction to be met by our solution.

2.2 Relationship Clustering

Clustering Web search results faces a number of challenges [6,10]. We find that most of these challenges are still to be met when dealing with relationships. Among others, we consider the following three challenges the most vital to our first attempt to relationship clustering.

Challenge 1: Similarity Measurement. Clustering requires a similarity measure. Although there has been extensive study on the similarity between texts, the measurement of the similarity between relationships should differ greatly because relationships are structured and associated not with ambiguous words but with symbolic labels (i.e. URIs) having machine-readable meanings. We expect to devise a measure that can exploit these features.

Challenge 2: Cluster Labeling. Assigning a representative, distinctive, and meaningful label to each cluster is a primary challenge to search result clustering. When dealing with texts, one heuristic is to select one or several words or phrases that best characterize the centroid of the cluster. However, defining the centroid of a group of path-shaped relationships is not a trivial task, since we expect the label of a cluster to reflect not only the central topic of the constituent relationships but also their common structural characteristics.

Challenge 3: Intuitiveness of Sub-hierarchies. Counterintuitive hierarchies of clusters are often met due to complex similarity measures and inappropriate cluster labeling, which would confuse the user. We expect to provide the user with a hierarchy that explicitly—via cluster labeling—specifies the principle of dividing a cluster into sub-clusters. It correlates with the previous two challenges.

3 Approach

In this section, after giving the preliminaries, we illustrate the basic idea of our approach by a running example, and show how it meets the three challenges identified in Sect. 2.2. After that, we formally describe the clustering algorithm.

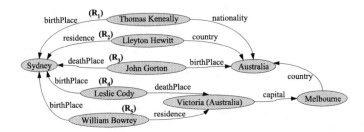

Fig. 1. A data graph containing five relationships from `Sydney` to `Melbourne`.

3.1   Preliminaries

At the schema level, let $\mathbb{C}$ and $\mathbb{P}$ be the sets of classes and properties, respectively. Classes are organized into a subsumptive hierarchy which characterizes the subclass-superclass relationship between them; properties are organized in a similar way. For simplicity, we focus on the properties that connect entities, but our approach can also be extended to handle literals. At the data level, let $\mathbb{E}$ be the set of entities. An entity may belong to zero or more classes.

   We formalize the data to be searched as a labeled digraph called *data graph*, denoted by an ordered 6-tuple $G = \langle V, A, s, t, l_V, l_A \rangle$ with

- V, a finite set of vertices,
- A, a finite set of directed edges,
- $s : A \rightarrow V$, assigning to each edge its source vertex,
- $t : A \rightarrow V$, assigning to each edge its target vertex,
- $l_V : V \rightarrow \mathbb{E}$, assigning to each vertex an entity as its label, and
- $l_A : A \rightarrow \mathbb{P}$, assigning to each edge a property as its label.

Figure 1 shows a real data graph extracted from DBpedia, which serves as a running example used in the paper.

   Given a start entity e_S and an end entity e_E, we define a *relationship of length n* from e_S to e_E as a simple path of length n connecting them in G which not necessarily follows the directions of the edges, or in other words, it is an alternating sequence of vertices and edges, $v_0 a_1 v_1 a_2 v_2 \cdots a_n v_n$, such that

- for $1 \leq i \leq n$, either $s(a_i) = v_{i-1}$, $t(a_i) = v_i$, or $s(a_i) = v_i$, $t(a_i) = v_{i-1}$,
- for $0 \leq i < j \leq n$, $v_i \neq v_j$, and
- $l_V(v_0) = e_S$, $l_V(v_n) = e_E$.

We only consider simple paths where no vertices (and thus no edges) are repeated, since otherwise there may be an infinite number of relationships. Edges in a relationship are not required to go the same direction, because the inverse of a property also has meaning for human readers. Figure 1 illustrates five relationships, R_1–R_5, from `Sydney` to `Melbourne`.

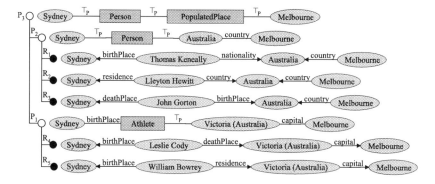

Fig. 2. A hierarchical clustering of five relationships from `Sydney` to `Melbourne`.

3.2   An Illustrative Example

Before formally describing the clustering algorithm, we illustrate its basic idea by using R_1–R_5 in Fig. 1, all of which are of length three. Our algorithm assumes that the relationships to be clustered are of the same length; otherwise we can group the relationships by length and then apply our algorithm to each group.

Our clustering algorithm follows an agglomerative manner [24]; that is, each relationship starts in its own cluster, and a hierarchy of clusters is built by progressively merging the most similar pair of clusters.

To explain our similarity measure, we need to firstly introduce how we assign a meaningful and representative label to each cluster. The **label of a singleton cluster** is just the unique relationship it contains; for instance, R_4 labels the singleton cluster $\{R_4\}$. The **label of the union of two clusters** is a *relationship pattern*, which is a high-level abstraction of relationships where vertices can be not only entities but also classes, and the directions of edges are omitted; we will formalize this concept in the next subsection. For instance, P_1 in Fig. 2 is a relationship pattern that labels the cluster $\{R_4, R_5\}$, where $\top_{\mathbb{P}}$ represents a pseudo-property that is a superproperty of all the properties. We call P_1 a *super-pattern* of R_4 and R_5 in the sense that for each entity, class, and property in R_4 and R_5, the element in the corresponding position in P_1 is either the same or its type, superclass, and superproperty, respectively; in particular, P_1 comprises the *least common elements* among the possibilities. For instance, `Leslie Cody` in R_4 and `William Bowrey` in R_5 are both instances of `Person` and `Athlete`, and we choose `Athlete` in P_1 because it is the least common type given `Athlete` being a subclass of `Person` and thus contains the most *information content* as discussed in [17]; `deathPlace` in R_4 and `residence` in R_5 share no common superproperty, so we use $\top_{\mathbb{P}}$ in P_1; other elements in the corresponding positions are the same and thus are retained in P_1.

We define the **similarity between two clusters** as *the information content associated with the relationship pattern that would label their union*, which indicates how many commonalities the two clusters share. We approximate the information content associated with a relationship pattern by the sum of the information contents of its elements. So, at the beginning of the clustering process in our running example, $\{R_4\}$ and $\{R_5\}$ are merged before $\{R_1\}$ and $\{R_2\}$ because the label of $\{R_4, R_5\}$, which would be P_1, contains more information content than P_2 which would label $\{R_1, R_2\}$, mainly because P_2 contains one more top property, the information content of which is trivially zero. We will formalize our similarity measure in the next subsection.

In addition, after successively forming $\{R_4, R_5\}$ labeled with P_1 and $\{R_1, R_2\}$ labeled with P_2, we will immediately merge $\{R_1, R_2\}$ and $\{R_3\}$ into $\{R_1, R_2, R_3\}$, still labeled with P_2, because R_3 also matches this pattern. Finally, only two clusters remain, labeled with P_1 and P_2; they are merged into the root cluster labeled with P_3, and a hierarchy is formed.

Referring to the **three challenges** identified in Sect. 2.2, our approach has met all of them as follows. Firstly. we measure the similarity between relationships by exploiting their schematic patterns and using information theory. Secondly, we label a non-singleton cluster with a structured relationship pattern that holds the most common semantic characteristics shared by the constituent relationships. Thirdly, the division of a cluster into sub-clusters is based on different sub-patterns, which are exposed to the user via cluster labeling.

3.3   Clustering Algorithm

Overview. We formalize the aforementioned solution in Algorithm 1. Initially, line 2–4 assign each relationship to a singleton cluster. Line 5–17 repeatedly merge the most similar pair of clusters and finally build a hierarchy. In particular, after merging two clusters, line 12–17 continue to merge the result with the remaining clusters that also match the relationship pattern just obtained. We will elaborate on line 6 and 13 later. Basically, given m relationships, an agglomerative clustering using priority queues runs in $O(m^2 \log m)$ time. In practice, the running time also depends on the measurement of similarity between $O(m^2)$ pairs of clusters. In addition, line 13 compares $O(m^2)$ pairs of relationship patterns, and influences the number of iterations of the `while` loop.

Relationship Pattern. The label of a cluster is a *relationship pattern*, which we formalize as an alternating sequence of classes and properties, $c_0 p_1 c_1 p_2 c_2 \cdots p_n c_n$. To allow an entity to appear as a "vertex" in a relationship pattern (and thus to cover both non-singleton and singleton clusters), we introduce a pseudo-class $I(e) = \{e\}$ for each entity e such that it is a subclass of all of e's types. A relationship pattern $c_0 p_1 c_1 p_2 c_2 \cdots p_n c_n$ is a *sub-pattern* of (denoted by $\sqsubseteq$) another $c'_0 p'_1 c'_1 p'_2 c'_2 \cdots p'_n c'_n$ when the following conditions hold:

- $c_i \sqsubseteq_C c'_i$, for $0 \leq i \leq n$, and
- $p_i \sqsubseteq_P p'_i$, for $1 \leq i \leq n$,

where $\sqsubseteq_C$ and $\sqsubseteq_P$ denote the class and property inclusion relations, respectively. In particular, we introduce $\top_C$ to be a pseudo-class that is a superclass of all the classes; similarly, $\top_P$ denotes the pseudo-top of the property hierarchy.

Algorithm 1: Hierarchical Relationship Clustering

Data: A set $\mathbb{R}$ of relationships of the same length
Result: A hierarchy H of clusters

1 $C \longleftarrow \emptyset$; $//C$ holds the remaining clusters.
2 **for** $R \in \mathbb{R}$ **do**
3      $C \longleftarrow C \cup \{\{R\}\}$;
4   H.addVertex($\{R\}$);
5 **while** $|C| > 1$ **do**
6      $(C_i, C_j) = \arg\max_{(C_k,C_l)\in C \times C, k \neq l} \mathrm{sim}(C_k, C_l)$;
7      $C_{ij} \longleftarrow C_i \cup C_j$;
8      $C \longleftarrow C \setminus \{C_i, C_j\}$;
9   H.addVertex(C_{ij});
10   H.addEdge(C_{ij}, C_i);
11   H.addEdge(C_{ij}, C_j);
12   **for** $C_k \in C$ **do**
13   **if** $\mathrm{label}C_k \sqsubseteq \mathrm{label}C_{ij}$ **then**
14             $C_{ij} \longleftarrow C_{ij} \cup C_k$;
15             $C \longleftarrow C \setminus \{C_k\}$;
16   H.addEdge(C_{ij}, C_k);
17     $C \longleftarrow C \cup \{C_{ij}\}$;
18 **return** H

Label of a Cluster. The `label` function in Algorithm 1 returns a relationship pattern as the label of a cluster. For a singleton cluster $\{R\}$ containing a unique relationship $R = v_0 a_1 v_1 a_2 v_2 \cdots a_n v_n$, we define its label as

$$\mathrm{label}(\{R\}) = I(l_V(v_0))\, l_A(a_1)\, I(l_V(v_1))\, l_A(a_2)\, I(l_V(v_2)) \cdots l_A(a_n)\, I(l_V(v_n)), \quad (1)$$

where l_V and l_A are defined in Sect. 3.1. The label of the union of two clusters C_i and C_j is defined as a super-pattern of their labels. When there is more than one super-pattern, we choose the one that holds the most commonalities shared by the two clusters, or in other words, the one that carries the largest amount of information about their commonalities:

$$\mathrm{label}(C_i \cup C_j) = \arg\max_P IC(P), \text{ s.t. } \mathrm{label}(C_i), \mathrm{label}(C_j) \sqsubseteq P, \quad (2)$$

where $IC(P)$ is the information content associated with the relationship pattern $P = c_0 p_1 c_1 p_2 c_2 \cdots p_n c_n$, which is given by a function of its probability according to information theory:

$$IC(P) = -\log p(c_0 p_1 c_1 p_2 c_2 \cdots p_n c_n). \quad (3)$$

The probability can be estimated by maximum likelihood based on the data graph G, namely in proportion to the number of relationships in G that match this pattern, which can be obtained by evaluating a structured query on G. However, we usually cannot afford the large number of such complex queries needed during the online execution of clustering. To make a trade-off, we approximate the probability by introducing a conditional independence assumption:

$$- \log p(c_0 p_1 c_1 p_2 c_2 \cdots p_n c_n) = - \log p(c_0) p(p_1) p(c_1) p(p_2) p(c_2) \cdots p(p_n) p(c_n) , \quad (4)$$

where $p(c_i)$ is the probability of observing an entity from c_i, which can be efficiently estimated by maximum likelihood based on G:

$$p(c_i) = \frac{|\{e \in \mathbb{E} : e \in c_i\}|}{|\mathbb{E}|} , \quad (5)$$

and $p(p_i)$ is the probability of observing the property p_i, which can be estimated in a similar way:

$$p(p_i) = \frac{|\{a \in A : l_A(a) = p_i\}|}{|A|} . \quad (6)$$

As a result, each "vertex" of $\texttt{label}(C_i \cup C_j)$ will be filled by a least common superclass of the classes in the corresponding positions in C_i and C_j, and an "edge" of $\texttt{label}(C_i \cup C_j)$ will be a least common superproperty.

Similarity Between Clusters. The $\texttt{sim}$ function in Algorithm 1 returns the similarity between two clusters C_k and C_l, which we define as the information content associated with the relationship pattern that would label the union of C_k and C_l, which exactly indicates how many commonalities the two clusters share:

$$\texttt{sim}(C_k, C_l) = IC(\texttt{label}(C_k \cup C_l)) . \quad (7)$$

4   User Study

We carried out a preliminary experiment involving human users to compare our approach with two baseline approaches.

4.1   Data Set

Our experiments were based on the English version of DBpedia 3.7 (http://wiki. dbpedia.org/Downloads37), which was the latest version at the time of experimentation. We derived a data graph covering a wide range of encyclopedic topics from the *Ontology Infobox Properties* data set, comprising 9,105,118 edges (i.e. RDF triples) after removing those containing literals. For cluster labeling and similarity measurement, we retrieved the types of each entity from the *Ontology Infobox Types* data set, and obtained the schema of the data graph from the *DBpedia Ontology*, which consists of hundreds of classes and more than one thousand kinds of properties.

4.2   Systems

Given two entities, our implementation searches the data graph in a bidirectional breadth-first manner for all the relationships between them of a limited length.

Fig. 3. A screenshot of **RClus**.

After that, we provided three user interfaces which present the results in different ways and support different sets of operations.

- **RClus** leverages our clustering-based approach to organize relationships into a expandable/collapsible hierarchy, as illustrated in Fig. 3. Each entity is prefixed by a thumbnail if available, provided by the *Images* data set of DBpedia; each class is prefixed by *some*; $\top_C$ is shown as *something*, and $\top_P$ is omitted. Relationships in a cluster are sorted alphabetically. Each non-singleton cluster is suffixed by its size in parentheses, and sibling clusters are sorted by their sizes decreasingly.
- **RList** simply lists all the relationships in alphabetical order; the way to present each relationship is in accordance with **RClus**.
- **RFacet** implements faceted categories by adding length, class, and property filters to **RList**. Each filter (i.e. facet value) is suffixed by the number of results to expect in parentheses. The values of a facet are sorted alphabetically. In accordance with RelFinder [11], by using these filters, users can hide the relationships of a specific length or containing a specific type of entity or property. We implemented **RFacet** as a representative of faceted relationship search instead of directly using RelFinder, in order to eliminate potential effects caused by other visualization and interaction techniques adopted by RelFinder as well as the differences between the data sets used.

4.3 Participants and Tasks

Since we lacked benchmark queries for evaluating relationship search, we decided to collect real queries from participants. However, we also intended to compare different systems based on the same set of search tasks, so we could not allow participants to submit arbitrary queries. Based on these considerations, among the seventeen participants having a background in computer science invited from a university and paid for participation, we asked two of them (called query providers) to provide real information needs, based on which search tasks were established; the other fifteen carried out these tasks by using different systems so that the systems were tested based on the same tasks.

Specifically, six pairs of entities were identified from the information needs received, covering a wide range of topics. To make a comparative study, three of them were reformulated as lookup tasks (T1–T3) and the other three as

Table 1. Tasks, settings, and results

	Task description (with start and end entities italicized)	Max. length	No. of results
T1:	Find the music genres that originated from both *Jazz* and *Blues*	2	92
T2:	Find the directors that have cooperated with both *Jackie Chan* and *Zhang Ziyi*	4	557
T3:	Find the prizes that have been awarded to both scientists working on *Graph theory* and scientists working on *Group theory*	4	113
T4:	Imagine you are preparing for a research paper on the connections between *Aristotle* and *Confucius*. Find several kinds of connections to discuss, and find an example for each type	3	62
T5:	Imagine you are preparing for a blog on the connections between *Ang Lee* and *David Lynch*. Find several kinds of connections to discuss, and find an example for each type	4	216
T6:	Imagine you are preparing for an article on the connections between *ExxonMobil* and *ConocoPhillips*. Find several kinds of connections to discuss, and find an example for each type	4	766

exploratory search tasks (T4–T6), as shown in Table 1. For each task, the maximum length of a relationship to return was tuned by the two query providers and fixed to ensure a return of a moderate number of interesting relationships.[2]

4.4   Procedure

Each participant used each of the three systems to continuously carry out a lookup task and an exploratory search task (in random order), so each of the six tasks was carried out exactly once by each user. The order of the systems to use and the assignment of the tasks were both randomly determined but counterbalanced across participants. Before experimenting with a system, the participant was given a tutorial about the system, after which five minutes was given to freely explore the system. Then, ten minutes was given to conduct each task and submit the results. After completing a task, the participant was asked to respond to a questionnaire comprising five 5-point Likert items, from 1 for strongly disagree to 5 for strongly agree.

- Q1: I am very familiar with these two entities.
- Q2: I found this task very difficult.
- Q3: I felt very confident about the results I submitted.
- Q4: I found the functions in this system very useful for performing this task.
- Q5: I thought the functions in this system sufficient to support this task.

---

[2] But in the online demo, this parameter is user-configurable.

After using a system and completing two tasks, the participant was asked to fill out an SUS [5] for usability testing. After using all the three systems and completing all the six tasks, comments were collected via a face-to-face interview.

Table 2. Questionnaire results (Avg.)

| | T1 | | | T2 | | | T3 | | |
	RList	RFacet	RClus	RList	RFacet	RClus	RList	RFacet	RClus
Q1	2.50	2.60	2.00	4.20	4.67	3.75	2.50	3.50	3.14
Q2	2.17	2.00	3.00	2.60	2.17	1.50	2.50	3.25	2.43
Q3	4.50	**4.60**	4.25	3.60	**4.17**	3.75	4.50	4.25	**4.57**
Q4	4.00	**4.60**	4.50	3.00	3.00	**3.75**	2.75	3.25	**4.57**
Q5	3.50	**4.60**	4.25	2.40	3.00	**3.25**	2.75	2.75	**4.00**

| | T4 | | | T5 | | | T6 | | |
	RList	RFacet	RClus	RList	RFacet	RClus	RList	RFacet	RClus
Q1	3.20	3.33	3.25	2.60	3.20	2.20	1.00	1.50	2.33
Q2	3.60	4.17	3.75	3.80	4.00	3.40	4.60	4.50	3.33
Q3	2.60	3.00	**4.75**	**3.60**	3.20	3.20	3.00	**3.50**	3.33
Q4	3.00	3.50	**4.00**	2.80	3.20	**4.20**	2.40	3.00	**3.83**
Q5	2.80	3.33	**3.50**	2.40	2.40	**4.00**	1.80	2.50	**3.17**

4.5 Results

Table 2 summarizes the questionnaire results. As given by the points of Q2, the three exploratory search tasks (T4–T6) were generally considered much more difficult than the lookup tasks (T1–T3).

Among the three lookup tasks, T1 was the easiest since, according to Table 1, it returned a relatively small number of relationships. In this task, all the three systems exhibited considerable effectiveness, as indicated by the points of Q4 and Q5. However, in T2 and T3 which returned a large number of relationships, **RList** became less effective, when **RClus** outperformed **RFacet**.

Among the three exploratory search tasks, **RList** performed consistently worse than the other two systems, when **RClus** considerably outperformed **RFacet**. However, it is worth noting that the participants only showed a medium level of confidence in their results in T5 and T6 when using **RClus** and all the other systems, as given by the points of Q3, which will be discussed later.

In addition, the SUS scores achieved by **RList**, **RFacet**, and **RClus** were 71.00, 63.50, and 72.83, respectively, showing that our **RClus** system (as well as the other two systems) is fairly usable, though it was not our primary focus in the experiments.

4.6 User Feedback and Discussion

The experimental results can be partially explained by the comments collected from the participants, as discussed in the following.

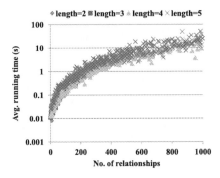

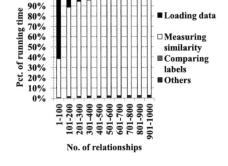

Fig. 4. Average time for a single run of our algorithm, under different numbers of relationships of different lengths.

Fig. 5. Percentages of running time for different steps of our algorithm, under different numbers of relationships.

Five participants considered **RList** easy to understand due to its simplicity. However, seven believed that without an effective ranking or an appropriate organization, this system is hardly useful when faced with a large result set.

Three participants favored the class filters provided by **RFacet** when carrying out lookup tasks. However, nine participants disliked filtering in a click-to-exclude manner, but preferred click-to-include. Two participants complained that there were too many facet values to browse and select. For these and other reasons, six participants never used any filter in exploratory search tasks. Besides, three participants suggested a hierarchical organization of class filters.

Five participants felt that the hierarchical clustering provided by **RClus** offered a useful overview to start exploration. However, four participants disliked the deep hierarchies met in the experiments, and two proposed to solve this by automatically merging fine-grained clusters.[3] Besides, four participants expected a more concise method for visualizing a hierarchy of relationship patterns.

Last but not least, most participants in T5 and T6 reported only a medium level of confidence in their answers because they were uncertain about the completeness given a large result set, regardless of the system used.

To summarize, presenting a large number of relationships as a flat list is clearly unsatisfactory. Providing faceted categories is beneficial in particular in lookup tasks, but requires careful design. In exploratory search tasks, clustering-based exploration is preferable, but deep hierarchies should be avoided.

5   Performance Testing

We also tested the scalability of our approach, which was implemented in Java on an IBM server with four Xeon E7440 and 4 GB memory for JVM. The data graph was stored on disk as an indexed triple table using MySQL 5.1.58.

---

[3] Following this suggestion, in the latest online demo we have merged very small clusters to flatten the hierarchies.

To show how the length and the number of relationships would influence the running time of our clustering algorithm, for each n from 2 to 5, we randomly selected and tested 2,000 pairs of entities between which at least one relationship of length n could be found. Figure 4 plots, on a log-lin scale, the average time for a single run of our algorithm under different numbers of relationships and different n. Not surprisingly, basically more time was needed for more or longer relationships. When $n = 3$ or 4, the algorithm processed roughly 200 relationships within one second. Given ten seconds, the algorithm processed around 700 relationships when $n = 4$, or 500 relationships when $n = 5$.

To find potential bottleneck points of our implementation, we identified three key steps: measuring similarity (i.e. line 6 of Algorithm 1), comparing labels (i.e. line 13 of Algorithm 1), and loading data (e.g. types of entities, precomputed probabilities) from database. As shown in Fig. 5, given no more than 100 relationships, data loading took most of the running time. For larger numbers of relationships, similarity measurement became dominating. Fortunately, this step could be easily parallelized to considerably improve the performance.

6   Related Work

6.1   Relationship Search

Anyanwu et al. in their seminal work [4] formalize several types of connections between entities called semantic associations, one of which is a path connecting the two entities, as studied in this paper. Among recently developed systems, RelFinder [11]provides several kinds of filters including length and the types of internal vertices (i.e. classes) and edges (i.e. properties) for refining the search. Viswanathan et al. [21] allow the user to specify the interval vertices through which a relationship to return should pass. Conkar [25] can find relationships under keyword-based constraints expressed using quantitative metrics including coverage and relevance. In general, all the three systems can be classified into exploratory relationship search and, more specifically, faceted categories-based methods, despite providing different kinds of facets including length, internal entities, types, and keywords. By comparison, our approach attempts to *implement exploratory relationship search by using a clustering-based approach*; from another point of view, *the filters we provide are structured schematic patterns*.

Besides, various extensions of the problem have been proposed. In REX [8], a relationship is not necessarily a path but conforms to a certain constrained graph pattern. The latest version of RelFinder [12] can simultaneously find relationships between more than two entities.

6.2   Path Finding and Relationship Ranking

Several subproblems of relationship search such as path finding and relationship ranking have attracted considerable attention.

Apart from executing SPARQL or other types of formal queries, relationship finding is usually solved by finding paths of a limited length between two vertices using graph search. BRAHMS [14] provides two main memory-based implementations: one based on depth-first search (DFS) and the other using bidirectional breadth-first search (bi-BFS). Experiments show that bi-BFS is up to a few orders of magnitude faster than DFS, but comes with high memory requirements. Gubichev et al. [9] present a disk-based implementation which extends the RDF-3X RDF database engine in several directions, e.g. adopting an optimized strategy for assigning IDs to vertices, and combining multiple lookups of neighbors of vertices into one single join operation. Following [14], our online demo uses bi-BFS and runs concurrently in multiple threads. However, *path finding is orthogonal to the contribution of this paper.*

To rank relationships, Aleman-Meza et al. [1,2] consider several kinds of weights, including the depths of the types of vertices and edges in the hierarchy, the length of a path, users' interests in types, rarity, popularity, etc. Viswanathan et al. [22] follow [1] but learn a user's interest from the Web browsing history. The SemRank model [3] measures informativeness, query relevance, etc. Chen et al. [7] employ learning-to-rank to capture a user's preferences. However, *relationship ranking is not the focus of this paper.*

6.3   Exploratory Search

The concept of exploratory search has been implemented in many Semantic Web applications other than relationship search, in particular in entity search. The primary method adopted there is faceted categories. In a typical faceted entity search system like mSpace [19] and Magnet [20], the user is provided with a collection of properties as well as their values to filter entities. When there are too many properties to show, only the more important and useful ones will be selected, e.g. those occurring frequently and having a moderate number of values to choose from [16]. When the number of values is large, organizing them into a hierarchy based on their types and providing keyword search featuring autocomplete are both beneficial to users [13]. Whereas all these approaches treat each property as a facet, a more powerful solution uses sequences of properties [23] and organizes them into a tree instead of a flat list. Clustering also finds applications in entity search. In Noadster [18], entities are hierarchically clustered based on their common property values, which are also used for cluster labeling. Compared with Noadster, *our approach follows a similar paradigm but solves a different problem by using different techniques.*

6.4   Clustering

A large body of work has been devoted to agglomerative clustering [24]. One major feature of our clustering algorithm lies in its measure of similarity between clusters. Different from wide-used measures like single linkage and complete linkage which combine the values of similarity between constituent elements, we *directly measure, by using information theory, how many commonalities the two clusters share in terms of their characteristic schematic patterns.* However, an empirical comparison of different clustering algorithms is left as future work.

7   Conclusion

In this paper, we have practiced clustering-based exploratory relationship search. To meet the three challenges faced in this novel problem, when our approach progressively groups similar relationships into larger clusters, it labels each cluster with a human-readable relationship pattern, which best reflects the semantic commonalities shared by the constituent relationships; the similarity between clusters is measured based on their corresponding relationship patterns by using information theory. As a result, users interact with the generated hierarchical clustering as if they start with an overview of the results and continuously, according to their needs, refine or coarsen a path query under semantic constraints at different levels of granularity, which is represented by a hierarchy of relationship patterns conforming with human intuition.

We have empirically compared our clustering-based solution with faceted categories via a user study. In general, both approaches have proven to be more effective than trivially listing all the results. In lookup tasks, they have their own advantages and suit different tasks to different degrees, which motivates us to consider an integration of the two methods. In exploratory search tasks, our clustering-based solution is preferable, although there is still room for improvement. In future work, we will primarily explore and empirically compare different clustering algorithms. Another challenge would be to visualize a hierarchical clustering of relationships in an appropriate manner.

Acknowledgments. This work was supported in part by the NSFC under Grant 61100040, 61170068, and 61223003, and in part by the JSNSF under Grant BK2012723. We thank the anonymous reviewers for their suggestions.

References

1. Aleman-Meza, B., Halaschek-Wiener, C., Arpinar, I.B., Ramakrishnan, C., Sheth, A.P.: Ranking complex relationships on the Semantic Web. IEEE Internet Comput. **9**(3), 37–44 (2005)
2. Aleman-Meza, B., Halaschek-Wiener, C., Arpinar, I.B., Sheth, A.: Context-aware semantic association ranking. In: 1st International Workshop on Semantic Web and Databases, pp. 33–50 (2003)
3. Anyanwu, K., Maduko, A., Sheth, A.: SemRank: ranking complex relationship search results on the Semantic Web. In: 14th International Conference on World Wide Web, pp. 117–127. ACM, New York (2005)
4. Anyanwu, K., Sheth, A.: ρ-Queries: enabling querying for semantic associations on the Semantic Web. In: 12th International Conference on World Wide Web, pp. 690–699. ACM, New York (2003)
5. Brooke, J.: SUS: a 'quick and dirty' usability scale. In: Jordan, P.W., Thomas, B., Weerdmeester, B.A., McClelland, I.L. (eds.) Usability Evaluation in Industry, pp. 189–194. Taylor & Francis, London (1996)
6. Carpineto, C., Osiński, S., Romano, G., Weiss, D.: A survey of web clustering engines. ACM Comput. Surv. **41**(3), 17 (2009)

7. Chen, N., Prasanna, V.K.: Learning to rank complex semantic relationships. Int. J. Semant. Web Inf. Syst. **8**(4), 1–19 (2012)

8. Fang, L., Das Sarma, A., Yu, C., Bohannon, P.: REX: explaining relationships between entity pairs. Proc. VLDB Endow. **5**(3), 241–252 (2011)

9. Gubichev, A., Neumann, T.: Path query processing on very large RDF graphs. In: 14th International Workshop on the Web and Databases (2011)

10. Hearst, M.A.: Clustering versus faceted categories for information exploration. Comm. ACM **49**(4), 59–61 (2006)

11. Heim, P., Hellmann, S., Lehmann, J., Lohmann, S., Stegemann, T.: RelFinder: revealing relationships in RDF knowledge bases. In: Chua, T.-S., Kompatsiaris, Y., Mérialdo, B., Haas, W., Thallinger, G., Bailer, W. (eds.) SAMT 2009. LNCS, vol. 5887, pp. 182–187. Springer, Heidelberg (2009)

12. Heim, Philipp, Lohmann, Steffen, Stegemann, Timo: Interactive relationship discovery via the Semantic Web. In: Aroyo, Lora, Antoniou, Grigoris, Hyvönen, Eero, ten Teije, Annette, Stuckenschmidt, Heiner, Cabral, Liliana, Tudorache, Tania (eds.) ESWC 2010, Part I. LNCS, vol. 6088, pp. 303–317. Springer, Heidelberg (2010)

13. Hildebrand, M., van Ossenbruggen, J., Hardman, L.: /facet: a browser for heterogeneous Semantic Web repositories. In: Cruz, I., Decker, S., Allemang, D., Preist, C., Schwabe, D., Mika, P., Uschold, M., Aroyo, L.M. (eds.) ISWC 2006. LNCS, vol. 4273, pp. 272–285. Springer, Heidelberg (2006)

14. Janik, M., Kochut, K.J.: BRAHMS: a workbench RDF store and high performance memory system for semantic association discovery. In: Gil, Y., Motta, E., Benjamins, V.R., Musen, M.A. (eds.) ISWC 2005. LNCS, vol. 3729, pp. 431–445. Springer, Heidelberg (2005)

15. Marchionini, G.: Exploratory search: from finding to understanding. Comm. ACM **49**(4), 41–46 (2006)

16. Oren, E., Delbru, R., Decker, S.: Extending faceted navigation for RDF data. In: Cruz, I., Decker, S., Allemang, D., Preist, C., Schwabe, D., Mika, P., Uschold, M., Aroyo, L.M. (eds.) ISWC 2006. LNCS, vol. 4273, pp. 559–572. Springer, Heidelberg (2006)

17. Resnik, P.: Using information content to evaluate semantic similarity in a taxonomy. In: 14th International Joint Conference on Artificial Intelligence, vol. 1, pp. 448–453. Morgan Kaufmann, San Francisco (1995)

18. Rutledge, L., van Ossenbruggen, J., Hardman, L.: Making RDF presentable: integrated global and local Semantic Web browsing. In: 14th International Conference on World Wide Web, pp. 199–206. ACM, New York (2005)

19. Schraefel, M.C., Wilson, M., Russell, A., Smith, D.A.: mSpace: improving information access to multimedia domains with multimodal exploratory search. Comm. ACM **49**(4), 47–49 (2006)

20. Sinha, V., Karger, D.R.: Magnet: supporting navigation in semistructured data environments. In: 2005 ACM SIGMOD International Conference on Management of Data, pp. 97–106. ACM, New York (2005)

21. Viswanathan, V., Ilango, K.: Finding relevant semantic association paths through user-specific intermediate entities. Hum.-centric Comput. Inf. Sci. **2**, 9 (2012)

22. Viswanathan, V., Ilango, K.: Ranking semantic relationships between two entities using personalization in context specification. Inf. Sci. **207**, 35–49 (2012)

23. Wagner, A., Ladwig, G., Tran, T.: Browsing-oriented semantic faceted search. In: Hameurlain, A., Liddle, S.W., Schewe, K.-D., Zhou, X. (eds.) DEXA 2011, Part I. LNCS, vol. 6860, pp. 303–319. Springer, Heidelberg (2011)

24. Xu, R., Wunsch II, D.: Survey of clustering algorithms. IEEE Trans. Neural Netw. **16**(3), 645–678 (2005)
25. Zhou, M., Pan, Y., Wu, Y.: Conkar: constraint keyword-based association discovery. In: 20th ACM International Conference on Information and Knowledge Management, pp. 2553–2556. ACM, New York (2011)

XML Multi-core Query Optimization Based on Task Preemption and Data Partition

Pingfang Tian[1,3]($\boxtimes$), Dan Luo[2], Yaoyao Li[1,3], and Jinguang Gu[1,3]

[1] College of Computer Science and Technology,
Wuhan University of Science and Technology, Wuhan 430065, China
tianpf@ontoweb.wust.edu.cn
[2] School of Computer Science and Technology,
Huazhong University of Science and Technology, Wuhan 430074, China
[3] Hubei Province Key Laboratory of Intelligent Information Processing
and Real-time Industrial System, Wuhan 430065, China

Abstract. In XML query optimization, most algorithms still use the traditional serial mode, so they can hardly take full advantage of multi-core resources. This paper proposes data partition for the XML database so that the new approach reaches load balance between different cores. Each thread processes the sub-regional data independently to reduce synchronization and communication overhead between cores. This paper also discusses the usage of task preemption in multi-core querying. According to experiments, our strategy requires less time consumption and gains better workload balance than both NBP and SBP.

Keywords: XML Query Optimization · Multi-core · Data partition · Task preemption

1 Introduction

XML has emerged as a standard of data representation and exchange on the Internet for a long time. It plays an important role in many application areas; it is also the technological basis of Interoperability for Web Service, documents and relation or object databases. However, the XML data query process has become the bottleneck of system performance. Therefore, the demand for efficient and fast XML processing grows rapidly. From the view of hardware, multi-core system is currently the dominant trend in computer processors [1]; however, most of XML processing algorithms are still using traditional serial processing modes, thus the advantages of multi-core resource can not be fully exploited. In order to solve the problems analyzed above, this paper designs an XML multi-core optimization strategy based on task preemption and data partition, and this strategy reduces the response time of user querying, mitigates the network loading, improves the querying efficiency and responsiveness of database.

The rest of this paper is organized as follows: Sect. 2 describes some related works of XML multi-core querying; Sect. 3 describes the XML query optimization problems in multi-core environment, including XQuery decomposition, data

W. Kim et al. (Eds.): JIST 2013, LNCS 8388, pp. 294–305, 2014.
DOI: 10.1007/978-3-319-06826-8_22, © Springer International Publishing Switzerland 2014

partition etc.; Sect. 4 describes a XML multi-core optimization strategy based on task preemption and data partition (Imp-NBPP), and discusses related algorithm; Sect. 5 evaluates and analyzes the performance of NBP, SBP and Imp-NBPP through the experiments, the result shows that the strategy of Imp-NBPP is much more better than the other two strategies in economizing query time and keeping load balance. Finally, Sect. 6 summarizes our work and discusses the future research.

2   Related Works

2.1   XML Multi-core Query Optimization

Multi-core is gradually growing popular, and it also influence the trend of computing; Traditional XPath and XQuery cannot take advantages of multi-core environment effectively, because existing algorithms are based on single core, thus it's hard to get obvious effects when putting them into multi-core environment directly. If the advantages of multi-core can be fully taken, query processing will get better performance. When considering the parallel algorithm of XML query processing (such as whole branches connection), the key is to partition XML data effectively, and to attribute them to different CPU cores with an appropriate method so as to reduce the synchronization overhead between the CPU cores. In addition, the multi-core processors face several additional challenges such as the use of memory and cache optimization techniques, data locality and data prefetching techniques to improve performance greatly [2]. The extra challenges prohibits developers making full use of the multi-core platform to optimize applications, to achieve the load balancing on CPU cores and to improve the performance effectively.

Some previous researches focus on improving performance of XML querying, such as XML pattern parse and hardware acceleration. The performance of parsing XML improved by concurrent method naturally inspires us applying pipeline. In this method, XML parse is divided into several stages, each stage will be executed by different thread. The method brings in acceleration, but due to the overhead produced by synchronization, load balancing and memory access, it's hard to implement on software pipeline. On the other hand, threads and cores can build up relationships through mapping, the more using of threads, yet the greater possibility of data access conflicts. Another possible method is data parallelism, it divides the XML file into one or more than one data blocks, each thread processes the data blocks independently, after processing each data partition, then merges and builds the final result.

This paper discusses two main problems in multi-core environment. One is the XML data partition method, the other one is unbalanced thread of querying processing, in which some threads are "busy", but some are "idle". Light task threads need wait for the heavy task thread, which affects the output of final result, and easily leads to "cask effect". In conclusion, the major target is to exploit the advantage of multi-core environment to improve XML query process performance.

2.2   Related Work About XML Multi-core Query Processing

Query optimization of XML database has been deeply researched by more and more scholars, the main studies are as follows:

Paper [3] proposes an XML parallel parsing method, and provides the design in details and their implementation. This method firstly uses incipient pre-parsing stage to determine the logical structure of XML document tree; then distributes the documents to each processing block with the complete, parallel parsing. In multi-thread environment, the strategy implements like this: when a thread is idle, it will steal half of other threads' tasks to process, when all the tasks are completed, the whole process of XML can be regarded DONE. However, the granularity of task stealing is a question which requires further discussion.

The parallel query process algorithms of XPath and XQuery is a even more critical issue in the effective use of multi-core CPU computing resource, paper [4] made a good try in twig parallel query algorithms of XML document. Aiming at twig parallel query, a parallel path stack algorithm called P-PathStack was proposed, which can skip many unnecessary nodes and avoid huge intermediate results. Obviously the P-PathStack is more effective in shared-memory multi-core system. The strategy focuses on skipping more unnecessary nodes, exploring more effective data partition algorithm, along with load balancing, designing more efficient parallel computing model to handle XPath/XQuery process.

To solve the problems currently in XML multi-core query strategy, the paper proposes a XML multi-core optimization strategy based on task preemption and data partition (Improved Node Based Partition and Preemption, Imp-NBPP). The strategy firstly partitions the XML document data to order to achieve load balancing, and then each thread deals with different data partition to reduce the overhead of synchronization and communication between threads. The paper also discusses the application of task preemption strategy in multi-core query.

3   XML Query Optimization in Multi-core Environment

3.1   XQuery Query Decomposition

Query decomposition is one of popular technology to deal with traditional distributed SQL query [5,6]. A global query can be decomposed into several sub-queries, meanwhile these sub-queries are sent to different parts of the process. Therefore, it's easy to come up with the idea of taking decomposition technique of SQL query into the process of heterogeneous data. XQuery query can be represented with some special expressions (e.g. FOR clause) to solve the problems of semi-structured data processing, whereas those expressions make the global XQuery decomposition a little difficult. The multi-document XQuery shown as Fig. 1.

Figure 1 mainly shows the query to item.xml and seller.xml. In Q2 and Q3, tick border ellipse, description, URI and address, represent the nodes need to

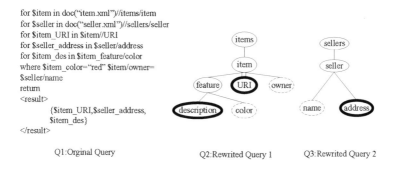

```
for $item in doc("item.xml")//items/item
for $seller in doc("seller.xml")//sellers/seller
for $item_URI in $item//URI
for $seller_address in $seller/address
for $item_des in $item_feature/color
where $item_color="red" $item/owner=
$seller/name
return
<result>
        {$item_URI,$seller_address,
        $item_des}
</result>
```

Q1:Orginal Query   Q2:Rewrited Query 1   Q3:Rewrited Query 2

Fig. 1. Multi-document XML query

return; dotted border ellipse, color, name and owner, represent predicated conditions and the nodes need to connect. For the query decomposition of XQuery, query is firstly decomposed into several combined twig query plans on each data stream, followed by multi-join and finally the twig query. To the XQuery with FWR sentence, it can be decomposed into three parts:

1. Structured to filter input stream;
2. Process the predicate;
3. Structured to filter the result.

Assume that XQuery preprocessor will parse XQuery expression with (FWR), and generate a query plan. In the predicate processing stage, filter is based on value, the same way as process the connection between input streams. Paper [7] focused on twig'n join query processing. A complex query contains '*','//'or '[]', can be divided into several simple paths or predicate expressions, known as partition component, it can be encoded into interval. Then, component connects the wildcard '*', descendant '//'or predicate '[]'. Each complex query can be divided into several parts. Paper [8] proposed an effective technology for multi-query processing in dynamic environment. The basic idea is to use XML compression technology to process the query as a whole data compression, rather than process queries and on-demand decomposition. From a technological perspective, using inclusion relationship to process XML data between queries can reduce bandwidth, and develop efficient data structures. If the method can make full use of semantic cache technology, providing the support for compress and sharing the data got from client by XML, it will be more efficient. XML query rewriting in a distributed computing environment has been highly valued. Paper [9] discussed the query rewriting in one client with many views. It designed two data structures to manage multiple XML views: MPTree builds the main path of XML view to generate candidate query rewriting program; PPLattice is built by predicate sub-tree of XPath to confirm the candidate query rewriting program. Through MPtree and PPLattice , we can reduce not only the research space but also the overhead furthest.

3.2 XML Data Partition

Due to the complexity of semi-structured data, many articles have suggested various XML data partitioning methods in parallel environment. Usually they store the XML nodes in form of ancestor and descendant nodes, and use the region coding partition method to handle the parallel structure connection. The key point of parallel query processing is to design a good data partition method, therefore it is obviously important to solve the problems of physical fragmentation in XML database data. Because XML database and traditional relational database are different: XML data is a kind of semi-structured data with flexible manifestation, whereas the data model in relational database is structured, so when relational database to storage XML data are put into real practise, it's hard to represent the relationship between each node [10]. Different from structured table structure, the complexity of XML lays on its complex nested structure, when we use the way of relational data table to store and manage XML data, the following two shortcomings comes out:

1. Large number of data redundancy;
2. The representation of the relationship between the nodes is also a difficult aspect.

Given an ancestor set $AList = a_1, a_2, \ldots, a_m$ and a descendant set $DList = d_1, d_2, \ldots, d_n$, where both $AList$ and $DList$ consist of nodes from the XML document tree. A structure join of $AList$ and $DList$, denoted as $AList \bullet DList$, returns all tuple pairs (a_i, d_j), where $a_i \in AList$ and $d_j \in DList$, such that a_i is an ancestor of d_j, and symbol $\bullet$ denotes structure join [4,11]. The $DList$ is partitioned into different blocks $DList_i(i = 0, 1, \ldots, n_b)$, except the last block, the size of each block is fixed, represented by b_s. Then partition $Alist$ to different block $Alist_i(i = 0, 1, \ldots, n_b)$ based on $Dlist$. For any e element contained in $AList$, e belongs to $AList_i$ if and only if it has one or more than one descendant in $DList_i$. In most cases, this partition method can guarantee all the elements in different blocks have no parent-child relationship, or ancestor-descendant. In other words, $AList_i \bullet DList_j = \varnothing$ (when $i \neq j$). Even in some cases $AList_i \bullet DList_j \neq \varnothing$, the method also can provides a number of partition conditions $AList_i \bullet DList_j \subseteq AList_j \bullet DList_j$. Therefore, it only needs to calculate $AList_i \bullet DList_i(i = 0, 1, n_b)$ and finally merge their results into final result. It can naturally and statically divide XML document into a number of equal-sized trees. The advantage of static partition is that it can generate very good load balancing for each thread of execution to utilize the benefits of parallel. The static partition also has its disadvantages, there are two hidden defects:

1. Node in corresponding XML document may not be continuous.
2. Static partition algorithm must be sequentially executed before parallel parsing, thus the parallel advantage may easily be offset by the overhead generated by the algorithm.

If the structure of XML document which will be processed is well-formated, static partitions can provide a good parallel performance, because the structure can easily be divided into equal-sized pieces. Division is primarily based on left-to-right division, so that each partition block is a continuous XML document. The partition will achieve the following two goals: First, to balance the workload of each site; Second, to minimize the system response time of user's queries.

4 XML Multi-core Optimization Strategy Based on Task Preemption and Data Partition

4.1 The Technique of Load Balance and Preemption

In the case of shared-nothing parallel architecture with few processors, physical partition technique based on declustering strategy is more effective [12]. The purpose is to achieve I/O parallelism and operation parallelism. The declustering objects are the tuples belonging to a relationship or all objects in a class, and they are declustered to all processors based on the number of processors. According to the structure features of XML document tree, viewing from the root to the leaf, the number of nodes close to root is relatively small, but they would appear in most XML queries. Strategy in paper [13] is that: to sacrifice a small amount of disk space at cost, put forward the concept of root-tree replication and sub-tree partition, copy the nodes close to root to all processing sites. The sub-tree partition refers to those data which are large but not frequently accessed, and the data are divided into different sites to be stored. In each site, the two forms of the tree generate a partial document tree through combination. In order to ensure the partial integrity of XML data, the strategy reduces the overhead of transmission, so that each site is stored in the form of a data tree, thus it can take full advantage of the parallel.

In order to partition XML document to one or more blocks, we simply treat it as a sequence of characters, then divide it into equal-sized blocks, and each thread processes a data block. It means that each thread should start parsing from a random point of XML document, since the XML document serialization is traversed by data model of tree structure according to the order from left to right, depth-first. With this block method, analyzed result might be difficult to combine into a tree, also the overhead might eliminate the benefits of parallel parse. Obviously, partition in light of logical structure of an XML document, rather than equal physical size, is the key to the efficient parallel processing of XML data. The reason is that the more XML partitions, the bigger node redundancy will be. Usually it appears, to a certain extent, as the number of partitions increases, the execution time will increase, since generating partitions will lead to a part of overhead. Paper [14] proves that if the number of threads is more than the available CPU core number, the performance will not be improved.

Paper [3] takes a simple strategy to partition half of workload. Shown in Fig. 2.

In other words, in light of partition requirement, the running threads resolve the sibling nodes of current element node according to a order from left to right. Subtask distribution refers to how and when task thread will assign tasks

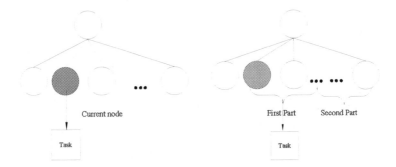

Fig. 2. A simple strategy

to request thread. In parallel XML parsing process, in order to maximize performance, parsing thread will parse XML data as many as possible without interruption, unless there are other threads idle or having task requirements. In addition, any thread can be the requester and distributor.

4.2 Region Coding

Usually XML documents use regional coding method, in XML document tree based on node encoding, it has the advantage of being able to determine the relationship between nodes is whether parent-child (pc) relationship or ancestor-descendant (ad) quickly without traversing the original XML document. In this paper, we use region coding <start, end, level> to encode XML document; the definition of pc and ad are as follows:

Definition 1. *Ancestors - Descendants relations: in XML document tree, for two nodes of n_1, n_2 , each of node coding is $n_1 : (D_1, S_1 : E_1, L_1)$ and $n_2 : (D_2, S_2 : E_2, L_2)$. n_1 is an ancestor node of n_2, if and only if the following conditions are satisfied:*

1. $D_1 = D_2$;
2. $S_1 < S_2$ and $E_2 < E_1$.

Definition 2. *Parent-Child relationships: in XML document tree, for two nodes of n_1, n_2 , each of their node coding is $n_1 : (D_1, S_1 : E_1, L_1)$ and $n_2 : (D_2, S_2 : E_2, L_2)$. n_1 is the father node of n_2, if and only if the following conditions are satisfied:*

1. $D_1 = D_2$;
2. $S_1 < S_2$ and $E_2 < E_1$;
3. $L_1 + 1 = L_2$; i.e. the layer interval of two nodes in XML tree is 1.

There are generally two different types of coding mode in the present encoding methods, and the region encoding method is the most widely used one, in this method, each element node is assigned a (start, end) encoding, start indicates the starting position of the element in the node tree, end indicates the end position in the node tree [15]. For any two different elements nodes u and v, the interval covered by their coding does not partly intersect, i.e. there are one of the following two relationships between u and v: completely contains or does not intersect [13].

4.3   Partitions and Query Processing Model

The existing parallel XML parsing algorithms often have the following problem: in order to partitions data, some kind of pretreatment operation is required to achieve the parallelization of fragmentation and analysis processing, however pretreatment often needs a lot of time, and the optimization of these operations is also relatively complex [16]. In order to obtain advantage in terms of performance, we extend two aspects of above data partition method:

1. create data block;
2. create a more fine-grained task in each data partition.

In the first level of the data partitions, we assign each CPU core a block. It's mainly to inherit the independence of primary characteristics and load balancing from data. It is important to store each block in a continuous independent memory space, to ensure that there is no shared address space and no synchronization. Of course to create more fine-grained partitions in each partition would reduce the XML nodes in XML query processing, and additional semantic cache can be taken into account for further optimizing. Therefore, they are useful to reduce latency and improve performance. The strategy to realize XML data partition has the following advantages:

1. Better data parallelism, because nodes are distributed to each CPU core evenly, path patterns are distributed evenly as well, it's easy to achieve better parallelism.
2. Since the data load processed by each CPU core is balanced, global database query response time can be greatly shortened.
3. The order of original data nodes is not destroyed, when build the final results, only natural connection is needed, it can avoid a large number of structural joining operations, while the connection of query result is time-consuming.

Query processing model and related processing algorithms are shown in Fig 3.

5   Evaluation and Analysis

To compare and analysis the query performance of the above three XML data partitioning strategies, we did some experiments in a Native XML database (Timber). Three XML documents are used in this experiment, namely mimi.xml (22 KB), pepople.xml (280 KB), africa.xml (1.39 M), along with a series of simple queries and complex queries. In the first step, we measure the performance improvement and the scalability of XML multi-core query. The strategy is used

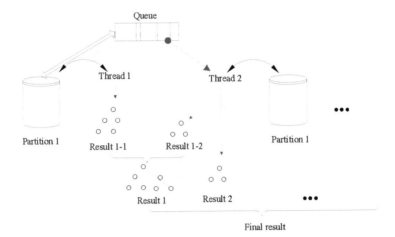

Fig. 3. Query processing model

to make a comparison among different sizes of XML documents. For each XML document, we prepare three groups of XQuery query with different partition strategiesm, so that they test the degree of query optimization and make comparisons between simple query and complex query. Each group contains two XQuery queries, one is simple XQuery, the other is complex XQuery (for example: contains nested sentence etc). The simple XQuery and complex XQuery examples are shown as following:

```
A Simple XQuery:
for $b in document("pepople.xml")//person
for $t in $b/address
where  $t/zipcode<25
return $b
```

```
A Complex XQuery:
for $b in document("pepople.xml")//person
for $t in $b/address
where  $t/country="United States" and $t/zipcode<25
return
<result>
  {$b/name }
{ for $k in document("pepople.xml")//person
where $k/name=$b/name and $k/profile/age<30
return
$k/homepage
}
</result>
```

Analyze simple XQuery and complex XQuery, the pattern trees according to different operations are as Fig. 4:

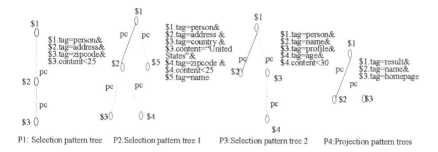

P1: Selection pattern tree   P2:Selection pattern tree 1   P3:Selection pattern tree 2   P4:Projection pattern trees

Fig. 4. Pattern trees according to simple query and complex query

In Fig 4, P1 is a selection pattern tree of simple XQuery, P2 and P3 are selection pattern trees corresponding to complex XQuery, which contain the join condition P2.name = P3.name, and P4 is the projection pattern tree of complex XQuery used to produce the final results. According to the partition and node coding method, the result is the nature connection of results produced by each thread processing. Our experiment ignore the connection of final result, because the construction of the final result is a natural connection of each partition result without any screening operation. We focus on effectiveness and feasibility of XML multi-core query optimizing strategy based on task preemption and data parallel.

The following Tables (Tables 1, 2 and 3) are the result of our experiment.

Table 1. XQuery queries on 22 K document(second)

Query Type	Single Core	Dual-Core			Quad-Core		
		SBP	NBP	Imp-NBPP	SBP	NBP	Imp-NBPP
Simple	0.125	0.109	0.109	0.109	0.093	0.094	0.093
Complex	0.235	0.219	0.218	0.204	0.203	0.203	0.203

Table 2. XQuery queries on 280 K document(second)

Query Type	Single Core	Dual-Core			Quad-Core		
		SBP	NBP	Imp-NBPP	SBP	NBP	Imp-NBPP
Simple	1.39	0.843	0.796	0.766	0.407	0.375	0.375
Complex	4.406	1.453	1.406	1.344	0.594	0.593	0.578

From the three tables above, it can be seen that: when queries differ in the sizes of XML documents, no matter in dual-core or quad-core environment, all of the query time have a significant reduction. In contrast, improved method costs less query time than strategy based on the number of nodes and strategy

Table 3. XQuery queries on 1.39 M document(second)

Query Type	Single Core	Dual-Core			Quad-Core		
		SBP	NBP	Imp-NBPP	SBP	NBP	Imp-NBPP
Simple	3.797	2.156	2.156	2.125	1.172	0.953	0.922
Complex	29.406	8.032	7.625	7.547	2.485	2.39	2.343

based on the size of document, so our strategy can save query time and improve the performance. In the aspect of time saving, it also figures out that the larger the XML document is, the more time will be saved. In addition, simple XQuery saves more time than complex XQuer. Viewing from analysis of load, improved method can gain the best load balance in three strategies. From the analysis of query time, we can see that, for relatively small documents, more time is needed to process the query and return the query results, whereas for larger documents, multi-core can effectively save query time. Query time of each core is likely to be a bottleneck in multi-core environment. Our strategy can effectively reduce the time interval to some extent, and make each query time close to each other.

6 Conclusions

This paper discusses the strategy based on task partition and data parallelism into XML multi-core query. The devised strategy can reduce query time, in comparison with NBP (Node Based Partition) and SBP (Size Based Partition). It also discusses several key technologies of XML data processing in multi-core environment, such as data partitioning strategy, and load balancing techniques, so it proposes XML multi-core optimization strategy based on task preemption and data partition (Improved Node Based Partition and Preemption, Imp-NBPP), the strategy can effectively avoid CPU idle through task preemption; Compared with the two data load balancing methods, experiment proves that the our strategy can save query time effectively, thus release network loads and improve query and responsiveness of the database efficiency.

The strategy discussed in the paper is suitable for other XML based dataset such as RDF/XML and pattern based query language such as SPARQL language, which will be our future work.

Acknowledgment. This work was partially supported by a grant from the NSF (Natural Science Foundation) of China under grant number 60803160, 61272110, 61100133 and 61303117, the Key Projects of National Social Science Foundation of China under grant number 11&ZD189. It was also partially supported by NSF of Hubei under grant number 2013CFB334, NSF of educational agency of Hubei Prov. under grant number Q20101110, and the State Key Lab of Software Engineering Open Foundation of Wuhan University under grant number SKLSE2012-09-07.

References

1. Zuo, W., Chen, Y., Heand, F., Chen, K.: Optimization strategy of top-down join enumeration on modern multi-core CPUs. J. Comp. **6**, 2004–2012 (2011)
2. Chen, S., Ailamaki, A., Gibbons, P.B., Mowry, T.C.: Improving hash join performance through prefetching. ACM Trans. Database Syst. **32**, 3–21 (2007)
3. Lu, W., Chiu, K., Pan, Y.: A parallel approach to XML parsing. In: 7th IEEE/ACM International Conference on Grid Computing, pp. 223–230 (2006)
4. Feng, J., Liu, L., Li, G., Li, J., Sun, Y.: An efficient parallel pathstack algorithm for processing XML twig queries on multi-core systems. In: 15th International Conference on Database Systems for Advanced Applications, pp. 277–291 (2010)
5. Kozankiewicz, H., Stencel, K., Subieta, K.: Distributed query optimization in the stack-based approach. In: Yang, L.T., Rana, O.F., Di Martino, B., Dongarra, J. (eds.) HPCC 2005. LNCS, vol. 3726, pp. 904–909. Springer, Heidelberg (2005)
6. Zhou, L., He, X., Li, K.: An improved approach for materialized view selection based on genetic algorithm. J. Comput. **7**, 1591–1598 (2012)
7. Tok, W.H., Bressan, S., Lee, M.L.: Twig'n join: Progressive query processing of multiple XML streams. In: 13th International Conference on Database Systems for Advanced Applications, pp. 546–553 (2008)
8. Wang, X., Zhou, A., He, J., Ng, W., Hung, P.: Multi-query evaluation over compressed XML data in daas. In: Service and Application Design Challenges in the Cloud, pp. 185–208 (2011)
9. Gao, J., Wang, T.J., Yang, D.: MQTree based query rewriting over multiple XML views. In: Wagner, R., Revell, N., Pernul, G. (eds.) DEXA 2007. LNCS, vol. 4653, pp. 562–571. Springer, Heidelberg (2007)
10. Tang, L., Yu, Y., Wang, G., Yu, G.: Design and implementation of a parallel data partitioning algorithm for XML data. Mini-Micro Syst. **25**, 1164–1169 (2004)
11. Wang, W., Jiang, H., Lu, H., Yu, J.X.: PBiTree coding and efficient processing of containment joins. In: Proceedings of the 19th International Conference on Data Engineering (ICDE03), pp. 391–402 (2003)
12. Wang, G., Tang, L., Yu, Y., Sun, B., Yu, G.: A data placement strategy for parallel XML databases. J. Softw. **17**, 770–781 (2007)
13. Wang, J., Meng, X., Wang, S.: Structural join of XML based on range partitioning. J. Softw. **15**, 720–729 (2004)
14. Machdi, I., Amagasa, T., Kitagawa, H.: Executing parallel twigstack algorithm on a multi-core system. In: Proceedings of the 11th International Conference on Information Integration and Web-based Applications and Services, pp. 176–184 (2009)
15. Zhang, C., Naughton, J., DeWitt, D., Luo, Q., Lohman, G.: On supporting containment queries in relational database management systems. In: Proceedings of the 2001 ACM SIGMOD International Conference on Management of Data, pp. 425–436 (2001)
16. Chen, R., Liao, H., Chen, W.: XML parsing schema based oil parallel sub-tree construction. Comput. Sci. **38**, 191–194 (2011)

Ranking the Results of DBpedia Retrieval with SPARQL Query

Shiori Ichinose[1], Ichiro Kobayashi[1($\boxtimes$)], Michiaki Iwazume[2], and Kouji Tanaka[2]

[1] Advanced Sciences, Graduate School of Humanities and Sciences,
Ochanomizu University, Tokyo, Japan
{ichinose.shiori,koba}@is.ocha.ac.jp
[2] Universal Communication Research Institute, National Institute of Information
and Communications Technology (NICT), Kyoto, Japan
{iwazume,tanaka}@nict.go.jp

Abstract. In recent years, a number of Semantic Web databases have been actively open to public as common resources on the Web by the effort of Linked Open Data community project. Due to this, we need a good method to search necessary data from those databases, depending on various purposes. In this study, we propose two methods to rank the results retrieved by a SPARQL query (especially, a SELECT query), using the information about the frequency of each property in a data set and the links between RDF resources. In order to evaluate our proposed methods, we set two cases for using SPARQL queries, and then rank the query results in each case. The usefulness of our proposed method has been confirmed by subject experiments.

Keywords: DBpedia · SPARQL · Ranking · PageRank algorithm · Semantic Web

1 Introduction

In recent years, the number of structured databases based on Linked Data [1] principles have been increasing and open to public. For example, some of them are DBpedia [2] which is a RDF data base; DBLP Bibliography Database[1] which is a database of major articles of computer science; Data.gov[2] which is a database of governmental statistical information, have been open to public, etc. A framework for connecting those structured databases is called Linked Data [1].

As a method to retrieve data from such databases, we can use a method to retrieve data with a SPARQL query, in addition to methods to simply browse data. A SPARQL query allows us to retrieve graph patterns whose minimum unit is an RDF triple consisting of three elements: subject, predicate and object. Using SPARQL queries, we can retrieve structured data, i.e., RDF triples, however, it is hard for us to tell which datum is important among the query results.

---

[1] http://dblp.uni-trier.de
[2] http://www.data.gov

W. Kim et al. (Eds.): JIST 2013, LNCS 8388, pp. 306–319, 2014.
DOI: 10.1007/978-3-319-06826-8_23, © Springer International Publishing Switzerland 2014

In this paper, using meta information of a data set such as link relation, frequency of properties, etc. and the structure of SELECT query in SPARQL, we propose a method to rank query results in terms of their importance. (The importance of data will be defined in Sect. 3.3.) As the target data, we use the DBpedia data set which works as a Hub database among Linked Open Data cloud. We set two cases for using SPARQL queries: (i) the case where RDF triples are retrieved, (ii) the case where resources which satisfy particular conditions are retrieved, and then rank the results of each case. Furthermore, we verify the usefulness of our method through subject experiments.

2   Related Studies

There are many studies to retrieve data from Semantic Web databases. Oren et al. [3] have proposed a method of indexing Semantic Web resources on the Web. Swoogle [4] is a retrieval engine which enables not only to collect Semantic Web document, but also to retrieve the documents with keywords.

Moreover, many methods to rank resources by weighting resources, utilizing the structure information of Semantic Web data. In particular, there are a lot of various studies to weight data using the algorithm to decide the importance of Web pages like the PageRank [5] and HITS [6] algorithms. In Swoogle [4], ReConRank [7], and ObjectRank [8], the algorithms employing the idea of the PargeRank algorithm to score structured data such as Semantic Web data have been proposed. Beagle [9] is a desktop data retrieval engine with ObjectRank algorithm. SemRank [10] employs a relation-based PageRank algorithm that uses the connections between resources provided by properties. In TripleRank [11], a method which ranks information using keys and facets based on 3-D tensor classification of the information of RDF triples.

Moreover, there are several studies to use DBpedia data set for ranking [12–15]. Meymandpour and Davis [12] rank higher class universities in the world by evaluating the RDF triples about the universities stored in the DBpedia data set. Mirizzi et al. [13] achieve to rank DBpedia resources based on the similarity between external resources and the DBpedia resources.

As well as our approach, Bamba and Mukherjea [14] and Mulay and Kumar [15] propose methods to rank the results retrieved with Semantic Web queries. Bamba and Mukherjea weight resources with a HITS-based algorithm and propose a method to rank the results based on the graph structure of the query results. Mulay and Kumar weight resources based on link-analysis using tri-strata structure consisting of data sets, triples, and resources. However, the information used in the method is limited to the information related to the property 'owl:sameAs[3]', they do not deal with the other kinds of properties.

In this study, we propose two methods to rank the results retrieved with a SPARQL query by weighting resources and properties taking account of the link relation in an RDF data set, referring to the algorithm of ranking the results retrieved by a Semantic Web query proposed by Bamba and Mukherjea [14].

---

[3] 'owl:' is the prefix for http://www.w3.org/2002/07/owl.rdf

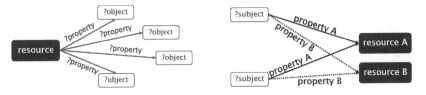

Fig. 1. Search RDF triples from a resource

Fig. 2. Search RDF triples from a triples

3 Framework for Proposed Ranking Method

We set the following two cases as the purpose for retrieving structured data.

- *The case of retrieving RDF triples through a resource:*
  As shown in Fig. 1, in this case, a user retrieves the most important RDF triple among multiple triples which have a particular subject within a range of specified resources. For example, from the resource of ':Tokyo[4]', the most valuable RDF triple which has ':Tokyo' as its subject is retrieved.
- *The case of retrieving resources from RDF triples:*
  As shown in Fig. 2, in this case, some RDF triples are used to narrow the candidates of the target resource. A user retrieves the most valuable resource, which satisfies retrieval conditions, within a range of specified triples. For example, resources, which belongs to a particular class, e.g., 'dbo:City[5]', are the targets to be retrieved.

These two cases are general in terms of the purpose of retrieving data with SPARQL queries. It is necessary for ranking query results retrieved by a SPARQL query to satisfy these requirements. In this study, we propose two methods to rank query results, referring to the query evaluation algorithm in the prior studies [14].

3.1 Use of SPARQL

SPARQL[6] is a query language to extract data from RDF graphs. The endpoint which works as an interface to accept SPARQL queries, is equipped in LOD database, therefore, we can obtain query results by providing SPARQL queries to the endpoint. In SPARQL 1.0, four kinds of query types: i.e., SELECT, ASK, CONSTRUCT, and DESCRIBE, are defined. SELECT query is used to retrieve a graph pattern whose minimum unit is an RDF triple. In SELECT clause, the variables representing the resources, literals and properties which we aim

---

[4] ':' is the prefix for http://dbpedia.org/resource/
[5] 'dbo:' is the prefix for http://dbpedia.org/ontology/
[6] http://www.w3.or/TR/rdf-sparql-query

to retrieve are provided. In WHERE clause, multiple constraints, which satisfy target resources, literals and properties, are described. In the case that we
retrieve information from a resource, in order to retrieve triples whose resource
is subject, a query includes ?property, ?object in SELECT clause, and then
is represented as follows:

```
    SELECT ?property ?object WHERE{
<resource> ?property ?object  .
}
```

In the case of retrieving resources from RDF triples, in order to define the
constraint on subject, in the query ?subject is included in SELECT clause. So,
the query is shown as follows:

```
    SELECT ?subject WHERE{
?subject <property> <object>.
}
```

There is a case where both property and object are provided in the query, or
a case where one of them is provided. In such cases, the variables not included
in SELECT clause can be any value — for instance, in the case that a resource
in DBpedia ':United_States' is provided, we can retrieve resources related to the
United States in terms of various factors, for example, people who were born in
the United States, books published in the United States, etc.

3.2   Graph Definition

The DBpedia data set we use in this study is open to public in the RDF format.
Here, we define a graph as a directed graph consisting of resources as vertices
and properties as edges between resources in the data set.

An RDF triple in DBpedia t is represented as (s, p, o). Here, s indicates
subject, p indicates predicate, i.e., property, and o indicates object. We define
all the sets of URI defined in the DBpedia data set as U, a set of RDF triples
is T , a set of vertices R, a set of edges is E, and then a graph, $G = (R, E)$,
consisting of resources is defined. $r \in R$ is either a subject s or an object o of an
RDF triple $t \in T$, and R is a subset of U. Moreover, as for $t \in T$, the relation
between resources $s \rightarrow o$ is defined as $e\{s, o\} \in E$, where $s, o \in R$.

The graph of resource G is a graph consisting of only relation between
resources in DBpedia data set, and the relation with the resources outside DBpedia is not considered.

3.3   Evaluation of Properties and Resources

We assume that the importance of an RDF triple is decided by evaluating the
weights of both subject and object and the property between the subject and the
object. In this section, we define the way of weighting properties and resources
to evaluate the query results.

Importance of Properties. Property is significantly important to rank query results. For example, in the case of retrieving writers and universities, the properties such as 'dbo:birthPlace' and 'dbo:influencedBy' for writers, 'dbprop:established[7]' and 'dbprop:campus' can be thought of as major information, respectively. Likewise, the importance of properties is decided based on the class of resources which is the subject of an RDF triple.

However, there are quite a few properties and classes used in DBpedia, so it is hard to manually decide the importance of each feature. To decide the importance of the features, we assume that important properties often appear in RDF triples whose subjects belongs to a particular class. So, we define Property Frequency - Inverted Property Frequency (PFIPF) as an evaluation index to decide the importance of a property based on the class to which its related resource belongs, referring to the idea of TF-IDF [16] often used an index to decide the importance of words.

We show the detail of the property frequency and the inverted property frequency as follows:

– *Property Frequency (PF)*
  PF is defined as the ratio of the number of resources which has some triples a particular property appears in, among all RDF resources in the target class. By PF, the properties which frequently appear in a particular class are highly regarded.
– *Inverted Property Frequency (IPF)*
  IPF is defined as the logarithm of the inverted value of the ratio of the number of classes in which a particular property appears among all classes.
  IPF works as a filter for general properties as well as inverted document frequency (IDF). The properties which appear in many classes are not regarded as valuable, but the ones which appear in few classes are regarded as valuable.

Importance of Resources. To evaluate resources, we use the PageRank algorithm [5] often used to evaluate the importance of Web pages. Now, we define the number of all resources in DBpedia as $|R|$; a set of resources x with the edges $\{x, r\} \in E$ where $r \in R$ as B_r, the number of edges from resource x as c_x, and then compute PageRank score, PR_r, based on Eq. (1).

$$PR_r = \frac{1 - d}{|R|} + d \sum_{x \in B_r} \frac{PR_x}{c_x} \tag{1}$$

Here, the reason for using the PageRank algorithm to decide the importance of resources in DBpedia is because it is generally used as an index to decide the importance of nodes in a directed graph, besides, it is also used as an algorithm to evaluate RDF data in the prior studies such as [4,7], therefore, we also employ it as the algorithm to evaluate the resources in DBpedia.

For a graph of resources, we adopt the power method to compute PageRank score. As the convergence condition, we employ the finite difference of iterative

---

[7] 'dbprop:' is the prefix for http://dbpedia.org/property/

calculation of PageRank vector. For stable PageRank score, we compute the score under the convergence condition of $|PR_r^k - PR_r^{k-1}| < 1E - X$ with $X = 5, 6, \ldots, 14$ which are the numbers obtained empirically as the ones effectively working to confirm convergence, and iterate the calculation until there is no ranking change between the case of $X = n$ and $X = n + 1$. As the data of DBpedia, we employ DBpedia 3.8. Figs. 3 and 4 show the number of iteration of the power method and necessary calculation time, and the ranking change to the value of X, respectively.

We see from Fig. 3 that the number of iteration of the power method is in proportion to necessary calculation time, and there is monotonic increase relation between them. On the other hand, we also see from Fig. 4 that the number of resources changing their ranking vastly decreases as X increases, and see that raking becomes stable around $X = 10$. In the case that the number of X is larger than 10, it is observed that the number of resources which change their ranking becomes mostly stable, rather the number of resources which change their ranking more than 1000 increases, therefore, we decide to employ the RageRank score of the case of $X = 10$ for ranking as the most stable value for the PageRank calculation convergence.

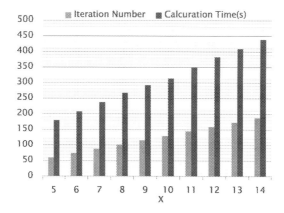

Fig. 3. The number of iteration of the power method and its calculation time

3.4   Ranking Algorithm

In this section, we explain our proposed ranking algorithms for the retrieval results by SPARQL queries. Each algorithm was developed by referring to the algorithm developed by Bamba and Mukherjea [14]. The results retrieved by a SPARQL query consists of partial graphs of the resource graph G. For example, as shown in Fig. 2, in the case of retrieving resources under particular constraints on RDF triples, the graph is constructed for each query result as shown in Fig. 5.

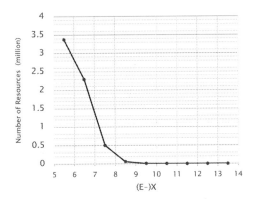

Fig. 4. Resource ranking changes corresponding to the value of X

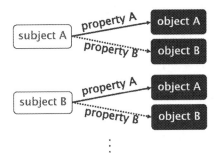

Fig. 5. A partial graph of results by resource retrieval

The query result is evaluated by the total score of the weights of both resources and properties in the retrieved graph.

Referring to the algorithm by Bamba and Mukherjea [14], we decide the factors to evaluate the importance of query results as follows:

- *Importance of resources and properties:*
  In the case that the weights of resources and properties are large, the weight of the result is larger than the case that the weight is small.
- *Size of a graph:*
  In the case that the graph of query results is large, the matching degree between the results and the query should be high, because much more information related to the given query should be included rather than the case where the graph is small.
- *Strength of user's interest:*
  The parameters used in SELECT clause are the items which a user is interested in. Whereas, regarding the items far from those items, i.e., resources and properties, on the graph, a user's interest can be getting decreased as the distance between the items increases.

Basically taking account of the factors mentioned above, we explain the algorithm of the prior study which has become the basis of our algorithm, and then propose two methods.

Algorithm by Bamba and Mukherjea [14] (Baseline Algorithm). In Bamba's method, the below algorithm is applied to evaluate each query result graph r. $decay$ is a parameter to adjust the weight of nodes and edges, depending on the strength of user's interest. $ScoreNode$ and $ScoreEdge$ are the score of nodes and edges, respectively.

Baseline Algorithm:

1. Let $decay = 1.0, scoreEdge(r) = 0.0, scoreNode(r) = 0.0$
2. Let Adj be a set initially consisting of the nodes of interest to the user as determined from SELECT clause of the query.
3. while Adj is not Empty
   (a) Let $Edges$ be the set of edges from the nodes in Adj.
   (b) $scoreNode(r)+ = \sum_{n \in Adj} Imp[n] * decay$ where Imp[n] is the importance of the node n. (It should be noted that if a node is not a resource, for example a literal like the label of a resource, its importance will be 0).
   (c) $scoreEdge(r)+ = \sum_{e \in Edges} IPF[e] * decay$ where IPF[e] is the Inverted Property Frequency of the edge e in the $Edges$ set.
   (d) $decay* = decayFactor$ where $decayFactor$ is a constant less than 1.0.
   (e) Reinitialize Adj to all the nodes that have not yet been visited and are adjacent to the previous nodes in the set.

After evaluating each query result graph r, both $scoreNode$ and $scoreEdge$ are normalized, and then the final score is computed by Eq. (2). Here, in Eq. (2), $nodeWt$ and $edgeWt$ indicate the weight of $nodeScore$ and $edgeScore$, respectively, and $nodeWt + edgeWt = 1.0$.

$$FinalScore = NormalizedScoreNode(r) \times nodeWt$$
$$+ NormalizedScoreEdge(r) \times edgeWt \qquad (2)$$

Query Evaluation Based on the Importance of Resources and Properties (Algorithm 1). The algorithm shown below is the algorithm to change IPF part in Bamba's algorithm with PFIPF. Depending on the class to which the subject of a RDF triple belongs, the importance of a property is changed.

Algorithm 1:

1. Let $decay = 1.0, scoreEdge = 0.0, scoreNode = 0.0$
2. Let Adj be a set initially consisting of the nodes of interest to the user as determined from SELECT clause of the query.
3. while Adj is not Empty
   (a) $LetClassedEdges$ be the set of edges from the nodes in Adj and the classes they belong to, i.e., (c, e).

(b) $scoreNode(r)+ = \sum_{n \in Adj} Imp[n] * decay$
where Imp[n] is the importance of the node n.(If a node is not a resource, for example a literal like the label of a resource, its importance will be 0.

(c) $scoreEdge(r)+ = \sum_{e \in ClassedEdges} PFIPF[c,e] * decay$
where PFIPF[c,e] is the value of Property Frequency (PF) multiplied by Inverted Property Frequency (IPF) at the edge e in the $Edges$ set for a Class c.

(d) $decay* = decayFactor$ where $decayFactor$ is a constant less than 1.0.

(e) Reinitialize Adj to all the nodes that have not yet been visited and are adjacent to the previous nodes in the set.

The difference between the baseline algorithm and Algorithm 1 is that at 3-a. When $Edges$ are updated, the class to which the subject of edge in the Edge set is updated at the same time. By using this set, at 3-c, $scoreEdge$ is updated by following Eq. (3).

$$scoreEdge(r)+ = \sum_{e \in Edges} PFIPF[c,e] \times decay \qquad (3)$$

The reason for changing Bamba's algorithm is because properties which are not often used are also included in LOD data set, in general, we cannot simply tell that the properties with low frequency are regarded as important. Furthermore, as described in the definition of the importance of properties, in order to regard the properties frequently appearing in a particular class as important, we change evaluation so as to focus on the class of resource's subject.

Query Evaluation Based on RDF Triple Unit (Algorithm 2). An RDF triple consisting subject, property and object is a unit of information. In the Algorithm 2, we evaluate an RDF triple as a whole as a unit of information, unlike evaluate a resource and a property individually. As for the baseline algorithm and the Algorithm 1, although resources and properties are individually evaluated, as for Algorithm 2, we regard an RDF triple as the unit of evaluation and then introduce a new score to evaluate it (see, Eq. (4)).

$$TripleScore = \frac{Imp(C_s) \times PFIPF(n_s, edge) \times Imp(n_o)}{linkNum(n_s) + linkNum(n_o) - 1} \qquad (4)$$

n_s and n_o indicate subject node and object node of an RDF triple, respectively. C_s represents the class to which subject s belongs. linkNum(n) indicates the number of edges from node n. The reason why the score is divided by the number of links is to take account of the possibility that the same resource is evaluated multiple times. In the case that n_o is not a resource but a literal, the score of a literal always becomes 0 in the use of the PageRank algorithm, because of this, TripleScore also comes to 0. To avoid this, the mean value of the all resource score is provided to each literal as its weight.

Algorithm 2:
1. Let $decay = 1.0$, $score = 0.0$
2. Let Adj be a set initially consisting of the triples content subject or object nodes of interest to the user as determined from SELECT clause of the query.
3. while Adj is not Empty
   (a) $LetClassedEdges$ be the set of edges from the nodes in Adj and the classes they belong to, i.e., (c, e).
   (b) $score(r) += \sum_{t \in Adj} TripleScore[t] * decay$
   (c) $decay* = decayFactor$ where $decayFactor$ is a constant less than 1.0.
   (d) Reinitialize Adj to all the triples that have not yet been visited and are adjacent to the previous triples in the set.

---

4 Experiments

We have set two types of SPARQL queries for the two purposes mentioned above, and rank the query results with the baseline algorithm, Algorithm 1, and Algorithm 2. However, in Bamba and Mukherjea [14], they have introduced their own original algorithm to evaluate resources, referring to the idea of HITS algorithm, besides, they have not shown the detail about parameter settings in the paper [14]. On the other hand, we decide the importance of properties with the PageRank algorithm. So, it is difficult to completely follow their experimental settings. Therefore, in our study we employ the PageRank algorithm in the baseline algorithm to decide the importance of resources and properties as well as Algorithm 1 and 2.

We use the same parameter settings in all algorithms, i.e., $decayFacotor = 0.5$, $nodeWeight = 0.5$, and $edgeWeight = 0.5$.

As the experimental environment, we downloaded whole data set of DBpedia 3.8, built a database with Virtuoso[8], and prepared an endpoint environment in a local machine. In Table 1, we show the specification of the experimental environment.

Table 1. Experimental environment and the data used in the experiment

Items	Values
CPU	Intel Core i7-3770K
Memory	32GB
OS	Ubuntu 12.10
data set	DBpedia 3.8
Total number of resources	9440897
Total number of RDF triples	158373972

---

[8] http://virtuoso.openlinksw.com/

In the experiments, in the case of retrieving information from resources, we use "Kyoto" and "University of Tokyo" as resources, besides, in the case of retrieving resources from RDF triples, we provide the following conditions: "Resources belong to the class of University" and "Resources belong to the class of City". Table 2 shows actual queries used in the experiments.

Queries 1 and 2 are adopted for the cases where RDF triples are retrieved by a resource, on the other hand, queries 3 and 4 are adopted for the cases where resources are retrieved by RDF triples. In the case that an RDF triple is retrieved by a resource, we use ':Kyoto' and ':University_of_Tokyo' as resources. In the case that a resource is retrieved by an RDF triple, we provide resources which belong to 'dbo:University' and resources which belong to 'dbo:City', respectively as each retrieval condition.

Table 2. Queries used in the experiments

ID	Content
Query 1	SELECT ?property ?object WHERE{ <http://dbpedia.org/resource/Kyoto> ?property ?object.}
Query 2	SELECT ?property ?object WHERE{ <http://dbpedia.org/resource/University_of_Tokyo> ?property ?object.}
Query 3	SELECT ?subject WHERE{ ?subject rdf:type[a] <http://dbpedia.org/ontology/City>. ?subject ?property ?object.}
Query 4	SELECT ?subject WHERE{ ?subject rdf:type <http://dbpedia.org/ontology/University>. ?subject ?property ?object.}

[a]'rdf:' is the prefix for http://www.w3.org/1999/02/22-rdf-syntax-ns.rdf

4.1   Subject Experiment

We have conducted subject experiments to evaluate the results of ranking by the three algorithms, i.e., Bamba's algorithm regarded as the baseline algorithm and the proposed algorithms, i.e., Algorithm 1 and 2. Twelve subjects evaluated the results of each algorithm with five grades. The average of evaluation scores of the cases where triples are retrieved and resources are retrieved are shown in Tables 3 and 4, respectively. Moreover, as an example of the results, the ranking results for "Kyoto" of each algorithm are shown in Tables 5, 6 and 7.

As for the retrieval of RDF triples, in the case that "Kyoto" is set as the retrieval target, Algorithm 1 provides highest score, whereas in the case that "University of Tokyo" is set as the retrieval target, Algorithm 2 provides the highest score. As for Algorithm 1, the evaluation for two queries is almost the same. On the other hand, as for Algorithm 2, the evaluation of the query for "Kyoto" is low. So, it is said that evaluation differs by each query.

Table 3. Average of evaluation: triple retrieval

Resource	Baseline	Algorithm 1	Algorithm 2
Kyoto (query 1)	2.38	**3.38**	2.79
University of Tokyo (query 2)	2.58	3.58	**3.63**

Table 4. Average of evaluation: resource retrieval

Class	Baseline	Algorithm 1	Algorithm 2
City (query 3)	2.63	2.88	**3.83**
University (query 4)	2.42	2.54	**3.79**

As for resource retrieval, Algorithm 2 works well. The evaluation of both baseline algorithm and Algorithm 2 is almost the same for each query, but the score of Algorithm 1 is a bit higher than the baseline algorithm.

4.2 Discussions

As for the subject experiments, the result with Algorithm 1 exceeds the that with the baseline algorithm. In particular, as for ranking RDF triples by query 1 and 2, we observe higher score than the other methods (see, Table 3). It can therefore be said that the change of properties' weight by Algorithm 1 provides good influence on ranking the results retrieved by SPARQL queries. As for resource retrieval by query 3 and 4, Algorithm 2 provides higher score than the other algorithms (see, Table 4). As evaluation of the retrieval results, in Algorithm 2, the scoring based on TripleScore has been introduced. We think that the scoring works well for these queries. However, as for retrieving RDF triples with query 1 and 2, the score provided by Algorithm 2 is either almost the same as that by Algorithm 1 or lower (see, Table 3). The top 10 ranking results retrieved by query 1, whose score is lower than the Algorithm 1, is shown in Tables 5, 6, and 7. As for the evaluation of the retrieval with query 1, it can be thought that the experiment subjects evaluated which RDF triple is strongly related to "Kyoto", which is the subject of the RDF triple. As for the ranking with Algorithm 1 shown in Table 5 and the ranking with Algorithm 2 shown in Table 7

Table 5. Ranking of the information about "Kyoto"(Bamba's algorithm)

rank	?property	?object	Score
1	http://dbpedia.org/ontology/wikiPageWikiLink	http://dbpedia.org/resource/United_State	12.4
2	http://dbpedia.org/ontology/wikiPageWikiLink	http://dbpedia.org/resource/France	3.92
3	http://dbpedia.org/property/stateParty	http://dbpedia.org/resource/Japan	3.52
4	http://dbpedia.org/ontology/wikiPageWikiLink	http://dbpedia.org/resource/Germany	3.00
5	http://dbpedia.org/ontology/country	http://dbpedia.org/resource/Japan	3.00
6	http://dbpedia.org/property/subdivisionName	http://dbpedia.org/resource/Japan	2.96
7	http://dbpedia.org/property/subdivisionType	http://dbpedia.org/resource/List_of_sovereign_states	2.66
8	http://dbpedia.org/ontology/wikiPageWikiLink	http://dbpedia.org/resource/World_War_II	2.33
9	http://dbpedia.org/property/coordinatesRegion	JP@en	2.33
10	http://dbpedia.org/property/populationDensityKm	auto@en	2.33

Table 6. Ranking of the information about "Kyoto"(Algorithm 1)

rank	?property	?object	Score
1	http://dbpedia.org/ontology/wikiPageWikiLink	http://dbpedia.org/resource/United_States	12.6
2	http://dbpedia.org/property/subdivisionName	http://dbpedia.org/resource/Japan	5.39
3	http://dbpedia.org/property/subdivisionType	http://dbpedia.org/resource/List_of_sovereign_states	5.33
4	http://dbpedia.org/ontology/isPartOf	http://dbpedia.org/resource/Kansai_region	4.23
5	http://dbpedia.org/ontology/country	http://dbpedia.org/resource/Japan	4.21
6	http://dbpedia.org/ontology/isPartOf	http://dbpedia.org/resource/Kyoto_Prefecture	4.21
7	http://dbpedia.org/ontology/isPartOf	http://dbpedia.org/resource/Kansai	4.19
8	http://dbpedia.org/ontology/wikiPageWikiLink	http://dbpedia.org/resource/France	4.16
9	http://dbpedia.org/ontology/timeZone	http://dbpedia.org/resource/Japan_Standard_Time	4.15
10	http://dbpedia.org/ontology/wikiPageWikiLink	http://dbpedia.org/resource/Germany	3.76

Table 7. Ranking of the information about "Kyoto"(Algorithm 2)

rank	?property	?object	Score
1	http://dbpedia.org/property/subdivisionName	http://dbpedia.org/resource/Japan	19.0
2	http://dbpedia.org/property/subdivisionType	http://dbpedia.org/resource/List_of_sovereign_states	18.8
3	http://dbpedia.org/ontology/country	http://dbpedia.org/resource/Japan	12.1
4	http://dbpedia.org/ontology/wikiPageWikiLink	http://dbpedia.org/resource/United_States	9.36
5	http://dbpedia.org/ontology/wikiPageWikiLink	http://dbpedia.org/resource/France	3.02
6	http://dbpedia.org/ontology/wikiPageWikiLink	http://dbpedia.org/resource/Germany	2.72
7	http://dbpedia.org/ontology/wikiPageWikiLink	http://dbpedia.org/resource/World_War_II	2.07
8	http://dbpedia.org/ontology/wikiPageWikiLink	http://dbpedia.org/resource/Japan	1.76
9	http://dbpedia.org/ontology/wikiPageWikiLink	http://dbpedia.org/resource/List_of_sovereign_states	1.75
10	http://dbpedia.org/ontology/wikiPageWikiLink	http://dbpedia.org/resource/Italy	1.71

which the experiment subjects rated lower, the information about foreign countries such as United States and France is ranked higher than the information about Japan to which Kyoto should be more related. However, in the case of the ranking with Algorithm 2, the information about Japan is included in the top 3 ranking. In the ranking with Algorithm 1, which the experiment subjects marked high score, shown in Table 6, much information about the area of Kyoto such as ':Kansai-region' is ranked higher. As for the ranking with Algorithm 2, TripleScore is introduced to evaluate an RDF triple itself, therefore, the weight of each element of an RDF triple influences each other rather than the case of using Algorithm 1. Therefore, the result of being strongly influenced by the importance of properties or objects reflects ranking, so it can be thought that we could not obtain stable results in ranking RDF triples.

5   Conclusions

We have proposed two ranking algorithms for the results retrieved by SPARQL queries. As the importance of properties, by introducing the idea of PFIPF based on the class of the subject of an RDF triple, we have become able to select important properties which are different in each class. It is easily said that the number of Linked Open Data will increase and methods to retrieve the data will be more important. The methods to rank the results retrieved by SPARQL queries we have proposed in this paper are also an approach to structured data retrieval. The method to rank resources using the link relation in data sets is a useful method in terms of both simple and costless.

As future work, as for evaluating resources, we will improve the unstable results in the experiments by adjusting scores, introducing another information, e.g., semantic information, etc. Furthermore, we aim to apply our method to other LOD data sets and improve it so as it will be more flexible method for any RDF database.

References

1. Bizer, C., Heath, T., Idehen, K., Berners-Lee, T.: Linked data on the web. In: Proceedings of the 11th IEEE International Symposium on Multimedia (ISM2009), Washington, DC, USA (2009)
2. Bizer, C., Lehmann, J., Kobilarov, G., Auer, S., Becler, C., Cyganiak, R.: DBpedia - a crystallization point for the web of data. Web Semant. **7**, 154–165 (2009)
3. Oren, E., Delbu, R., Catasta, M., Cyganiak, R., Stenzhorn, H., Tummarello, G.: Sindice. com: a document-oriented lookup index for open linked data. Int. J. Metadata Semant. Ontol. **3**(1), 37–52 (2008)
4. Ding, L., Pan, R., Finin, T.W., Joshi, A., Peng, Y., Kolari, P.: Finding and ranking knowledge on the semantic web. In: Gil, Y., Motta, E., Benjamins, V.R., Musen, M.A. (eds.) ISWC 2005. LNCS, vol. 3729, pp. 156–170. Springer, Heidelberg (2005)
5. Page, L., Brin, S., Motwani, R., Winograd, T.: The pagerank citation ranking: bringing order to the web. Technical report, Stanford University (1998)
6. Kleinberg, J.M.: Authoritative sources in a hyperlinked environment. J. ACM **46**(5), 604–632 (1999)
7. Hogan, A. Harth, A., Decker, S.: Reconrank: a scalable ranking method for semantic web data with context. In: Second International Workshop on Scalable Semantic Web Knowledge Base Systems (2006)
8. Balmin, A., Hristidis, V., Papakonstantinou, Y.: Objectrank: authority-based keyword search in databases. In: Proceedings of the Thirtieth International Conference on Very Large Data Bases (VLDB), Toronto, Canada, vol. 30, pp. 564–575 (2004)
9. Chirita, P.-A., Costache, S., Nejdl, W., Paiu, R.: Beagle[++]: semantically enhanced searching and ranking on the desktop. In: Sure, Y., Domingue, J. (eds.) ESWC 2006. LNCS, vol. 4011, pp. 348–362. Springer, Heidelberg (2006)
10. Anyanwu, K., Maduko, A., Sheth, A.: Semrank: ranking complex relationship search results of the semantic web. In: WWW'05, pp. 117–127 (2005)
11. Franz, T., Schultz, A., Sizov, S., Staab, S.: Triplerank: ranking semantic web data by tensor decomposition. In: Bernstein, A., Karger, D.R., Heath, T., Feigenbaum, L., Maynard, D., Motta, E., Thirunarayan, K. (eds.) ISWC 2009. LNCS, vol. 5823, pp. 213–228. Springer, Heidelberg (2009)
12. Meymandpour, R., Davis, J.G.: Ranking Universities Using Linked Open Data. In: Proceedings of the WWW2013 Workshop on Linked Data on the Web (LDOW2013), Rio de Janeiro, Brazil, CEUR Workshop Proceedings, vol. 996–09 (2013)
13. Mirizzi, R., Ragone, A., Di Noia, T., Di Sciascio, E.: Ranking the linked data: the case of DBpedia. In: Benatallah, B., Casati, F., Kappel, G., Rossi, G. (eds.) ICWE 2010. LNCS, vol. 6189, pp. 337–354. Springer, Heidelberg (2010)
14. Bamba, B., Mukherjea, S.: Utilizing resource importance for ranking semantic web query results. In: Bussler, C., Tannen, V., Fundulaki, I. (eds.) SWDB 2004. LNCS, vol. 3372, pp. 185–198. Springer, Heidelberg (2005)
15. Mulay, K., Kumar, P.S.: SPRING: ranking the results of SPARQL queries on Linked Data. In: Proceedings of the 17th International Conference on Management of Data (COMAD), Bangalore, India (2011)
16. Salton, G., McGill, M.: Introduction to Modern Information Retrieval. McGrawHill Book Company, New York (1984)

Personalized Search System
Based on User Profile

Yanhua Cai, Yiyeon Yoon, and Wooju Kim[✉]

Department of Information Industrial Engineering,
University of Yonsei, Seoul, Republic of Korea
yeonhcai@gmail.com, pryiyeon@nate.com,
wkim@yonsei.ac.kr

Abstract. With the development of Web technologies and the improvement of information technology standards, Internet has entered an age of information explosion. However, extraneous information is displayed on the top of the search results and the user interest in the search results in a text match without or seldom taking into account the search intents of the users. For most of the search engines, they either cannot become aware of the user interest properly or cannot find the information which users need efficiently. In our study, we solve these problems. We store users' search history in the user profile, and relocate the results of search history by the particular subject. The proposed method can provide a personalized search service that imparts higher priority to the user documentation saw, which is positioned at the top of the search results. On the basis of the proposed method, we developed a system with which the corresponding experiment has been performed to verify our proposed method. The experiment result shows the validity of our proposed method and the importance of personalized search.

Keywords: Personalized search · User profile · Data mining

1 Introduction

As the rapid growth of Internet, the search results from uncountable amount of web documents are enormous and relatively useless if there's no any corresponding optimization mechanism Hence, people have focused on how to provided a quick search to users, e.g., through the accurate information that fits the users' requirements.

Consequently, in order to increase the accuracy of the search, expanded classification of information has been introduced and a variety of search engines has been developed that focus on the keyword, such as through the provision of tag information to improve the satisfaction of search service various methods for attempts have been made [7]. However, the keyword-based search system [8], has been proposed to with the purpose to provide all the information that matches with the user query. However, because it is regardless of the search intent of the user, search results does not meet the user's intention. In some cases, because the keywords in the system are not present, it occurs that there is no desired result can be found.

W. Kim et al. (Eds.): JIST 2013, LNCS 8388, pp. 320–328, 2014.
DOI: 10.1007/978-3-319-06826-8_24, © Springer International Publishing Switzerland 2014

In particular, the search service of the digital library, the ability to provide accurate information appropriate to the field of topics related to users who are looking for professional materials by accessing the electronic library is insufficient, due to this, it takes a lots time to look for data. For example, if you take a lot of time to look for material that is not enough accurate information tailored to function and related topics of the Standing Committee to assist team stand with legislators must initiative a bill many are.

As a result, in this paper, after taking advantage of the search log of the parliament digital library, we construct a profile of the Standing Committee in a digital library. It can be re-arranged to apply a profile the results of a search by subject for particular words it is intended to that by, so that the viewed as priority publications related to the corresponding Standing Committee of the user.

2   Related Work

The personalized search based on query expansion were researched in [9, 10] two papers. But paper focused on the research about query expansion. They expand input queries by using guide words extracted from user profiles. However, the search engines still cannot find the desired information in case document does not contain the extended query.

Search logs using digital libraries can be used to build user profiles. User profiles retrieved document vector and vector cosine similarity measure's method has been proposed in [3]. To reorder the retrieved documents through which the proposed methodology, but the calculation consumption of the vector-built user profile similarity will be over-load, if the value is too large.

3   Proposed Personalized Search Service

3.1   Overview of Personalized Search Service

The flowchart of the proposed method is as follows (Fig. 1).

A Standing Committee is connected to a digital library system. Committee enters a keyword to data searching. Search results obtained from the current system of the search engine are re-ranked by applying the profiles of the Standing Committee. Finally, the re-ranked results will be sent to the Commission.

3.2   Building User Profile

In step 1, the associated log data is collected on a standing basis. Data that member of committee has read document from January 2011 to December. The collection the titles, keywords and contents with respect to ISBN are extracted from the reading of the original information (Fig. 2).

In step 2, by using the data obtained in the first step, e.g., the titles, keywords and etc. with respect to ISBN, morphological analysis is performed.

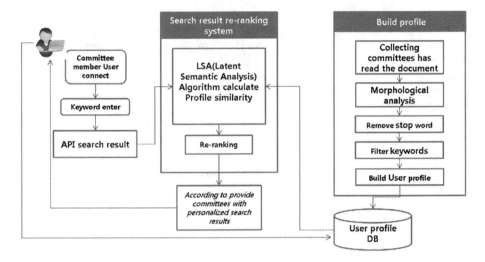

Fig. 1. Methodology flowchart presented

Fig. 2. User profile contract flowchart

In step 3, the treatment should stop words. Do not use the word as a search term, articles, prepositions, research, and conjunctions such a large role in the sentence does not indicate what the words are pre-stored in the table stop words. Stop words, stemming the words in the table if the process is to delete the term.

In step 4 filter-related keywords are calculation af-icf based on user. Af-icf is word weight by committee. Calculation of af-icf ideas from the tf-idf, tf-idf is weight that word in the document. Detailed calculation formula is as follows (Fig. 4).

$$af-icf_{ij} = \log af \times \left(\log \frac{P}{cf} + 1 \right)$$

- *af*: average term frequency by user *i*.
- *cf*: the number of users that have the appearance of a particular term *j*.
- *P*: the total number of users.

– Given a constant af-icf Threshold is selected by user-related index terms.
– User-specific values af-icf convergence point (knee) can be found through the L-method [11].
– The knee with a value greater than words to build a user profile with a small value terms, stop words are treated as a user-by-user basis (Fig. 3).

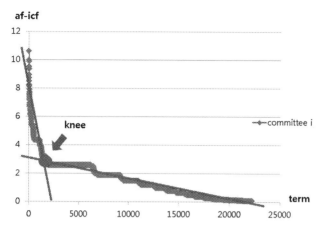

Fig. 3. Using L-Method filter term

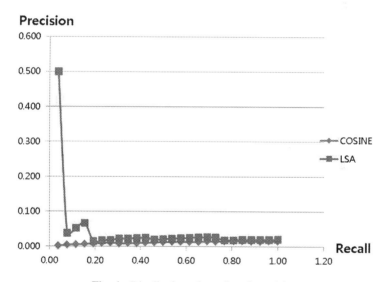

Fig. 4. Distribution of recall and precision

In step 5, term of af-icf is greater than a specific value (knee) to build user profiles, the Commission's profiles are made

3.3   Search Result Re-ranking System

- **Latent Semantic Analysis.**
  - *LSA uses the Singular Value Decomposition (SVD) to analyze the statistical relationships among words in a collection of text.*

- **A matrix X represents the relation of documents and words. The row of X means documents, and the column means words.**

	word 1	word 2	word 3	word 4	word 5	word 6
Document 1	1	0	1	0	0	0
Document 1	0	1	0	0	0	0
Document 2	1	1	0	0	0	0
Document 3	1	0	0	1	1	0
Document 5	0	0	0	1	0	1

- **The next step is to apply SVD to X, to decompose X into a product of three matrices ULA^T.**

	dim1	dim2	dim3	dim4	dim5
Document 1	-0.44	-0.3	0.57	0.56	0.25
Document 2	-0.13	-0.33	-0.59	0	0.73
Document 3	-0.48	-0.51	-0.37	0	-0.61
Document 4	-0.7	0.35	0.15	-0.58	0.16
Document 5	-0.26	0.65	-0.41	0.58	-0.09

X

$$\begin{pmatrix} 2.16 & 0 & 0 & 0 & 0 \\ 0 & 1.59 & 0 & 0 & 0 \\ 0 & 0 & 1.26 & 0 & 0 \\ 0 & 0 & 0 & 1 & 0 \\ 0 & 0 & 0 & 0 & 0.39 \end{pmatrix}$$

X

	Word 1	Word 2	Word 3	Word 4	Word 5	Word 6
dim1	-0.75	-0.28	-0.2	-0.45	-0.33	-0.12
dim2	-0.29	-0.53	-0.19	0.63	0.22	0.41
dim3	0.28	-0.75	0.45	-0.2	0.12	-0.33
dim4	0	0	0.58	0	-0.58	0.58
dim5	-0.53	0.29	0.63	0.19	0.41	-0.22

- **If X is of rank r, then L is also of rank r. Let L_k, where $k < r$, be the matrix produced by removing from L the $r - k$ columns and row, and let U_k and A_k be the matrices produced by removing the corresponding columns from U and A.**

	dim1	dim2
Document 1	-0.44	-0.3
Document 2	-0.13	-0.33
Document 3	-0.48	-0.51
Document 4	-0.7	0.35
Document 5	-0.26	0.65

X

$$\begin{pmatrix} 2.16 & 0 \\ 0 & 1.59 \end{pmatrix}$$

X

	Word 1	Word 2	Word 3	Word 4	Word 5	Word 6
dim1	-0.75	-0.28	-0.2	-0.45	-0.33	-0.12
dim2	-0.29	-0.53	-0.19	0.63	0.22	0.41

- **LSA works by measuring the similarity of documents using this $U_k \cdot L_k A_k^T$, instead of original matrix.** The similarity of two documents is measured by the cosine of the angle between their corresponding row vectors.

Cosine similarity.

$$\cos \theta = \frac{A \times B}{|A| \times |B|} = \frac{\sum_{i=1}^{n} A_i \times B_i}{\sqrt{\sum_{i=1}^{n} (A_i)^2} \times \sqrt{\sum_{i=1}^{n} (B_i)^2}}$$

	dim1	dim2
profile	-0.95	-0.48
Document 1	-0.28	-0.52
Document 2	-1.04	-0.81
Document 3	-1.51	0.56
Document 4	-0.56	1.03

	profile
Document 1	0.82
Document 2	0.98
Document 3	0.68
Document 4	0.03

The LSA algorithm, the Committee of the retrieved documents and the actual induction profiles are compared. The Committee of the document similarity with a higher profile and a high ranking in the search results are placed in the low similarity

Table 1. Current digital library rankings, recall and precision of the data

No	Query	Compare Data	Digital library rank	Recall	Precision
1	parliamentary report	PAMP1000030784	447	0.04	0.0022
2	parliamentary report	PAMP1000030668	453	0.08	0.0044
3	parliamentary report	PAMP1000028549	472	0.12	0.0064
4	parliamentary report	PAMP1000030697	568	0.15	0.0070
5	parliamentary report	PAMP1000030630	574	0.19	0.0087
6	parliamentary report	PAMP1000025608	599	0.23	0.0100
7	parliamentary report	PAMP1000025604	600	0.27	0.0117
8	parliamentary report	PAMP1000023149	926	0.31	0.0086
9	parliamentary report	PAMP1000030621	986	0.35	0.0091
10	parliamentary report	PAMP1000027995	1000	0.38	0.0100
11	parliamentary report	PAMP1000027960	1011	0.42	0.0109
12	parliamentary report	PAMP1000028476	1016	0.46	0.0118
13	parliamentary report	PAMP1000028568	1023	0.50	0.0127
14	parliamentary report	PAMP1000028591	1036	0.54	0.0135
15	parliamentary report	PAMP1000028489	1054	0.58	0.0142
16	parliamentary report	PAMP1000030785	1215	0.62	0.0132
17	parliamentary report	PAMP1000030776	1219	0.65	0.0139
18	parliamentary report	PAMP1000027954	1245	0.69	0.0145
19	parliamentary report	PAMP1000028528	1280	0.73	0.0148
20	parliamentary report	PAMP1000028526	1281	0.77	0.0156
21	parliamentary report	PAMP1000025684	1287	0.81	0.0163
22	parliamentary report	PAMP1000025583	1318	0.85	0.0167
23	parliamentary report	PAMP1000030797	1587	0.88	0.0145
24	parliamentary report	PAMP1000025605	1594	0.92	0.0151
25	parliamentary report	PAMP1000030771	1600	0.96	0.0156
26	parliamentary report	PAMP1000027986	1604	1.00	0.0162

Table 2. Standing Committee applying the digital library profile position, recall and precision of the data

No	Keyword	Compare Data	Committee profile rank	Recall	Precision
1	parliamentary report	PAMP1000030771	2	0.04	0.500
2	parliamentary report	PAMP1000030797	52	0.08	0.038
3	parliamentary report	PAMP1000025605	57	0.12	0.053
4	parliamentary report	PAMP1000027986	60	0.15	0.067
5	parliamentary report	PAMP1000025583	323	0.19	0.015
6	parliamentary report	PAMP1000025684	354	0.23	0.017
7	parliamentary report	PAMP1000028526	360	0.27	0.019
8	parliamentary report	PAMP1000028528	361	0.31	0.022
9	parliamentary report	PAMP1000027954	396	0.35	0.023
10	parliamentary report	PAMP1000030776	422	0.38	0.024
11	parliamentary report	PAMP1000030785	426	0.42	0.026
12	parliamentary report	PAMP1000028489	587	0.46	0.020
13	parliamentary report	PAMP1000028591	605	0.50	0.021
14	parliamentary report	PAMP1000028568	618	0.54	0.023
15	parliamentary report	PAMP1000028476	625	0.58	0.024
16	parliamentary report	PAMP1000027960	630	0.62	0.025
17	parliamentary report	PAMP1000027995	641	0.65	0.027
18	parliamentary report	PAMP1000030621	655	0.69	0.027
19	parliamentary report	PAMP1000023149	715	0.73	0.027
20	parliamentary report	PAMP1000025604	1098	0.77	0.018
21	parliamentary report	PAMP1000025608	1099	0.81	0.019
22	parliamentary report	PAMP1000030630	1124	0.85	0.020
23	parliamentary report	PAMP1000030697	1130	0.88	0.020
24	parliamentary report	PAMP1000028549	1226	0.92	0.020
25	parliamentary report	PAMP1000030668	1244	0.96	0.020
26	parliamentary report	PAMP1000030784	1250	1.00	0.021

results are placed in the lower ranks. After reordering the search results, the user can get a new search result. By the experiment in next section, it confirms the verification feasibility and superiority of the proposed method.

4   Experimental Evaluation

In this section, the evaluation of our proposed method is performed. All experimental data are from the searching result when the user is Land Infrastructure and Transport Committee member with the keyword "parliamentary report".

- Evaluation

  - 11 Point average precision

$$P_{11-pt} = \frac{1}{11} \sum_{j=0}^{10} \tilde{P}(r_j)$$

Define 11 standard recall points $r_j = \frac{j}{10} : r_0 = 0,\ r_0 = 0.1, \ldots, r_{10} = 1$.
$\tilde{P}(r_j)$: i.e. the precision at our recall points.

As shown in Fig. 4, which is based on the data in Tables 1 and 2, for both recall and precision, the performance of our proposed method is better than that of the related method.

5   Conclusion

In this paper, a personalized search service is proposed that using the digital library of user profile search results and the reorder of the priorities of the relevant data to the user.

The electronic library Parliament Standing profile is constructed by analyzing the search log. A profile similarity analysis system has been developed and an experiment has been performed whose results demonstrate the feasibility of the proposed method. Specifically, the standing experiments applying the profile priority ranking is more suitable than the existing digital library users in terms of displaying the search results.

One of the most important contributions of our proposed personalized search services is the consideration of users intents and the requirements of most important information. Our future work is to collect the Standing Committee to improve the retrieve of the log by establishing a keyword system. Besides, the research of data dimension reduction in LSA algorithms, and the value of dimension k of data comparing experimental research, will be continued.

References

1. Landauer, T.K., Dumais, S.T.: A solution to Plato's problem: the latent semantic analysis theory of acquisition, induction, and representation of knowledge. Psychol. Rev. **104**, 211–240 (1997)
2. Turney, P.: Mining the web for synonyms: PMI-IR versus LSA on TOEFL. In: Proceedings of the Twelfth European Conference on Machine Learning (2001)
3. Sin J.Y.: Digital library using search log data, personalized search service, one study. Master thesis, Industrial Management, Graduate School of Engineering, Yonsei University (2012)
4. Kim, J.-H.: Life term to search for legal information and legal terminology correspondence between the search methodology. Master thesis, Information Industrial Engineering, Yonsei University (2011)
5. Jeong S.T.: Life term semantic-based legal information retrieval methodology study. Master's thesis, Industrial Management, School of Business, Yonsei University (2011)
6. Shin, D.-H.: Latent semantic analysis using a content-based information retrieval system. MS thesis, Seoul National University (1999)
7. IT as a promising future strategy excavation report item, Institute for Information Technology (2006) 12

8. Kim, J.-T., Kim, Y.S., former Wu.: Network based u-Health services promoted trends. Institute for Information and Communication Technology Trends 1321 Conference call informationweek
9. Liu, F., Yu, C., Meng, W.: Personalized web search for improving retrieval effectiveness. IEEE Trans. Knowl. Data Eng. **16**(1), 28–40 (2004)
10. Shen, D., Sun, J., Yang, Q., Chen, Z.: Building bridges for web query classification. In: Proceeding of the 29th Annual International ACM SIGIR Conference on Research and Development in Information Retrieval (SIGIR'06), pp. 131–138 (2006)
11. Salvador, S., Chan, P.: Determining the number of clusters/segments in hierarchical clustering/segmentation algorithms. In: Proceedings of the 16th IEEE International Conference on Tools with Artificial Intelligence (2004)

Ontology Mapping

Utilizing Weighted Ontology Mappings on Federated SPARQL Querying

Takahisa Fujino$^{(\boxtimes)}$ and Naoki Fukuta$^{(\boxtimes)}$

Graduate School of Informatics, Shizuoka University, Hamamatsu, Japan
{gs12033@s,fukuta@cs}.inf.shizuoka.ac.jp

Abstract. Technologies to allow the heterogeneity of ontologies on the web of data have a key role on the emergence of Linked Open Data. Much research exists on generating better ontology mappings, and also they produce 'weights' (i.e. confidence) of each generated mapping. Although the semantics of such weighted ontology mappings has been discussed, how they can effectively be used in querying remains as a crucial issue. In this paper we show how such weighted ontology mappings can effectively be used in SPARQL-based querying on heterogeneous data sources by slightly extending the syntax of SPARQL query language. We show how such an extended query can be translated to a standard SPARQL query and how the extended query can customize the behavior of processing the query by reflecting the user's demands.

Keywords: SPARQL · Weighted ontology mapping · Heterogeneous ontologies

1 Introduction

There are many SPARQL endpoints (e.g., DBpedia [4]) that allow us to retrieve and see the data by queries that are written in SPARQL query language. We can retrieve data not only from one endpoint but also from several endpoints using federated queries [11,12,15]. However, it is not always easy to understand the ontology used at an endpoint well enough to write a query based on it. The main reason could be that, often an ontology is constructed with some specific terms that were not used in others. Furthermore, it often lacks detailed documentations.

The use of ontology mappings, alignments, and matching techniques [7,8,14] could be useful for writing queries when it allows us to use any ontologies in the queries unless those ontologies are not directly used in the endpoint [9]. Bizer et al. [5] proposed a framework to find appropriate mappings to exchange data. Rivero et al. [16] proposed an approach to convert the mappings to directly be executable on SPARQL queries. Makris et al. [13] proposed a framework for utilizing such ontology mappings by applying query rewriting techniques. However, due to the limited information about the concepts to be aligned (i.e.,

---

Takahisa Fujino – currently with Denso Corporation.

W. Kim et al. (Eds.): JIST 2013, LNCS 8388, pp. 331–347, 2014.
DOI: 10.1007/978-3-319-06826-8_25, © Springer International Publishing Switzerland 2014

lack of sufficient background knowledge) or the existence of significant semantic gaps among the ontologies that came from their background conceptual models themselves, defining clear and crisp mappings between two ontologies is often difficult without a loss of semantics.

Ontology mappings often include some weights to represent the confidence or the expected semantic similarity for each generated mapping relation, and it is called weighted ontology mappings [3]. When we could utilize these confidence values in mappings to make an effective query to the SPARQL endpoints, we could also enjoy querying to any SPARQL endpoints without deeply understanding their used ontologies. Even when we know the used ontologies for the target endpoints, explicitly using such confidence values (i.e., 'weights') in mappings in a SPARQL query allows us to eliminate the chance to retrieve and transfer unwanted data and hide them from retrieved data on different endpoints in federated querying.

When we would like to utilize an ontology mapping in a query, and also when the mapping allows to map a concept to multiple concepts, we can prepare a query that gathers wider results (i.e., it obtains higher recalls) by looking up to all possible candidates based on the mapping. However, it will also produce numerous unwanted results in executing the query, and sometimes the query cannot be completed within a feasible time. To avoid spending much time on executing queries at an endpoint, an engine that can decompose queries into simple subqueries for faster execution on each endpoint was proposed [1]. Schwarte et al. [17] proposed an improvement method in query performance over federated query engines. However, even when we could use such techniques to increase the query processing speed, there is a difficulty to extract expected results from numerous unwanted results. Furthermore, to effectively limit the obtained results and their search spaces, such federated querying with unweighted multi-mappings cannot be easily processed since there is no hint in the standard SPARQL queries to know which part is more important than the others while executing them.

In this paper, we show how weighted ontology mappings can effectively be used in SPARQL-based querying on heterogeneous data sources by slightly extending the syntax of SPARQL query language. We show how such an extended query can be translated to a standard SPARQL query and how the extended query can customize the behavior of processing the query by reflecting the user's demands. Also we present a system that can translate the extended queries into standard SPARQL queries, run them, and generate corresponding code fragments to be used to develop applications that have access to the web of data.

2 Utilizing Heterogeneous Weighted Ontology Mappings

2.1 Weighted Ontology Mappings

Ontology mapping/alignment/matching, which is one of the most active areas of ontology research, is realized by many methods [2]. For example, Duan et al. [7] proposed a clustering-based approach to effectively reuse alignments provided by experts. Ontology Alignment Evaluation Initiative (OAEI)[1] proposed the

---

[1] http://oaei.ontologymatching.org/

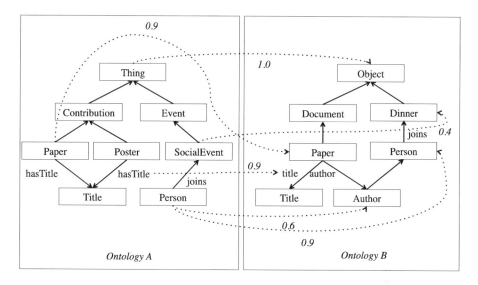

Fig. 1. Example of weighted ontology mappings between ontologies A and B

Alignment API[2] for better interoperability among ontology alignment mechanisms and applications that use such alignment mechanisms.

Most automatic ontology matching tools generate mappings with weights, to show the confidence of the mapping relation or its expected semantic similarity for each mapping relation, and such mappings are called *weighted ontology mappings*. Atencia et al. [3] presented a formal semantics for weighted ontology mappings to interpret the confidence value associated with a mapping.

Figure 1 shows an example of weighted ontology mappings and the basic model of mappings used in this paper. We assume that the classes or properties in a source ontology (e.g., ontology A) can be mapped to several classes or properties in the target ontology (e.g., ontology B). The confidence values, typically represented as real values among 0 and 1, can be included in each mapping relation definition, and a specific format can be defined to represent it (e.g., the alignment format[3] used in the Alignment API). In this paper, we assume that a mapping has such weights to represent its reliability degrees.

2.2 Heterogeneous Ontology Mappings and Reliability Issues

Ontology alignment methods often utilize different kind of measures to calculate matching degrees among entities. Euzenat et al. [8] classified ontology matching techniques, e.g., strings-based (terminological) approaches, data instance-based (extensional) approaches, and others with different characteristics. Several ontology matching systems use mixed measures [2] to improve their alignment quality.

There are some ontology alignment methods to produce precise and crisp mappings among their entities. However, in some cases, producing a precise

---

[2] http://alignapi.gforge.inria.fr/
[3] http://alignapi.gforge.inria.fr/format.html

mapping is very difficult [7,14], especially when the two ontologies have semantic gaps. Alignment methods generally require much time to produce an alignment when the two ontologies are huge. Hence, an efficient way to get an alignment has also been proposed (e.g., [18]).

To utilize these ontology mappings that have a variety of confidence values, it is important to explicitly control their effective use in a query, since without such control a query can produce so much unwanted data in the process of its execution so that they cannot be processed. However, with SPARQL, almost no method can explicitly control how to process queries with such 'somewhat unreliable' weighted ontology mappings. In this paper, we show that the SPAR-QLoid, a small extension to SPARQL that allows us to utilize such weighted ontology mapping effectively, can effectively handle such situations.

Our initial motivation was to realize a mechanism that utilizes such hetero-geneous ontology mappings that were produced using different methods with different characteristics. This means that several alignments may be prepared by different methods, and each alignment has several possible mappings among entities with their matching confidence values.

Ontology alignment methods are based on several measures, which are only effective and applicable in certain situations [19]. For example, a string-based matching technique is generally applicable only when the languages are the same or very close. An instance-based matching technique is applicable only when identical instances are shared between two ontologies. However, the instance data cannot always be recognized as identical when the URIs are different. When a matching method is based on mixed measures, it is sometimes difficult to choose the best parameters to produce the results. Therefore, we argue that ontology alignment methods must be selected depending on the situation, i.e., the time of writing the query itself.

SPARQLoid enables users to control their utilization of different ontology mappings as well as confidence values in a query. They can choose ontology mappings to get the data in the target endpoint and specify the criterion for sorting the results and cutting output data that are far from accurate.

3   Syntax and Translation Algorithm

Syntax. Listing 1.1 shows the BNF-style definition of the syntax used in a SPARQLoid query. Here, we show a typical SPARQLoid query in listing 1.2. Notice that the SPARQLoid query in listing 1.2 obtains the matched instances as ?a, ?b, and ?c using the vocabularies in the conference namespace. How-ever, each target endpoint to be accessed by a URI (*targetEndpoint* or *federated-QueryEndpoint* in listing 1.2) is expected to be constructed using each different ontologies. For example, the class *conference:Regular_author* and the property *conference:has_the_first_name* require appropriate mappings to run the queries in listing 1.2 on the endpoints. In listing 1.2, *conference:Regular_author* was specified as more important than *conference:has_the_first_name* by the ranking clause. In case of a federated SPARQL querying, the weights of mappings are merged in the specified ways. In listing 1.2, some threshold values are specified to cut the irrelevant results that come from lower-weight mappings. For example, the mappings for *conference:Regular_author* whose confidence values are higher than 0.3 are used to execute the query on the target endpoint. In listing 1.3, we used another threshold syntax. Here, the query specified that the top-5 or top-7 mappings for *conference:Conference_contribution* are used to execute the query.

```
SPARQLoidQuery::=Prologue SelectQuery
SelectQuery::=SelectClause DatasetClause* WhereClause
 SolutionModifier
WhereClause::='WHERE'? GroupGraphPattern AdditionalClause
AdditionalClause::=RankingClause* ThresholdClause*
RankingClause::='RANKING''<'IRI'>''{'RankingExpression'}'
RankingExpression::=RankingTerm|RankingExpression'+'
 RankingTerm|RankingExpression'-'RankingTerm
RankingTerm::=IRI|RankingTerm'*'RankingFact|
 RankingTerm'/'RankingFact
RankingFact::=(RankingExpression)|DOUBLE|FLOAT|INTEGER
ThresholdClause::='THRESHOLD''<'IRI'>''{'ThresholdExpression'}'
ThresholdExpression::=(ValueExpression|CountExpression)
ValueExpression::=(IRI'='FLOAT)*
CountExpression::=(IRI':'INTEGER)*
```

Listing 1.1. A part of SPARQLoid syntax

```
SELECT ?a ?b ?c
WHERE
{  ?a rdf:type conference:Regular_author .
   ?a conference:has_the_last_name ?b .
   { SERVICE < federatedQueryEndpoint >
      { ?a conference:has_the_first_name ?c . }
   }
   RANKING < targetEndpoint >
    {conference:Regular_author*0.5+conference:has_the_last_name*0.2}
   THRESHOLD < targetEndpoint >
    {conference:Regular_author=0.3,conference:has_the_last_name=0.3}
   RANKING < federatedQueryEndpoint >
    {conference:has_the_first_name*0.3}
   THRESHOLD < federatedQueryEndpoint >
    {conference:has_the_first_name=0.2}
}
```

Listing 1.2. Example of federated SPARQLoid query

```
SELECT ?a
WHERE
{  { ?a rdf:type conference:Conference_contribution . }
   UNION
   { SERVICE < federatedQueryEndpoint >
     { ?a rdf:type conference:Conference_contribution . }
   }
   RANKING < targetEndpoint >
    { conference:Conference_contribution*0.5 }
   THRESHOLD < targetEndpoint >
    { conference:Conference_contribution:5 }
   RANKING < federatedQueryEndpoint >
    { conference:Conference_contribution*0.5 }
   THRESHOLD < federatedQueryEndpoint >
    { conference:Conference_contribution:7 }
}
```

Listing 1.3. Another example of federated SPARQLoid query

Query Translation Algorithm. The query translator engine translates a SPARQLoid query into a standard SPARQL query that can directly be executed on the target endpoint. The following is the query translation algorithm[4]:

Algorithm 1. Translate a SPARQLoid query into a standard executable query

---

Input: Q : SPARQLoid query, M : specified onto1, onto2, target endpoint, alignment endpoint, and method name, J : Set of M
Output: Q' : translated query
  if not translationRequired(Q) **then**
    return Q
  else
    $QueryComponents$;
    for each M in J **do**
      $ReferringMappingOperations \Leftarrow$ createCondition(M);
      $sourceEntities \Leftarrow$ extractSourceEntities(Q, M);
      $thresholdClause \Leftarrow$ extractThreshold(Q, M);
      $rankingClause \Leftarrow$ extractRanking(Q, M);
      for each $sourceEntity$ in $sourceEntities$ **do**
        if $thresholdClause.isValueType$ **then**
          $value \Leftarrow thresholdClause.\text{getValue}(sourceEntity)$;
        else if $thresholdClause.isCountType$ **then**
          $count \Leftarrow thresholdClause.\text{getCount}(sourceEntity)$;
          $value \Leftarrow$ findValidValue($sourceEntity, count, M$);
        end if
        $ReferringMappingOperations \Leftarrow$ createReferringOp($sourceEntity, value$);
      end for
      $QueryComponents \Leftarrow ReferringMappingOperations$;
    end for
    $QueryComponents \Leftarrow$ createMainbodyOp(Q);
    $QueryComponents \Leftarrow$ createRankingOp($rankingClause$);
    $Q' \Leftarrow$ createSPARQL($QueryComponents$);
    return Q'
  end if

A translated query contains three major processing parts: *referringMapping*, *mainbody*, and *ranking*. In this algorithm, each major part is recursively extracted from the source query and appropriate query code-fragments are inserted into the translated query. The extraction and code-fragment generation are based on a pattern-based code generation technique. An example of a translated query is shown in Fig. 2, which is transformed from the query shown in listing 1.2. In the process of translating the *referringMapping*, the URIs for the specified source

---

[4] A preliminary idea of this query translation mechanism was previously presented in [9]. However, the previously presented approach only considers the use of two ontologies with mappings to access a single data source but no consideration to access two or more data sources by federated querying. Furthermore, in this paper, we extended the syntax and the algorithm to cover federated querying situations and also we added evaluation about effectiveness of controlling queries utilizing weighted ontology mappings in querying.

ontology, the target ontologies, and the endpoints of the alignment providers are used to make the translated query accessible to the obtained weighted mappings. Here, the used alignment algorithms can be specified as the names of methods that implement specific alignment algorithms. In the process, appropriate code fragments are generated to obtain mappings for each entity to process the *referringMapping* in the query. In the *mainbody*, the translated query retrieves the data in the target endpoint using the mappings that were obtained in the code fragments in *referringMapping*. Finally, in the generated code-fragments for the *ranking*, the reliability degrees for the retrieved data are calculated based on the control of the ranking clause in the query. Based on the combinations of these operations, a source SPARQLoid query is successfully translated into a standard SPARQL query that can be executed on standard SPARQL endpoints.

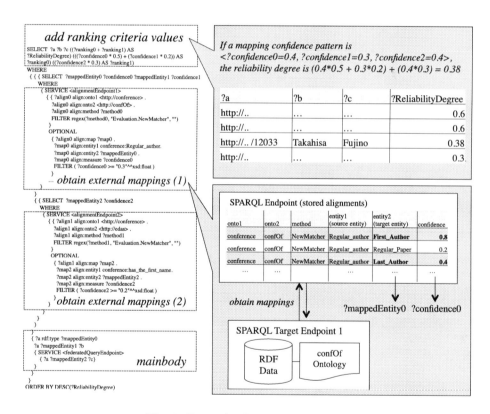

Fig. 2. Example of a translated query

4   Implementation

We implemented the SPARQLoid[5] engine that allows users to use the SPARQLoid syntax queries by providing functions to translate the corresponding standard

---

[5] A preliminary idea about SPARQLoid has been demonstrated in [10].

SPARQL queries, execute them, and also produce the corresponding code-fragments in several programming languages that can be embed into applications to effectively control query processing with weighted ontology mappings.

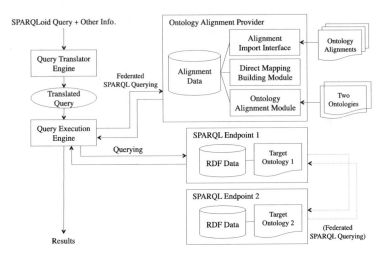

Fig. 3. Processing model of SPARQLoid

To shorten the generated query strings and to avoid storing a large amount of mapping data in user applications, the system can generate a query that refers to an external endpoint that stores necessary ontology mappings. When the mapping relations to be used are compact enough, the SPARQLoid engine can also embed the mapping data into the translated query to avoid frequent access to external endpoints that provide mappings.

Figure 3 shows the core processing model of SPARQLoid queries. We assume that potential users are developers who want to build an application that uses SPARQL queries to retrieve semantic web data from heterogeneous resources. Users prepare a special SPARQL query that uses special primitives; the query can also be based on the user's (i.e., well-understood) ontology. The specified extended form of the SPARQL query is translated by the query translator engine, which uses heterogeneous ontology mapping data that can be produced by some ontology alignment methods or some specific pre-generated mappings stored in SPARQLoid engine or other sufficient endpoints. Next the translator engine produces a translated query that can be executed against more than two endpoints if the user tries to obtain data against the endpoints. SPARQLoid engine users can see the obtained data or use the translated query in their applications. The translated query works with SPARQL 1.1[6].

Figure 4 shows a running example of SPARQLoid engine. In step 1, users input the basic information for query translation: target endpoints, target ontologies, source ontology, alignment endpoints, and alignment methods. In step 2,

---

[6] http://www.w3.org/TR/sparql11-query/

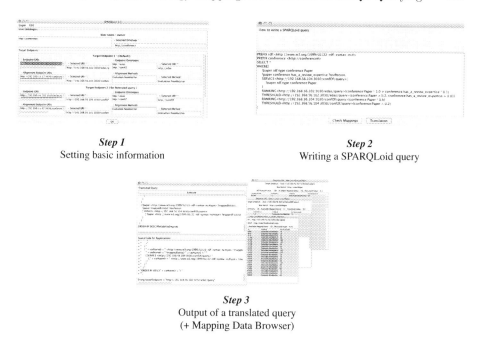

Step 1
Setting basic information

Step 2
Writing a SPARQLoid query

Step 3
Output of a translated query
(+ Mapping Data Browser)

Fig. 4. Running example of SPARQLoid

users write a SPARQLoid query. In step 3, they can see the translated query and get programming-friendly code in Java. If they want to change the threshold values for each entity, they can confirm the weighted ontology mapping information in this step. After performing these steps, they can see the results obtained by a translated query or embed the code in their applications.

5   Evaluation

To validate the effectiveness of using weighted ontology mappings in federated SPARQL queries, we conduct experiments on datasets that have weighted ontology mappings.

5.1   Datasets and Queries

There are no standard testing datasets and queries to evaluate queries with weighted ontology mappings. To calculate precision, recall, and F-measure which are often used on information retrieval benchmarking, the correct answers (i.e., which should be obtained from the prepared queries) must be defined beforehand. In this paper, we use datasets from a conference track[7] on OAEI, whose domain is well understandable. In the experiment, we used one of the ontologies in the

---

[7] http://oaei.ontologymatching.org/2012/conference/index.html

dataset as the ontology for querying. Other ontologies are used as the ontologies used in the target endpoints. Since the exact mappings in some OAEI datasets can be defined[8], the alignments we used for the two ontologies were prepared precisely. In our experiment, we assume that the results obtained from the golden master SPARQL queries, which have already resolved the mappings based on the correct mappings prepared in the experiment, are correct answers. With these answers, we calculate the precision, recall, and traditional F-measure of the results obtained by the translated queries with weighted ontology mappings.

Although the content stored in the T-Box (i.e., classes, properties, etc.) could be prepared from the datasets used in the ontology matching tracks in OAEI, the content in the A-Box, which is often retrieved by SPARQL queries, is not specially included in most ontologies used in the datasets. Therefore, we randomly generated ten instances for each class and added them based on the property's domain and range in a target ontology.

```
SELECT ?a
WHERE
  { ?a rdf:type confOf:Contribution
    { SERVICE <FederatedEndpointURI>
      { ?a rdf:type edas:RejectedPaper}
    }
  }
```

```
SELECT ?a ?b
WHERE
  { ?a rdf:type confOf:Conference
    OPTIONAL
      { SERVICE <FederatedEndpointURI >
        { ?a edas:hasFirstName ?b}
      }
  }
```

```
SELECT ?a
WHERE
  { { ?a rdf:type confOf:Contribution }
    UNION
    { SERVICE <FederatedEndpointURI >
      { ?a rdf:type edas:Document}
    }
  }
```

(a) Join   (b) Optional   (c) Union

Fig. 5. Three types of examined federated SPARQL queries: Join, Optional, Union (These queries are examples of the golden master queries whose mappings have been correctly resolved. We generated the experimented queries based on different ontologies that are directly used in the golden master queries. Also the experimented queries contain extended SPARQLoid notations to specify how those mapping-problem should be resolved.)

A target-executable federated SPARQL query over a target endpoint was translated into a SPARQLoid query based on precise ontology alignment. Some threshold values in SPARQLoid queries were varied in the evaluation process. We prepared three simple federated query patterns: a SERVICE graph pattern in a series of joins (we call these queries 'join queries') that has ten sub-patterns, a SERVICE graph pattern in a series of optional values (optional queries) that has ten sub-patterns, and a SERVICE graph pattern in a series of matching alternatives (union queries) that has three sub-patterns. We prepared a query generator that generates queries for the experiments based on the above three patterns. Example queries based on the three query patterns are shown in Fig. 5. Sub-patterns are generated by slightly changing them (e.g., *rdf:type* to an object property in the union queries).

---

[8] It is because they are generated among the identical ontologies with different names, etc.

5.2   Preparation of Datasets and Mappings

The OAEI datasets used in our evaluation are Sofsem (conference) as a source ontology in the queries and ConfTool (confOf) and Edas (edas) ontologies as target ontologies prepared in the conference tracks. ConfTool has 38 classes, 13 object properties, and 23 datatype properties. Sofsem has 60 classes, 46 object properties, and 18 datatype properties. Edas has 104 classes, 30 object properties, and 20 datatype properties. We added ten instances for each class and additional links among the instances in the three ontologies. There are 11,041 triples in the confOf endpoint and 9,388 triples in the edas endpoint in our evaluation.

To confirm how our approach can be applied to process queries on heterogeneous ontology conditions even when generated mappings contain inaccurate mappings, we used for our evaluation a simple ontology matcher that was based on *levenshtein distance*, which is an example of string metrics prepared in Alignment API. In the metrics, *conference:Paper* corresponds to *confOf:Paper* with weight 1.0, and *conference:Paper* corresponds to *confOf:Poster* with weight 0.5. The ontology matcher produces many-to-many mappings with some weights. In our evaluation, all source classes, object properties, and datatype properties are mapped to all target classes, object properties, and datatype properties, respectively. In the prepared reference alignment on the OAEI datasets, there are 15 correct mappings in conference-confOf and 17 correct mappings in conference-edas.

5.3   Experimental Settings

We performed our experiment on a computer running Mac OS X 10.8.3 on Intel Core i7 CPU at 2.3 GHz and 16 GB RAM. The edas endpoint was built in Fuseki 0.2.5[9] on a virtual machine running Ubuntu 12.04 on 1536 MB RAM and the confOf endpoint was built in Fuseki 0.2.5 on a virtual machine running Ubuntu 10.04 on 3027 MB RAM. The conference-confOf alignment endpoint was built in Fuseki 0.2.5 on a virtual machine running Ubuntu 10.04 on 888 MB RAM and the conference-edas alignment endpoint was built in Fuseki 0.2.5 on a virtual machine running Ubuntu 10.04 on 888 MB RAM. SPARQLoid was built on Mac OS X by using Jena 2.10.1[10] and ARQ 2.10.1 in Jena. The ontology matcher in our evaluation was implemented with Alignment API 4.4.

5.4   Results

In the conference-confOf*edas problem, 407 executable queries for the target endpoint were generated. It included 30 join queries, 216 optional queries, and 161 union queries. In the conference-edas*confOf problem, 438 executable queries for the target endpoint were generated. It included 40 join queries, 237 optional

---

[9] http://jena.apache.org/documentation/serving_data/

[10] http://jena.apache.org/

Table 1. Avg. F-values on different mapping lookup thresholds (edas*confOf) (Since a small amount of queries have failed in their executions, the value has been slightly varied in Tables 1 and 2).

	Join F-value(precision/recall)	Optional F-value(precision/recall)	Union F-value(precision/recall)	Overall F-value(precision/recall)
Uniform 1	0.64(0.63/0.71)	0.54(0.53/0.74)	0.65(0.62/0.88)	0.59(0.57/0.79)
Uniform 2	0.54(0.48/0.75)	0.48(0.42/0.74)	0.59(0.51/0.9)	0.52(0.46/0.8)
Uniform 3	0.53(0.45/0.76)	0.41(0.34/0.8)	0.56(0.45/0.95)	0.48(0.39/0.85)
Uniform 4	0.56(0.47/0.83)	0.37(0.31/0.82)	*0.49(0.39/0.93)	0.43(0.35/0.86)
Uniform 5	0.47(0.39/0.85)	*0.33(0.27/0.81)	0.48(0.38/0.95)	0.4(0.32/0.87)
Unibest	0.78(0.74/0.88)	0.67(0.63/0.84)	0.82(0.76/0.95)	0.74(0.69/0.89)
Optimum	0.82(0.79/0.88)	0.76(0.74/0.85)	0.87(0.83/0.95)	0.81(0.78/0.89)

Table 2. Avg. F-values on different mapping lookup thresholds (confOf*edas)

	Join F-value(precision/recall)	Optional F-value(precision/recall)	Union F-value(precision/recall)	Overall F-value(precision/recall)
Uniform 1	0.55(0.53/0.66)	0.68(0.67/0.85)	0.62(0.58/0.88)	0.65(0.62/0.85)
Uniform 2	0.58(0.53/0.73)	0.53(0.46/0.89)	0.57(0.49/0.9)	0.55(0.48/0.88)
Uniform 3	0.61(0.55/0.83)	0.49(0.41/0.9)	0.55(0.46/0.92)	0.52(0.44/0.9)
Uniform 4	*0.56(0.48/0.82)	0.44(0.35/0.91)	0.51(0.42/0.94)	0.48(0.39/0.91)
Uniform 5	0.54(0.46/0.89)	0.42(0.35/0.91)	0.49(0.39/0.94)	0.45(0.37/0.92)
Unibest	0.81(0.77/0.89)	0.82(0.8/0.91)	0.79(0.73/0.95)	0.81(0.77/0.92)
Optimum	0.88(0.86/0.92)	0.87(0.85/0.91)	0.88(0.83/0.95)	0.87(0.85/0.93)

queries, and 161 union queries. We conducted based on these SPARQLoid queries, each of which has two thresholds, to specify how many mappings will be looked up for each corresponding entity (i.e., class or property) in the range from 1 to 5. Those thresholds are specified as follows:

$$THRESHOLD{<}target1{>}\{conference{:}Conference\_contribution{:}3\}$$
$$THRESHOLD{<}target2{>}\{conference{:}Rejected\_contribution{:}4\}$$

In this case, we denote it as $< 3, 4 >$. Totally we conducted the experiment with 10,175 (=407*5*5) SPARQLoid queries in confOf*edas problem and 10,950 (=438*5*5) SPARQLoid queries in edas*confOf problem.

Tables 1 and 2 show the average F-values of the returned results with uniform lookup depth for each mapping candidates and average F-values on optimal threshold values for each queries. These values were averaged for each query types (i.e., Join, Optional, and Union). Overall values indicate the average F-values of all SPARQLoid queries that meet the conditions. "Uniform 1" means that their thresholds values are always set to $< 1, 1 >$. These uniform values show the baseline performance when we look up mapping candidates uniformly in all query. In "Unibest", we calculated average of the best F-values for each query when the two threshold values are equal, i.e., one of $< 1, 1 >< 2, 2 >< 3, 3 ><$ $4, 4 >< 5, 5 >$. These unibest values show the effectiveness of the approach that the number of utilizing mappings allowed to be adjusted for each query. In the "Optimum", we calculated average of the best F-values in all parameter patterns (i.e., 25 patterns). These optimum values show the effectiveness of our approach that the number of utilizing mappings can be adjusted for each entity in a query. We can see that specifying appropriate thresholds for each entities that have mappings in each query produced higher F-values in both problems.

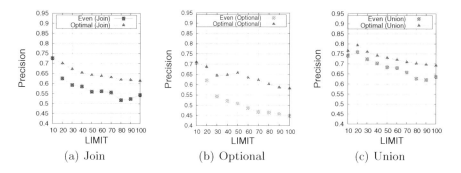

Fig. 6. Performance on different weighs and limits (edas*confOf)

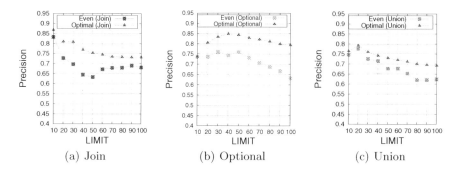

Fig. 7. Performance on different weighs and limits (confOf*edas)

From these results, we can see that there are cases that we should set the optimal threshold values for each query content even in the same query type. From Table 2, we obtained relatively higher F-values when threshold values are '3' on the join queries and threshold values are '1' on the optional and union queries. The join queries needed relatively a large number of mapping candidates to be looked up since these join queries could produce meaningful results only when the all mapped entities are meaningful (i.e., the results are produced only when the correct set of mapping combinations are used, otherwise they just return no result). If we can put appropriate threshold values for each query, we can get higher F-values. Here, for example, the best F-value with uniform lookup depth for each mapping candidate is 0.65, while the best F-value with appropriate threshold values for each query is 0.81 in confOf*edas problem. Furthermore, we obtained higher F-values when we put optimal threshold values for each entity in each query, and in that case, the value is 0.87.

To evaluate effectiveness of the use of ranking clause that we also proposed, we conducted another experiment which is analyzed with ranking parameters in the range from 0.1 to 0.9 (and from 0.9 to 0.1 to another) for each query. Also the LIMIT clause is used whose value is in the range from 10 to 100 for each query. Totally we conducted 39,420 (=438*9*10) SPARQLoid queries in edas*confOf

Table 3. Comparison of features

	SPARQLoid	Mosto[16]	SPARQL-RW[13]	R2R[5]
Ability to generate SPARQL-compatible queries	✓		✓	
Use of weighted ontology mappings	✓			✓
Control syntax for weighted ontology mappings	✓			
Ability to translate data		✓		✓
Ability to integrate data		✓		✓
Ability to generate mappings	*	✓		
Support for complex mappings		✓	✓	✓
Ability to publish mappings	**			✓
Support for interlinking mappings				✓
Application embeddability	✓		✓	✓

✓ implemented, * partially implemented, ** under development

problem and 36,630 (=407*9*10) in confOf*edas problem. We conducted based on these SPARQLoid queries, each of which has two ranking parameters to be set, to specify the weights for each entity. Here, we used the threshold parameter $< 5, 5 >$. An example of query which has ranking parameter $<< 0.2, 0.8 >>$ can be described as follows:

$$RANKING<target1>\{conference:Conference\_contribution*0.2\}$$
$$RANKING<target2>\{conference:Rejected\_contribution*0.8\}$$

Figures 6 and 7 show the average precision for each top k results where the k was specified by LIMIT clause for the two experimental datasets that we used above. These values were averaged for each query types (i.e., Join, Optional, and Union). For each chart, the horizontal axis indicates the specified LIMIT values and the vertical axis indicates the average precision value. "Even" means the results when we set the even weights for the used mappings in ranking the results, i.e., $<< 0.5, 0.5 >>$. These "Even" values show the baseline performance that we are not aware of importance of each entity in ranking the query results. "Optimal" shows averaged precision values when optimal ranking parameters are set for each queries. We obtained higher precision values when the ranking parameters are set optimally.

6   Related Work

In this section, we show a qualitative comparison among our proposed SPAR-QLoid and other related approaches. Table 3 shows a summary of the comparison. Mosto [16] is a tool to perform the data translation using automatically generated SPARQL executable mappings. The approach in Mosto is different from our approach in the point that, they realized a translator that can translate data to be accessed via mappings in a SPARQL query that can run directly on a SPARQL endpoint, while our approach may refer external mappings in the translated query. SPARQL-RW [13] is a framework which provides transparent query access over mapped RDF datasets. The approach in SPARQL-RW is query rewriting, with respect to a set of predefined mappings between ontology schemas. Our system can utilize not only predefined mappings but also weighted ontology mappings while SPARQL-RW does not consider weighted ontology mappings. The R2R framework [5] is a framework to automatically discover and publish mappings among Linked Data sources for better integration of them. The approach in the R2R framework is presenting heuristic mapping generation for specified sources, rather than the utilization of the generated mappings with weights. Our SPARQLoid focuses on aggressively utilizing weighted ontology mappings in federated querying.

7   Conclusion and Future Work

In this paper, we presented an approach to utilize weighted ontology mappings on federated SPARQL-based querying and evaluated the effectiveness of the approach. To overcome issues in the use of heterogeneous ontology mappings and weighted ontology mappings, our proposed approach enables users to control detailed processing in queries to obtain preferable results. In our evaluation, we confirmed that the extended primitives can effectively control the processing of queries when there are weighted ontology mappings. By utilizing weighted ontology mappings in our system, we could get higher F-value in the obtained results without deeply understanding target ontologies. Also, we could get higher precision in top k results by enabling a ranking mechanism based on the mapping-weights of the obtained entities, as well as specifying priority for each entity as the ranking criteria.

Our future work includes utilizing complex correspondences or other ontology mapping relations. In this paper we assume there are mappings which define property and concept equivalence. Other works (e.g., EDOAL [6]) provide functionality for representing complex correspondences (i.e. more structured correspondences) between the entities of different ontologies. However, in the current situation, most of automatic ontology matchers do not return complex correspondences. Our future work includes the utilization of complex correspondences as the mappings to be used. Although we utilize property and concept equivalent relation (i.e. $\equiv$), a formal semantics for weighted ontology mappings [3] also includes subsumption ($\sqsubseteq$ and $\sqsupseteq$), equivalence ($\equiv$) and disjointness ($\perp$) relations. In our current approach, such relations are not fully utilized. Although there is a way to transform two ontology mapping relation (e.g., $\sqsubseteq$ and $\sqsupseteq$) into one equivalent-relation mapping [3], there is no standard way to apply such approaches to weighted ontology mappings. To effectively handle such situations in our approach is our future work.

References

1. Acosta, M., Vidal, M.-E., Lampo, T., Castillo, J., Ruckhaus, E.: ANAPSID: an adaptive query processing engine for SPARQL endpoints. In: Aroyo, L., Welty, C., Alani, H., Taylor, J., Bernstein, A., Kagal, L., Noy, N., Blomqvist, E. (eds.) ISWC 2011, Part I. LNCS, vol. 7031, pp. 18–34. Springer, Heidelberg (2011)
2. Aguirre, J.L., Eckert, K., Euzenat, J., Ferrara, A., van Hage, W.R., Hollink, L., Meilicke, C., Nikolov, A., Ritze, D., Scharffe, F., Shvaiko, P., Šváb Zamazal, O., Trojahn, C., Jiménez-Ruiz, E., Grau, B.C., Zapilko, B.: Results of the ontology alignment evaluation initiative 2012. In: Proceedings of the 7th Ontology Matching, Workshop (OM2012) (2012)
3. Atencia, M., Borgida, A., Euzenat, J., Ghidini, C., Serafini, L.: A formal semantics for weighted ontology mappings. In: Bernstein, A., et al. (eds.) ISWC 2012, Part I. LNCS, vol. 7649, pp. 17–33. Springer, Heidelberg (2012)
4. Auer, S., Bizer, C., Kobilarov, G., Lehmann, J., Cyganiak, R., Ives, Z.: DBpedia: a nucleus for a web of open data. In: Aberer, K., et al. (eds.) ISWC/ASWC 2007. LNCS, vol. 4825, pp. 722–735. Springer, Heidelberg (2007)
5. Bizer, C., Schultz, A.: The R2R framework: publishing and discovering mappings on the web. In: Proceedings of the 1st International Workshop on Consuming Linked Data (COLD2010) (2010)
6. David, J., Euzenat, J., Scharffe, F., Trojahn, C.: The alignment API 4.0. Semant. Web **2**(1), 3–10 (2011)
7. Duan, S., Fokoue, A., Srinivas, K., Byrne, B.: A clustering-based approach to ontology alignment. In: Aroyo, L., Welty, C., Alani, H., Taylor, J., Bernstein, A., Kagal, L., Noy, N., Blomqvist, E. (eds.) ISWC 2011, Part I. LNCS, vol. 7031, pp. 146–161. Springer, Heidelberg (2011)
8. Euzenat, J., Shvaiko, P.: Ontology Matching. Springer, Berlin (2007)
9. Fujino, T., Fukuta, N.: A SPARQL query rewriting approach on heterogeneous ontologies with mapping reliability. In: Proceedings of the 3rd IIAI International Conference on e-Services and Knowledge Management (IIAI-ESKM2012), pp. 230–235 (2012)
10. Fujino, T., Fukuta, N.: SPARQLoid - a querying system using own ontology and ontology mappings with reliability. In: Proceedings of the 11th International Semantic Web Conference (Posters & Demos) (ISWC2012) (2012)
11. Ladwig, G., Tran, T.: Linked data query processing strategies. In: Patel-Schneider, P.F., Pan, Y., Hitzler, P., Mika, P., Zhang, L., Pan, J.Z., Horrocks, I., Glimm, B. (eds.) ISWC 2010, Part I. LNCS, vol. 6496, pp. 453–469. Springer, Heidelberg (2010)
12. Ladwig, G., Tran, T.: SIHJoin: querying remote and local linked data. In: Antoniou, G., Grobelnik, M., Simperl, E., Parsia, B., Plexousakis, D., De Leenheer, P., Pan, J. (eds.) ESWC 2011, Part I. LNCS, vol. 6643, pp. 139–153. Springer, Heidelberg (2011)

13. Makris, K., Bikakis, N., Gioldasis, N., Christodoulakis, S.: SPARQL-RW: transparent query access over mapped RDF data sources. In: Proceedings of the15th International Conference on Extending Database Technology (EDBT2012), pp. 610–613 (2012)
14. Noy, N.F.: Ontology mapping. In: Steffen, S., Rudi, S. (eds.) Handbook on Ontologies, pp. 573–590. Springer, Berlin (2009)
15. Quilitz, B., Leser, U.: Querying distributed RDF data sources with SPARQL. In: Bechhofer, S., Hauswirth, M., Hoffmann, J., Koubarakis, M. (eds.) ESWC 2008. LNCS, vol. 5021, pp. 524–538. Springer, Heidelberg (2008)
16. Rivero, C.R., Hernández, I., Ruiz, D., Corchuelo, R.: MostoDE: a tool to exchange data amongst semantic-web ontologies. J. Syst. Softw. **86**(6), 1517–1529 (2013)
17. Schwarte, A., Haase, P., Hose, K., Schenkel, R., Schmidt, M.: FedX: optimization techniques for federated query processing on linked data. In: Aroyo, L., Welty, C., Alani, H., Taylor, J., Bernstein, A., Kagal, L., Noy, N., Blomqvist, E. (eds.) ISWC 2011, Part I. LNCS, vol. 7031, pp. 601–616. Springer, Heidelberg (2011)
18. Seddiqui, M.H., Aono, M.: An efficient and scalable algorithm for segmented alignment of ontologies of arbitrary size. J. Web Semant. **7**(4), 344–356 (2009)
19. Tordai, A., Ghazvinian, A., van Ossenbruggen, J., Musen, M.A., Noy, N.F.: Lost in translation? empirical analysis of mapping compositions for large ontologies. In: Proceedings of the 5th International Workshop on Ontology Matching (OM2010) (2010)

MAPSOM: User Involvement in Ontology Matching

Václav Jirkovský[1,2(✉)] and Ryutaro Ichise[3]

[1] Czech Technical University in Prague, Zikova 1903/4, Prague, Czech Republic
[2] Rockwell Automation Research and Development Center, Pekařská 695/10a,
Prague, Czech Republic
vjirkovsky@ra.rockwell.com
[3] National Institute of Informatics, Tokyo 101-8430, Japan
ichise@nii.ac.jp

Abstract. This paper presents a semi-automatic similarity aggregating system for ontology matching problem. The system consists of two main parts. The first part is aggregation of similarity measures with the help of self-organizing map. The second part incorporates user feedback for refining self-organizing map outcomes. The system calculates different similarity measures (e.g., string-based similarity measure, WordNet-based similarity measure...) to cover different causes of semantic heterogeneity. The next step is similarity aggregation by means of the self-organizing map and the ward clustering. The final step is the active learning phase for results tuning. We implemented this idea as MAPSOM framework. Our experimental results show that MAPSOM framework can be used for problems where the highest precision is needed.

1 Introduction

Nowadays, information is very important in everyday life. It is important in making every informed decision that we are faced. Correct informed decisions are essential for top management in factories as well as for everyday personal life. Therefore we need complete access to available information, which is distributed and heterogeneous. The problem of information sharing can be divided into two steps. The first step consists in identifying information sources with suitable data for given task. This problem is addressed in the areas of information retrieval and filtering [1]. After identifying information sources, access to the data has to be established. This problem of integration heterogeneous and distributed computer systems is known as interoperability problem [2].

Interoperability problems (on a technical and on an information level) dealing with heterogeneous data were already identified and discussed within a domain of the distributed database systems - structural heterogeneity [3] and semantic heterogeneity [4]. Structural heterogeneity lies in the fact that different information systems use various data structures. On the other hand, semantic heterogeneity means variance in the contents of information and intended meaning. It may

W. Kim et al. (Eds.): JIST 2013, LNCS 8388, pp. 348–363, 2014.
DOI: 10.1007/978-3-319-06826-8_26, © Springer International Publishing Switzerland 2014

cause application unusability and dealing with this type of heterogeneity is crucial in every extensive knowledge system. There are three main causations of the semantic heterogeneity [5]:

- *Confounding conflicts* occur when information items seem to have the same meaning, but differ in reality, e.g., owing to different temporal contexts.
- *Scaling conflicts* occur when different reference systems are used to measure a value. Examples are different currencies.
- *Naming conflicts* occur when naming schemes of information differ significantly. A frequent phenomenon is the presence of homonyms and synonyms.

Ontologies came to the forefront because developers are interested in the reusing or sharing knowledge across systems and they were promising solution how to deal with semantic heterogeneity. The popular ontology definition in AI community is expressed as: Ontology is an explicit specification of conceptualization [6]. In other words, ontologies could be used for establishing explicit formal vocabulary to share among applications. However, the utilization of ontologies did not fulfill completely desire to overcome heterogeneity problem and therefore there are needs to find correspondences among ontologies - ontology matching problem.

Hence, there are many different similarity measures based on e.g., natural language processing, structural similarities between ontologies, and similarities among concept names and/or labels for solving ontology matching. However, these similarity measures are not sufficient for matching complicated ontologies. Therefore these basic similarity measures can be considered as building blocks for complex matching solutions, e.g., similarity measure aggregation. The problem of similarity measure aggregation is a complex task and it is difficult to develop a universal matching tool. Requirements are different for a different application domain. An automatic solution suitable for a semantic web application is inadequate for the manufacturing domain. The best precision of a system is needed in manufacturing or medical solutions and therefore manual or semi-automatic approaches are more convenient.

In response to the above challenge, we have developed a semi-automatic similarity aggregating system for ontology matching. The system consists of two main parts. The first part is aggregation of similarity measures with the help of self-organizing map. The second part incorporates user for refining self-organizing map outcomes. This part is built on active learning techniques.

This paper is organized as follows: after the general introduction, we give an overview of different approaches to semantic integration and existing solutions. Next, we describe our solution based on self-organizing map for ontology matching with a user involvement with the help of active learning approach. Then, we present experimental evaluation of our method. Finally, we conclude the paper with summarizing the future visions.

2   Ontology Matching Problem

In this section, we introduce the ontology matching problem [7]. We assume in our use that the ontologies present a concept hierarchy to classify information (instances of objects). The ontology used for our paper can be defined as follows: *The ontology O contains a set of concepts, $C_1, C_2, \ldots, C_n$ that are organized into a hierarchy. Each concept is labeled by string and can contain instances* [8].

The goal of ontology matching is to find the relations between entities expressed in different ontologies [7]. The basic relation between the elements is one-to-one relation, e.g., "Liberty maps to Freedom". There are also more complex types of semantic relation, e.g., "Courses maps to Undergrad-Courses and Grad-Courses". In this paper, we only treat the problem of one-to-one matching.

Most of the ontology matching systems are used in the Semantic Web domain for integration of a lot of large different ontologies, and therefore the systems has to be fully automatic. On the other hand, a system with the highest possible precision and recall is needed for communication among experts from different domains, e.g., in manufacturing or/and in medicine. One of the possible examples for such a problem is described in [9]. The semi-automatic or manual systems have normally better outcomes but they are more time consuming.

The problem of finding relations between entities defined in ontologies can be expressed as a problem of finding the most similar entities. There are many different already implemented similarity measures for finding out similarity of entities. In the following paragraphs, details about similarity measures, similarity measures aggregation, and user involvement in ontology matching are presented.

2.1   Basic Similarity Measures

String-based techniques are methods comparing strings. They are applied to the name, the label or the comments of entities (e.g., the class names or URIs). A *prefix or suffix similarity* can be defined as method which is testing if one string is a prefix or suffix of another. The next very popular similarity is *n-gram*. This method computes the number of n-grams, i.e., sequences of n characters, between them. It is efficient if only some characters are missing. There are many string-based similarity measures and therefore we are not able to describe all of them as well as all of the following kinds of similarity measures. String-based methods are unsuitable if synonyms with different structure are used.

On the other hand, **language-based techniques** rely on using Natural Language Processing methods to help extract meaningful terms. It can be divided into intrinsic methods (i.e., linguistic normalization) and extrinsic. Extrinsic methods use external resources. Example of an extrinsic resource is *WordNet* [10]. WordNet is an electronic lexical database for English, based on the notion of synsets or sets of synonyms. WordNet provides also hypernyms and meronyms. Several similarity measures have been developed for the WordNet - e.g., cosynonymy similarity or Resnik.

There are more characteristics to match than names of ontology entities. **Structure-based** similarity measures can compare the structure of entities that can be found in ontologies. The structure-based comparison can be divided into a comparison of the internal structure of an entity or the comparison of the entity with other entities to which it is related. The internal structure is mainly used database schema matching, while the relational structure is primarily exploited in matching formal ontologies and semantic networks. The example of structure based similarity measure is for example the structural topological dissimilarity on a hierarchy [11].

Extensional techniques can be exploited when individuals are available. The main idea of this approach consists in the fact that if two classes have the exactly the same set of individuals then there is a high probability that these classes represent the same entity. Jaccard similarity can be for example exploited for extensional matching.

Last but not least category of similarity measures is **semantic-based techniques**. These techniques belong to the deductive methods. Unfortunately, pure deductive methods do not perform very well when they are directly utilized for an essentially inductive task like ontology matching. Hence they need a preprocessing phase which provides "pre-alignments" - e.g., entities which are presupposed to be equivalent. The semantic-based techniques can be called verification or amplification of "pre-alignments". Examples of semantic techniques are propositional satisfiability, modal satisfiability techniques, or description logic based techniques.

2.2   Similarity Aggregation

Already mentioned similarity measures are basic measures and they are suitable for different dissimilarity. They can be considered as building blocks of complex solutions. The question is how to use these blocks for the best outcomes. One of the possibilities is to aggregate them. A construction of such a framework is a difficult task. What algorithm is suitable for aggregation of independent measures and how to balance the values between dimensions?

Following paragraphs describe common ways how to aggregate different similarity measures. These ways are related to our implemented solution which is presented in the Sect. 4.

Triangular Norm - Weighted Product. Triangular norms are commonly used as conjunction operators in uncertain calculi. The weighted product is one of triangular norms and could be used for ontology matching. The weighted product between two objects x, x' from set of objects o is as follows:

$$sim(x, x') = \prod_{i=1}^{n} sim_i(x, x')^{w_i}, \tag{1}$$

where $sim_i(x, x')$ is the ith similarity measure of objects and w_i is the corresponding weight. The main drawback of this triangular norm is the fact that if one of the dimensions is 0, then the result of aggregation is 0.

Multidimensional distances are suitable for independent basic similarity measures. Well known multidimensional distance is Minkowski distance:

$$sim(x, x') = \sqrt[p]{\sum_{i=1}^{n} sim_i(x, x')^p},$$
(2)

where $sim_i(x, x')$ is the ith similarity measure of objects x, x'. The distance is equal to the Euclidean distance for $p = 2$, to the Manhattan distance for $p = 1$, and to Chebishev distance for $p = +\infty$.

Weighted sum is the next simple linear similarity aggregation which can be considered as a generalization of the Manhattan distance with weighted dimensions.

$$sim(x, x') = \sum_{i=1}^{n} w_i \times sim_i(x, x'),$$
(3)

where $sim_i(x, x')$ is the ith similarity measure of objects and w_i is the corresponding weight. The weighted sum is called to be normalized, if $\sum_{i=1}^{n} w_i = 1$.

2.3   User Involvement in Ontology Matching

User involvement is one of the requirements in some applications. There are three areas in which users can be involved in a matching solution: (i) by providing initial alignments (and parameters) to the matchers, (ii) by dynamically combining matchers, and (iii) by providing feedback to the matchers in order for them to adapt their results [7].

1. The first area lies in providing system inputs and covers providing the ontologies to be matched, choosing system parameters, and providing initial alignments. These aspects offer an opportunity for users to control the algorithm behavior.
2. Dynamically combination of matchers is used for combining different methods to follow particular needs. Examples of such systems are Rondo [12] or COMA++ [13]. The next possibility is to combine different methods in opportunistic way - systems choose the next method in regard of the input data. This approach is limited implemented for example in Falcon-AO [14].
3. User feedback is usually used for setting improvement of parameters of a matcher, thresholds for filtering, and aggregation parameters. The feedback is normally utilized for the error computation - a distance between the feedback and the result of matching system.

Finally, the history of the prior matching actions is the last possibility, how to improve a matching quality. Taking the existing matches and the user action history into account makes the process of matching more interactive and personalized [15].

3   Related Work

A lot of systems have been developed to deal with the ontology matching problem and many innovative and interesting ideas are proposed. In this section, we would like to introduce some of these systems for ontology matching, especially for similarity aggregation, which are related to our proposed system.

As we already mentioned, automatic ontology matching usually cannot deliver high quality results in traditional applications, e.g., on large datasets [16]. Unfortunately, there are only few studies on how to involve users in ontology matching [17] and most of them are dedicated to design-time matcher interaction [18].

Recently, some frameworks had focus on manual designing, checking, and correcting alignments. The framework [19] involved cognitive studies for a graphical visualization of alignments. An environment [20] for manually designing alignments by usage of connected perspective that allows to quickly deemphasize non relevant aspects of ontologies while keeping the connections between relevant entities.

With the development of interactive approaches the issues of their usability will become very critical. The main characteristics are possibilities of visualization scalability [21] and better user interface usability in general. In the following paragraphs we will describe some frameworks for ontology matching, which are relating with our proposed system, i.e., similarity measure aggregation and utilization of machine learning approach.

n-Harmony [22] is based on harmony measure, which is an adaptive aggregation method that assigns higher weight to more reliable and important similarity measure. The n-Harmony measure conciders top-n values in similarity metrix to get more reliable weight for similarity measures.

X-SOM [23] is an ontology mapping and integration tool. This tool automatically combines several matching techniques by means of a multi-layer perceptron network. The X-SOM provides also a "human-intensive" behavior - uncertain mappings are submitted to the user for the verification.

Cluster-based Similarity Aggregation system [24] is an automatic aggregating system for ontology matching. The system calculates five different basic similarity measures to create five similarity matrixes and then these matrixes are combined into a final similarity matrix. At the end, the pruning process is applied.

Every system for ontology matching is suitable for different application domain. We can divide these systems into two main categories - automatic ontology matching systems and manual ontology matching systems. Fully automatic systems are primarily used for matching very large ontologies - e.g., Semantic web, and manual systems are used for matching ontologies in applications where the highest precision is needed and every fault can cause the loss of money. Of course, the border between these categories is not strictly determined. Our solution is compromise between these two categories and offer variability in level of precision and recall - if more active learning iterations are processed then we obtain higher precision and recall. In addition, user works with group of candidates (similarity pairs) to minimize the user effort.

4   MAPSOM - Ontology Matching Framework

Many researchers have usually effort to develop fully automatic systems for ontology matching. These systems have usually deficiencies in precision and recall caused by impossibility to find similar elements automatically. However, human is still more capable than computer and many applications require the best precision and recall more than quickness, e.g., in manufacturing or in medicine.

In following sections, we describe implemented ontology matching framework based on similarity measure aggregation by means of Self-Organizing Map. The framework combines automatic ontology matching approach as the first step and the second step is comprised of optimizing the first step outcomes.

4.1   Self-Organizing Map-Based Similarity Measures Aggregation

The self-organizing map (SOM) is neural network introduced by Teuvo Kohonen [25]. The SOM implements a characteristic nonlinear projection from the high-dimensional space onto a low-dimensional array of neurons and the mapping has inclination to preserve the topological relationships. Furthermore, the SOM has important applications in the visualization of high-dimensional systems and is possible to discover categories and abstractions from raw data.

The SOM usually consists of a two-dimensional regular grid of cells - neurons (see Fig. 1). Each neuron in the SOM is a d-dimensional weight vector (codebook vector) where d is equal to the dimension of the input vectors. The neurons are connected to adjacent neurons by a neighborhood relation, which determines the topology of the map, i.e. hexagonal or rectangular topology. The mapping is ensured by the SOM algorithm in the following way: assuming a general distance measure between input vector x and codebook m_i denoted $d(x, m_i)$, then the image $c(x)$ (winner) is defined as

$$c(x) = \arg \min_i d(x, m_i) \qquad (4)$$

The learning algorithm in the SOM is called competitive learning. The basic idea of the traditional SOM learning algorithm can be expressed as

$$m_i(t + 1) = m_i(t) + h_{c(x),i}(t)[x - m_i(t)] \qquad (5)$$

where t is the index of the time step, $m_i(t)$ is the weight vector of the neuron i in the time step t, $h_{c(x),i}$ is called the neighborhood function, and x is the input vector.

The SOM is used in our solution for similarity aggregation. Input vectors for SOM training are composed of different similarity measure values between concepts from a source and target ontology. The main benefit - pairs of concepts with the similar features are located in a nearby area of the SOM output layer

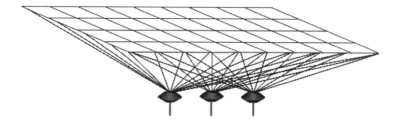

Fig. 1. The SOM with rectangular topology.

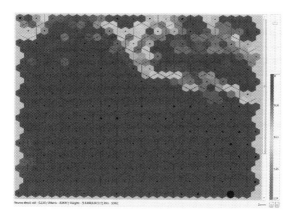

Fig. 2. Example of U-Matrix and hit histogram of hexagonal SOM

after training. Then, neurons from output layer are clustered by ward clustering and classified to be positive or negative. Now, the user can utilize information about neurons from clustering, visualization (see below), and initial classification to prove the classification or can tune the classification by means of active learning.

Visualization is needed for meaningful user interaction. There are three main different visualization possibilities implemented in MAPSOM - U-Matrix, Ward clustering, and hit histogram. These methods are described in following paragraphs.

The U-Matrix method (see Fig. 2 - an example of U-Matrix and hit histogram) [26] visualizes the distances among neurons in a SOM, and thus the U-Matrix is possible to show cluster structure of the SOM. High values indicate a cluster border and areas of low values indicate clusters themselves.

Ward Clustering is implemented for automatic cluster creation. It is possible to operate with whole group of neurons instead of a single neuron. The cluster method of Ward belongs to the hierarchical agglomerative cluster algorithms (i.e., every single neuron is a cluster by itself and the clusters with minimal distance are merged in every step). The distance measure characterizing Ward's

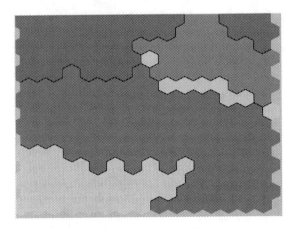

Fig. 3. Ward clustering

method is based on the variance criterion. This distance measure is called the Ward distance and is defined as follows:

$$d_{rs} := \frac{n_r \cdot n_s}{n_r + n_s} \cdot \|\bar{x}_r - \bar{x}_s\|^2, \tag{6}$$

where r and s denote two specific clusters, n_r and n_s denote the number of data points in the two cluster, and $\bar{x}_r$ and $\bar{x}_s$ denote the centers of gravity of the clusters. $\|.\|$ is the Euclidean norm. The number of clusters is variable and user can vary this number. Ward clustering of SOM with hexagonal topology and four clusters is depicted on Fig. 3.

Hit Histogram can be combined with ward clustering, U-Matrix, or with both of them. This implemented visualization method allows see, how many samples correspond to a neuron. U-Matrix and hit histogram combination of SOM with hexagonal topology and 15 neurons in every dimension is depicted on Fig. 2.

Initial Classification. In our solution, an initial classification of clusters is needed. *Boolean conjunctive classifier B* and *linear weighted classifier L* are used for this purpose.

Boolean conjunctive classifier **B** is defined as

$$F_B = \bigwedge_{i=1}^{n} (\sigma_i(s, t) \geq \tau_i), \tag{7}$$

where s and t are a certain pair of ontology entities, σ_i is a similarity function, and τ_i is a threshold of ith similarity function.

Linear weighted classifier **L** (derived from weighted sum - see Eq. 3) has the form

$$F_L = \sum_{i=1}^{n} \omega_i \sigma_i(s,t), \qquad (8)$$

where s and t are a certain pair of ontology entities, σ_i is a similarity function, and ω_i is a weight of ith similarity function.

The cluster classification is not computed for every single neuron but it is computed from center of mass of the cluster. This fact causes the situation that the classification of neurons may vary depending on the number of clusters.

From different point of view, somebody can wonder about advantages and suitability of the SOM utilization for ontology matching problem. Samuel Kaski describe in [26] that SOM is appropriate for data feature exploration. A usage of the SOM involves following advantages - possibility of visualization of high-dimensional data, clustering, and non-linear projection capability. These characteristics are essential for data exploration together with user involvement and data advanced visualization of high dimensional data.

4.2   User Involvement in Ontology Matching

A primary target is to classify pairs from ontologies with the highest accuracy initially. This effort lies in the best setting of classifier parameters. Unfortunately, there are some limitations specially based on domain dependency of ontology. User involvement in the process of semantic integration seems to be one possible approach to address this limitation and allows enhance this process.

Skilled user (domain expert) is able to correctly classify pairs in positive or negative matching assignments. In real applications, it is impossible to let the user control every sample in set of all possible mappings between ontologies. Hence, solution combining minimization of user effort and maximization of accuracy has to be developed. Machine learning, especially active learning, offers possibility how to cope with the problem.

Active Learning. Active learning (query learning) is a part of machine learning and is commonly used in speech recognition, information extraction, classification, and filtering. The main idea of active learning is that a machine learning approach can achieve better outcomes with fewer training labels than in usual systems if it is allowed to choose the data for learning. An active learner may poses queries, usually in the form of unlabeled data instances to be labeled by an oracle (e.g., a human annotator) [27].

For many real-world learning problems, large collections of unlabeled data can be gathered at once. This motivates pool-based sampling [28], where data are selectively drawn from the pool, which is usually non-changing (static). Samples are queried according to their informative contribution.

Active learning methods involve evaluating the informativeness of unlabeled samples, which are sampled from a given distribution or newly generated.

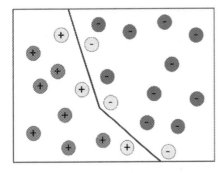

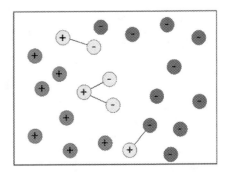

Fig. 4. Candidate selection - the smallest distance from the boundary

Fig. 5. Candidate selection - the smallest distance from different classes

There are many proposed formulations of query strategies. Typically, most commonly used query framework is uncertainty sampling. This framework selects samples with at least certain label assignment. For example, when using a probabilistic model for binary classification, uncertainty sampling simply queries the instances whose posterior probability of being positive is nearest 0.5 [29].

Candidate Selection. Certain number of candidates for user approval is selected in the first step of active learning approach. Selection of these candidates has to fulfill a selection criterion in order to ensure approval of the most uncertain classification. If we don't consider any additional domain knowledge for the selection then it is easy to see that these candidates are the closest samples to the boundary between positive and non-positive alignments.

There are two information sources for utilization in selection criterion for distance description between samples with different classification. The first source is classifier settings and the second source is composed of topological information from output SOM layer. Information from classifier allows finding precise boundary location. Hence, it is needed to find samples with the smallest distance from this boundary (Fig. 4). Main drawback of this approach could reside in necessity of updating classifier settings after every step. In the second case, candidates for user approval can be described as the closest samples which have different classification (Fig. 5). The second approach has its drawback in higher computational complexity. The second approach is preferred in our solution because iterative parameters update is difficult or impossible for some specific types of classifiers. Outcomes of these two selection algorithms may not be the same.

Important parameter of candidate selection is how many samples will be presented to user. Parameter value must be chosen with respect to minimize user effort.

It is good to mention the fact that more than one sample can be assign to a neuron. These candidates have similar value therefore we work with them as one candidate in active learning process to eliminate user involvement. On the other hand, except this batch processing user has also possibility to change assignment for certain sample.

Summarization. Finally, we can summary whole process of ontology matching with MAPSOM framework. First, the self-organizing map is defined and trained. Then, initial classification of neurons (of mapping candidates belonging to the neuron) is processed. After that, a user can explore data and initial classification with the help of SOM visualization (U-Matrix, Ward clustering, and hit histogram) - a user can change the assigned mapping in this step. Finally, a user can apply the active learning classification tuning for achieving better results.

5   Experiments

In this section, we analyze the experimental results to show functionality of the proposed semi-automatic framework. The experiments were conducted using benchmark datasets with the reference matching provided by Ontology Alignment Evaluation Initiative (OAEI) - 2013 Campaign. The benchmark dataset[1] offers a set of tests which are wide in feature coverage, progressive and stable. It serves the purpose of evaluating the strength and weakness of matchers and measuring the progress of matchers. The conducted tests with MAPSOM framework are divided into two parts and they are presented in the followings paragraphs. The first part shows how different initial parameters affect matching results and the second part presents MAPSOM functionality on dataset with real ontologies (dataset No. 302).

The first part of tests shows variability and dependencies between SOM parameters and results. The most important parameters are: a number of neurons of the SOM, types of similarity measures, and a type of the initial classifier. The used parameters of the SOM and learning algorithm were: 15 neurons in the both dimensions, hexagonal topology, sigma - 0.5, neighborhood size - 8, learning rate - 0.4, and 1500 iterations. We exploited following similarity measures: n-gram measure, Levenstein measure, Needleman-Wunsch measure [30], and Lin and Wu&Palmer for WordNet [7]. In this part, we tried different initial classifiers, i.e., the boolean conjunctive classifier (described in Eq. 7) with the threshold equal to 0.2 for every similarity measure and the linear weighted classifier (described in Eq. 8) with the weight equal to 0.2 for every similarity measure and threshold equal to 0.4. There are many suitable datasets in the OAEI 2013 Challenge for demonstrating MAPSOM functionality. We chose the dataset No. 222 from the OAEI benchmark (MAPSOM functionality is similar with the other datasets and this dataset is sufficient for our demonstration). Table 1. shows experimental results for previously mentioned parameters and for different number of clusters. It is possible to achieve the best precision and recall with any initial setting and the main difference is in the number of iterations.

---

[1] http://oaei.ontologymatching.org/2013/benchmarks/

Table 1. Dependencies between different SOM configuration and results (dataset No. 222)

Iteration	Bool. Conj. Class. 4 Clusters		Bool. Conj. Classifier 20 Clusters		Linear Weight. Class. 20 Clusters	
	Precision	Recall	Precision	Recall	Precision	Recall
1	0.38	1	0.49	1	1	0.74
2	0.39	1	0.51	1	1	0.81
3	0.4	1	0.54	1	1	1
4	0.41	1	0.63	1		
5	0.48	1	0.7	1		
6	0.61	1	0.7	1		
7	0.68	1	0.73	1		
8	0.84	1	1	1		
9	0.84	1				
10	0.97	1				
11	0.99	1				
12	0.99	1				
13	1	1				

Table 2. MAPSOM and a real ontology (dataset No. 302)

Iteration	1	2	3	4	5	6	7	8	9	10	11	12	13	14
Precision	0.72	0.88	0.96	0.96	1	1	1	0.83	0.83	0.83	0.83	0.84	0.81	0.85
Recall		0.46	0.46	0.46	0.46	0.46	0.46	0.46	0.5	0.5	0.5	0.5	0.52	0.54	0.61

The second part of experiments demonstrates MAPSOM functionality with real ontology and comparison with the best systems in OM2013. Dataset No. 302 is composed of finding alignments between reference ontology and real Bib-Tex/UMBC ontology. The result is verified by reference matching from OAEI. We used SOM with 20 neurons in the both dimensions, hexagonal topology, and following parameters of training algorithm: sigma - 0.5, neighborhood size - 10, learning rate - 0.5, and 1500 iterations. The similarity measures are n-gram measure, Levenstein measure, Needleman-Wunsch measure, and Lin and Wu&Palmer for WordNet. The initial classifier is boolean conjunctive classifier with thresholds equal to 0.2 for every similarity measure. The number of clusters before the active learning step is 15. The Table 2. shows the evolution of the precision and recall depending on the iterations.

Finally, we compare the results of our experiment with some of the best systems tested also on the dataset No. 302 (see Table 3.).

Discussion. The results from the experiments show the proposed system has the ability to produce appropriate alignments. In this paper, it is not enough space to present detailed experiments. Hence, we incorporated demonstrative experiments to prove MAPSOM functionality.

Table 3. MAPSOM and comparison with some of the best systems tested on dataset No. 302

ASMOV		Lily		n-Harmony		MAPSOM	
Precision	Recall	Precision	Recall	Precision	Recall	Precision	Recall
0.71	0.56	0.84	0.65	0.93	0.55	0.85	0.61

It is not reasonable to compare MAPSOM framework results only with other automatic matching systems, because our MAPSOM can work also in fully manual mode and when user has a lot of time it is possible to achieve better results than fully automatic systems can achieve. The presented experiments were conducted in semi-automatic mode. The first experiment shows results of dataset No. 222 depending on different initial settings. From these demonstrative results, we can see the number of clusters can rapidly change the time which is needed to achieve the best result. The MAPSOM architecture is also robust to the initial setting and results of algorithm are equal after adequate number of active learning iteration. On the other hand, there are many parameters of the SOM, learning algorithm, and active learning process. Therefore user need some experience to set the optimal MAPSOM parameters.

The second experiment was conducted to prove MAPSOM functionality with real ontology dataset. We can see the results of dataset No. 302 are comparable with the other systems.

6   Conclusions

The framework MAPSOM for the semi-automatic ontology matching was presented in this paper. The architecture combines the self-organizing map for aggregating similarity measures and the active learning tuning. We have demonstrated the MAPSOM is able to perform the ontology matching task with good precision and recall after only a few iteration of the active learning process. Hence, the proposed MAPSOM is suitable for the intended purpose, e.g., for manufacturing or medicine application.

Our MAPSOM is new and there is still chance to improve the proposed system. In future work, we plan to carry out more tests to verify the quality and the general usability of our framework.

Next step is to investigate thoroughly correlation between types of similarity measures and quality of output matching. In current version of the framework, we use limited number of similarity measures and we suppose the precision, recall, and speed of active learning convergence will be improved with more and carefully chosen similarity measures.

Finally, the last step is an exploration of the possibility, how the user feedback from previous MAPSOM executions can be applied for improving the initial settings of the framework. This fact can improve functionality and speed up convergence of results.

Acknowledgements. This research has been supported by the Grant Agency of the Czech Technical University in Prague, grant No. SGS12/188/OHK3/3T/13.

References

1. Belkin, N.J., Croft, W.B.: Information filtering and information retrieval: two sides of the same coin? Commun. ACM **35**(12), 29–38 (1992)
2. Wache, H., Voegele, T., Visser, U., Stuckenschmidt, H., Schuster, G., Neumann, H., Hbner, S.: Ontology-based integration of information-a survey of existing approaches. In: Proceedings of IJCAI Workshop on Ontologies and Information Sharing, pp. 108–117 (2001)
3. Kashyap, V., Sheth, A.: Semantic and schematic similarities between database objects: a context-based approach. Int. J. Very Large Data Bases **5**(4), 276–304 (1996)
4. Kim, W., Seo, J.: Classifying schematic and data heterogeneity in multidatabase systems. Computer **24**(12), 12–18 (1991)
5. Goh, C.H.: Representing and reasoning about semantic conflicts in heterogeneous information systems. Ph.D. thesis (1996)
6. Gruber, T.R.: Toward principles for the design of ontologies used for knowledge sharing? Int. J. Hum.-Comput. Stud. **43**(5), 907–928 (1995)
7. Euzenat, J., Shvaiko, P.: Ontology Matching. Springer, Heidelberg (2007)
8. Ichise, R.: Machine learning approach for ontology mapping using multiple concept similarity measures. In: Proceedings of the 7th IEEE/ACIS International Conference on Computer and Information Science, pp. 340–346 (2008)
9. Jirkovský, V., Obitko, M.: Ontology mapping approach for fault classification in multi-agent systems. In: Proceedings of the IFAC Conference on Manufacturing Modelling, Management, and Control, pp. 951–956 (2013)
10. Miller, G.A.: Wordnet: a lexical database for english. Commun. ACM **38**(11), 39–41 (1995)
11. Valtchev, P., Euzenat, J.: Dissimilarity measure for collections of objects and values. In: Liu, X., Cohen, P., Berthold, M. (eds.) IDA 1997. LNCS, vol. 1280, pp. 259–272. Springer, Heidelberg (1997)
12. Melnik, S., Rahm, E., Bernstein, P.A.: Rondo: a programming platform for generic model management. In: Proceedings of the ACM SIGMOD International Conference on Management of Data, pp. 193–204. ACM (2003)
13. Aumueller, D., Do, H.H., Massmann, S., Rahm, E.: Schema and ontology matching with coma. In: Proceedings of the ACM SIGMOD International Conference on Management of Data, pp. 906–908. ACM (2005)
14. Jian, N., Hu, W., Cheng, G., Qu, Y.: Falcon-ao: Aligning ontologies with falcon. In: Proceedings of K-CAP Workshop on Integrating Ontologies, pp. 85–91 (2005)
15. Bernstein, P.A., Melnik, S., Churchill, J.E.: Incremental schema matching. In: Proceedings of the 32nd International Conference on Very Large Data Bases, pp. 1167–1170 (2006)
16. Giunchiglia, F., Yatskevich, M., Avesani, P., Shvaiko, P.: A large dataset for the evaluation of ontology matching. Knowl. Eng. Rev. **24**(2), 137–157 (2009)
17. Shvaiko, P., Euzenat, J.: Ten challenges for ontology matching. In: Meersman, R., Tari, Z. (eds.) OTM 2008, Part II. LNCS, vol. 5332, pp. 1164–1182. Springer, Heidelberg (2008)
18. Do, H.H., Rahm, E.: Matching large schemas: approaches and evaluation. Inf. Syst. **32**(6), 857–885 (2007)

19. Falconer, S.M., Storey, M.-A.D.: A cognitive support framework for ontology mapping. In: Aberer, K., et al. (eds.) ISWC/ASWC 2007. LNCS, vol. 4825, pp. 114–127. Springer, Heidelberg (2007)
20. Mocan, A., Cimpian, E.: An ontology-based data mediation framework for semantic environments. Int. J. Seman. Web Inf. Syst. **3**(2), 69–98 (2007)
21. Robertson, G.G., Czerwinski, M.P., Churchill, J.E.: Visualization of mappings between schemas. In: Proceedings of the SIGCHI Conference on Human Factors in Computing System, pp. 431–439. ACM (2005)
22. Zhao, L., Ichise, R.: Aggregation of similarity measures in ontology matching. In: Proceedings of the 5th International Workshop on Ontology Matching, pp. 232–233 (2010)
23. Curino, C., Orsi, G., Tanca, L.: X-som: A flexible ontology mapper. In: Proceedings of the 18th International Workshop on Database and Expert Systems Applications, pp. 424–428. IEEE (2007)
24. Tran, Q.V., Ichise, R., Ho, B.Q.: Clusterbased similarity aggregation for ontology matching. In: Proceedings of the 6th International Workshop on Ontology Matching, pp. 142–147 (2011)
25. Kohonen, T.: The self-organizing map. Proc. IEEE **78**(9), 1464–1480 (1990)
26. Kaski, S., Kohonen, T.: Exploratory data analysis by the self-organizing map: Structures of welfare and poverty in the world. In: Proceedings of the 3rd International Conference on Neural Networks in the Capital Markets (1996)
27. Settles, B.: Active learning literature survey. Computer Sciences Technical Report 1648, University of Wisconsin-Madison (2009)
28. Lewis, D.D., Gale, W.A.: A sequential algorithm for training text classifiers. In: Croft, B.W., van Rijsbergen, C.J. (eds.) Proceedings of the 17th Annual International ACM SIGIR Conference on Research and Development in Information Retrieval, pp. 3–12. Springer, New York (1994)
29. Lewis, D.D., Catlett, J.: Heterogenous uncertainty sampling for supervised learning. In: Proceedings of the 11th International Conference on Machine Learning, pp. 148–156 (1994)
30. Needleman, S.B., Wunsch, C.D.: A general method applicable to the search for similarities in the amino acid sequence of two proteins. J. Mol. Biol. **48**(3), 443–453 (1970)

Automatic and Dynamic Book Category Assignment Using Concept-Based Book Ontology

Heeryon Cho and Hyun Jung Lee[✉]

Yonsei Institute of Convergence Technology, Yonsei University,
85, Songdogwahak-ro, Yeonsu-gu, Incheon 406-840, Republic of Korea
heeryon@yonsei.ac.kr, hjlee5249@gmail.com

Abstract. We propose concept-based book ontology for automatic and dynamic category assignment to books through collaborative filtering. It is general for authors or book systems to assign one or more categories to books, but determining book categories based on book reviews have long been neglected. Popularization of online reviews has generated abundant reviews, and it is valuable to additively consider these reviews for assigning relevant book categories. The proposed concept-based book ontology is constructed by conceptual categories that are extracted from the existing book category hierarchy using semantic relationships. Moreover, category-specific review words are constructed through collaborative filtering with the semantically related review words. We built an automatic and dynamic book category assignment prototype system using the concept-based book ontology with the Amazon book department data and confirmed the effectiveness of our approach through empirical evaluations.

Keywords: Ontology · Book category hierarchy · Book reviews · Conceptual category · Category assignment

1 Introduction

The World Wide Web is increasingly becoming an open platform for publishing one's own creations including books, videos, music, cartoons, etc. Marketing one's work is more effective and efficient by placing one's work within a concentrated repository of similar category of works, and this has lead to the establishment of online sales platform and marketplace which handle thousands of products in the unlimited storage space of cyberspace. Maintaining an organized marketplace of multifarious products however requires great human effort. The task of assigning relevant product categories to a given product is usually done mechanically by the product distributor under a fixed rule excluding much input from the consumers, for instance, in the form of consumer review reports. In the case of books, the book categories are determined subjectively by the author or the seller (i.e., individual expert), disregarding the opinions of the actual readers. For instance, the author might assume that his/her book is best represented in category 'A,' but readers might view the book to be belonging to category 'B.' This forces the book away from the ideal place in the online marketplace

W. Kim et al. (Eds.): JIST 2013, LNCS 8388, pp. 364–379, 2014.
DOI: 10.1007/978-3-319-06826-8_27, © Springer International Publishing Switzerland 2014

incurring disserve to both the author and readers. Fortunately, collecting readers' opinions via online book reviews is now possible, and if we can reflect readers' opinions in determining the book category through collaborative filtering, then more adequate categories can be assigned to the book. We aim to realize this by proposing concept-based book ontology for automatically assigning book categories based on the book reviews.

Unlike other manufactured products with limited number of distinct functions and attributes, books are especially difficult to categorize since they deal with vast range of ideas, concepts, and conceptually related categories. The number of book categories is immense and it is often the case for a book to relate to conceptually similar multiple categories. Therefore, in this research, we propose a prototype system which automatically and dynamically assigns book categories to books using concept-based book ontology.

The concept-based book ontology is constructed using semantic relationships between the existing book category hierarchy and reader-provided book reviews. The process is illustrated as follows. First, we obtain an existing book category hierarchy from an online bookstore. For this research we selected partial book category hierarchy from Amazon's online book department. Second, we extract semantically similar categories distributed across different category hierarchies. For example, '*Horror*' category in the Amazon book department is defined in at least two different hierarchies: '*Books > Literature & Fiction > Genre Fiction > Horror*' and '*Books > Teen & Young Adult > Horror.*' These semantically similar categories are extracted across different hierarchies under a shared concept, '*Horror.*' We will call such category *conceptual category*. Third, we retrieve existing books categorized under the conceptual categories, and the book reviews linked to these books are collected. The collected book reviews are tokenized and lemmatized to generate a list of book review words so that the words that represent the conceptual categories can be extracted. These review words are constructed for each book categories included in the conceptual categories. We will call these review words *category-specific review words*. In this research, we construct fifteen category-specific review words. Once the fifteen category-specific review words are constructed, stop words are removed from the review words, and the common review words appearing across different conceptual categories are also removed to delete words like "book" and "read". Finally, concept-based book ontology is constructed based on the existing book category hierarchies, conceptual categories, and category-specific review words.

Using the concept-based book ontology, the inferred categories based on a book review are automatically assigned to a book. To verify the effectiveness of our approach, we conducted an automatic book category assignment experiment. We borrowed the existing Amazon book category hierarchy and the Amazon's customer book reviews to construct our book ontology. Using this ontology, new book categories were automatically assigned. We evaluated the adequacy of the assigned categories and the F-measure was 73.6 %.

The rest of the paper is structured as follows: Sect. 2 explains the existing researches on collaborative filtering, conceptual ontology, and concept-based information retrieval. Section 3 illustrates how the book category is assigned to the book using the concept-based book ontology and book reviews. The concept-based book ontology is explained in Sect. 4. The evaluation of our prototype system is performed in Sect. 5, and conclusion and further works are given in Sect. 6.

2 Related Works

Our concept-based book ontology is based on collaborative filtering of book reviews and the extraction of conceptual categories for the ontology. Using the concept-based book ontology, the book categories are assigned to books by inferring the relevant category from book reviews. We review the existing works on collaborative filtering, conceptual ontology, and concept-based information retrieval.

2.1 Collaborative Filtering

Collaborative filtering is defined as the process of filtering or evaluating items that are based on the opinions of other people [4, 10]. The advent of the Web and the Internet has enabled the collection of community's opinion, and this has opened up the possibilities of additively filtering and applying these opinions to applications depending on users' requirements. For instance, in recommendation systems, users' preferences, behaviors, and opinions are collected, filtered, and analyzed to recommend a product they want; a group having the same preferences as the user can be inferred through collaborative filtering. If a new user visits a book store and is included in one of the groups, then the same kind of books are recommended [8]. For instance, Amazon.com proposed "item-to-item collaborative filtering" which used neighborhood matrix with items depending on the similarity of items. The similarity is defined by collaborative filtering of users' buying behaviors [8]. Collaborative filtering is applied to recommendation of soft goods like books, music, and movie soft goods recommendation as well as physical goods. For the recommendation, contents can be roughly classified with official content like user ratings and informal content using implicit ratings like email [10]. Since pure content-based recommendation is often inadequate to find goods a consumer wants [10], it is important to use collection and filtering of implicit contents for the recommendation; this improves the satisfaction value [11].

In this research, the conceptual category as implicit contents is extracted from the book category as explicit contents in Amazon.com using book reviews. The category-specific review words that are included in the conceptual category are extracted through collaborative filtering using book reviews that contain opinions of the readers.

2.2 Conceptual Ontology

To assign book categories to books, we propose concept-based book ontology to infer and update relevant book categories using the book reviews and conceptual categories defined in the book ontology. The ontology is constructed by inter-relationships among the extracted conceptual categories as concepts [6]. The ontology is applied to Amazon's book domain and concepts are conceptually multi-layered with semantic relationships among book category, conceptual category, and category-specific review words [1, 6]. Ontology is appropriate to infer solutions according to users' purposes under consideration of semantics among implicit concepts based on inter-relationships. Some studies use semantics and ontology for collaborative filtering and apply to recommendation of soft goods like music, book, movie, and so on [2, 7, 9, 12, 13]. In this research, to realize the dynamic assignment of book categories to books using collaborative filtering, we adopt ontology that is constructed with the extracted concepts from the explicitly represented book categories provided by book systems like Amazon.com.

2.3   Concept-Based Information Retrieval

Going beyond a simple keyword search, concept-based (or ontology-driven) information retrieval has been proposed and studied to realize effective search, retrieval, filtering, and presenting of relevant information. An overview of concept-based information retrieval techniques and software tools in the form of prototypes and commercial products are given in [5]. To establish definitions of concepts, it is necessary to first identify concepts inside the text and then classify found concepts according to the given conceptual structure. This could be done manually or automatically. In this research, we manually extract book categories across different book category hierarchies to create conceptual categories.

There are several ways to identify concepts present in the text, and one way is through categorization. In this research, we use existing book categories to extract conceptual categories that form the basis of our concept-based book ontology. Using conceptual categories is known to aid information retrieval. Cho et al. use first-level concept categories such as agent, matter, event, location, and time to group pictogram interpretations into high-level categories to enhance pictogram retrieval performance [3]. In this research, we use the conceptual categories to extract semantically relevant category-specific book review words, and improve the performance of book category assignment using the category-specific review words built from book reviews.

3   Book Category Assignment

In this section, we introduce the overall process of automatically and dynamically assigning book categories using concept-based book ontology and book reviews. The overall process is illustrated in Fig. 1.

To assign semantically relevant book categories to a book, concept-based book ontology is built based on *conceptual category* (*CC*) that is extracted from conceptually similar book categories across different book category hierarchies (Fig. 1 *Conceptual Category Collaborative Filtering*). Next, *category-specific review words* (*CRW*) are extracted from the book reviews linked to the books categorized under the conceptual categories (Fig. 1 *Category-Specific Review Words Extraction*). Finally, the conceptually relevant inferred book categories are assigned to a book by comparing a newly added book review (*NBR*) to *CRW* (Fig. 1 *Automatic & Dynamic Book Category Assignment*). We explain each of these three steps in more detail in the next sections.

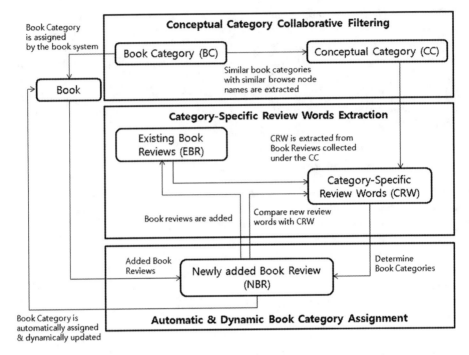

Fig. 1. Overall process of automatic and dynamic book category assignment.

3.1   Conceptual Category Collaborative Filtering

Conceptual category is extracted from the shared concepts among book categories. The shared concepts are conceptually similar across different hierarchies. Whereas book categories have vertical hierarchy among the categories, conceptual category is constructed using the shared concepts positioned across horizontal categories. The book category (BC) and the conceptual category (CC) are defined as follows.

$BC = \{bcv_i \mid bcv_i$ is i^{th} book category vector, $1 \le i \le m$ and m is a number of book category vectors$\}$; BC is a set of vectors of book categories.

$bcv_i = \{(bc_h) \mid 1 < h < l,\ h$ is h^{th} book category and l is a number of book category$\}$; i is i^{th} book category vector.

$CC = \{cc_j \mid cc_j$ is j^{th} shared category among book category vectors, $1 \le j \le n$ and n is a number of conceptual categories$\}$; CC is a set of conceptual categories. cc_j is defined as in formula (1).

$$cc_j = \cap_{i=1}^{k} bcv_i, \tag{1}$$

where $1 < k \le m$, k is a number of shared book category vectors. Figure 2 shows how to extract conceptual categories from the book category.

We give an example of book category and conceptual category. When a new book is registered to an online book store, the book is categorized under the book categories

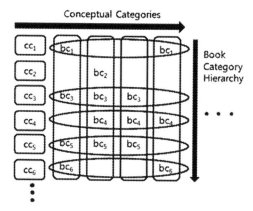

Fig. 2. Vertical Book Category (*BC*) hierarchy and horizontally extracted Conceptual Category (*CC*).

assigned by the author or a book system. Usually the registered book is assigned to one or more popular categories such as '*Horror*', '*Romance*', etc. However, the readers might have different notions about the book when searching for the book. For instance, a reader might be looking for a horror story for his teenage son, but if the book is only categorized under the '*Horror*' category targeted to adults, then the book will not be found even though the book contains adequate horror story for teenagers. To overcome this limitation, we extract conceptually similar meaning categories, in this case '*Horror*' categories, horizontally across different hierarchies to form a single conceptual category.

Here is a real-world example. When we look at the '*Horror*' category in the Amazon book category hierarchy, it is defined in at least two different hierarchies: '*Books > Literature & Fiction > Genre   Fiction > Horror*'   and   '*Books > Teen & Young Adult > Horror.*' We extract these similar meaning categories across two different vertical hierarchies to form a single conceptual category '*Horror*'. In this paper, we extracted six different conceptual categories.

3.2   Category-Specific Review Words Extraction

Once we extract conceptual categories, we generate *category-specific review words* (*CRW*) related to *conceptual category* (*CC*) from *Existing Book Reviews* (*EBR*) as in Fig. 1. We assume that the customer-provided book reviews are saved on the *EBR*. From the *EBR*, the book reviews are categorized under the conceptual category to form *CRW*. To construct *CRW* for each conceptual category, these book reviews are tokenized and lemmatized to form a set of review words as follows:

$CRW = \{(RW_i) \mid RW_i$ is review word set of cc_j, $1 \leq i \leq u\}$
$RW_i = \{(rw_{ij}) \mid rw_{ij}$ is j^{th} review word of RW_i, $1 \leq j \leq z\}$
$NRW = \{(nrw_k) \mid nrw_k$ is k^{th} new review word of a book, $1 \leq k \leq w\}$

In this paper, *CRW* are generated under six conceptual categories; a total of fifteen RW_i are generated. RW_i is a set of tokenized and lemmatized review words related to cc_j. The jth review word of RW_i is rw_{ij}. *NRW*, a set of new book review words, is compared with *CRW* to find the relevant book category. Figure 3 summarizes the relationship between cc_j and rw_{ij} of *CRW*.

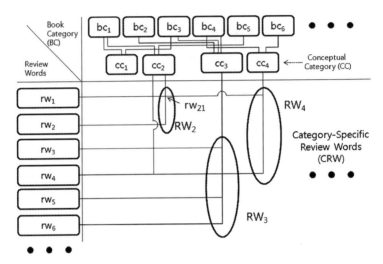

Fig. 3.   Category-specific review words (CRW).

3.3   Automatic & Dynamic Book Category Assignment

When a book is registered to the book system, it is general that the author or the book system to assign book categories. A book might be assigned to several categories as in formula (2).

$$b_q \in \left\{ \cup_{j=1}^{p} cc_j, \ 1 \leq p \leq n \right\}, \text{ where } b_q \text{ is a book } q. \tag{2}$$

Even if the book is assigned to several categories, the related book categories can be dynamically changed by referring to the book reviews of the book. Consequently, new book category can be added to the book through comparing *NRW* of b_q to *CRW*. We calculate the similarity between *NRW* and semantically related RW_i of *CRW* using the cosine similarity between *NRW* and RW_i as in formula (3).

$$Sim(NRW, RW_i) = \frac{NRW \cdot RW_i}{\|NRW\| \|RW_i\|}, \tag{3}$$

where i is i^{th} category-specific review words and $1 \leq i \leq u$. In this research, u is fifteen. Cosine similarity is a commonly used technique for calculating the similarity of two vectors, and we adopted cosine similarity to calculate the similarity of *NRW* and RW_i. There are, of course, other techniques for calculating similarity such as Euclidean distance, neighborhood similarity, etc., but we used one of the most general methods as a start to confirm the usefulness of the conceptual category proposed in this study. In the future, we plan to test other techniques such as Naive Bayes, Random Forest, Support Vector Machine, etc. to improve the similarity measurement.

Now let us assume that a horror novel, which might become popular among teenagers in the future, is newly registered to the book system with the author assigning a conceptual category '*Horror*' as the only relevant category. After a week, a reader uploads a book review about this new horror novel. Upon receiving the book review, the system then calculates the similarity between the new book review (*NRW*) and the existing RW_{Adult} and RW_{Teen} generated using the existing book reviews collected under the conceptual category '*Horror*'; based on the similarity values, a new book category '*Adult*' or '*Teen*' is inferred and assigned to the horror novel. In other words, the cosine similarity of the new review words and the two *RW* (one generated for the '*Adult*' category and another generated for the '*Teen*' category) of the *CRW* are calculated and compared, and the category with the greater cosine similarity is assigned to the horror novel.

4   Concept-Based Book Ontology

Previously, we composed *CRW* that is semantically relevant to *CC*; then, concept-based book ontology is used to infer relevant book categories from reader-provided book reviews using the semantic relationships. The concept-based book ontology is defined by classes *C* and properties *P* as follows.

$$O_b = \{C, P\}, \tag{4}$$

where *C* and *P* are a set of classes and properties, respectively. Classes are determined by *BC*, *CC*, *CRW*, and *RW*. *BC* is book category, *CC* is conceptual category, *CRW* is category-specific review words, and *RW* includes all review words.

$$C = \{BC, CC, CRW, RW\} \tag{5}$$

Properties that define relationships among classes are determined by object properties "*isKindOf*," "*isPartOf*," and "*hasWord*". Property "*isKindOf*" is a relationship between a *Book Domain Layer* and its corresponding *BC Layer*. "*isPartOf*" defines relationship between *BC Layer* and *CC Layer*. "*hasWord*" relationship represents a semantic relationship between *CC layer* and *CRW Layer* and *RW Layer*. Figure 4 shows conceptual hierarchy of concept-based book ontology.

Figure 5 displays a part of the constructed concept-based book ontology which is built by performing the aforementioned steps. The bridge between *BC* and *CRW* are formed under *CC*. *BC* has hierarchical structure among categories. *CC* is extracted from conceptually similar categories. The dotted line shows the conceptual relationships between *BC* and *CC*, and *CC* and *CRW*.

5   Experiment and Discussion

Amazon (http://www.amazon.com) is an online retail store that sells a wide range of products including books. A partial book category hierarchy was selected and the

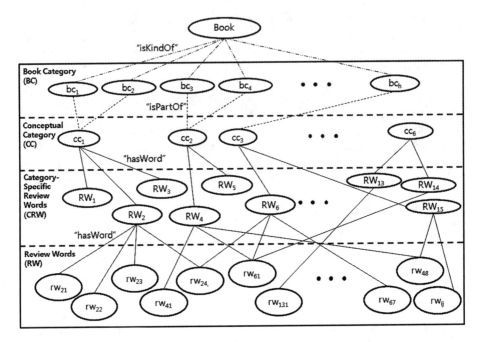

Fig. 4. Conceptual hierarchy of concept-based book ontology.

customer book reviews were collected from the Amazon book department to create a prototype system for the evaluation experiment.

5.1   Experiment

Figure 6 shows the partial book category selected from the Amazon book department. Each node indicates the individual book category, and the number inside the square bracket of each node indicates the 'Browse Node ID' given by Amazon. The dotted square at the top of Fig. 6 pointing to the '*Horror*' node illustrates the actual book category hierarchy displayed on the left side menu of Amazon webpage using the Browse Node ID (http://www.amazon.com/exec/obidos/tg/browse/-/17441).

We extracted six conceptual categories (*CC*) and generated fifteen category-specific review words (*CRW*) to perform fifteen book category assignment evaluation tasks. The books categorized under the specific book category (or Browse Node ID) were collected, and the customer book reviews linked to these books were processed to generate *CRW*. The details of the *CC* and *CRW* used in the experiment are summarized in Table 1.

When generating *CRW*, stop words were removed from the *CRW* and the shared words among the *CRW* within the same *CC* were removed. For example, the shared words between the *CRW* generated from the 119 reviews linked to the Browse Node ID 2967 and the *CRW* generated from the 312 reviews linked to the Browse Node ID 720360 were removed. Moreover, the shared words among all fifteen *CRW* were also removed. Such removed words included "book", "read", etc.

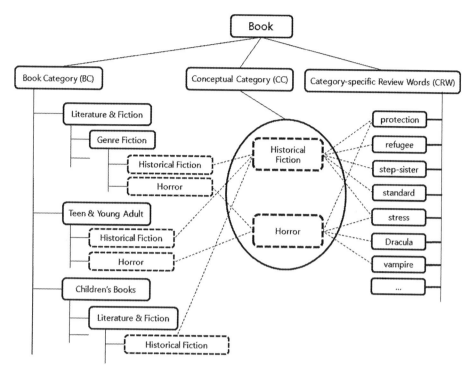

Fig. 5. Partial concept-based book ontology with *BC*, *CC*, and *CRW*.

The frequency of the review words in *CRW* was obtained and normalized for each *CRW* so that each review word was given a value between one and ten. Then, each *RW* in *CRW* were sorted according to the descending order of the normalized frequency (i.e., from large number to small number), and approximately top one thousand unique review words were extracted and used as *CRW* in the experiment. This was done because the number of book reviews collected for each *CRW* varied resulting in large variation in the number of unique review words and frequencies.

Once the *CRW* were generated, new book reviews with more than fifty words were freshly collected from Amazon to perform the evaluation experiment. The experiment consisted of fifteen book category assignment tasks:

① $T1_{2967}$: The book is '*Action & Adventure*'. Should the book be assigned a children (2967) category?

② $T1_{720360}$: The book is '*Action & Adventure*'. Should the book be assigned an adult (720360) category?

③ $T2_{2926}$: The book is '*Historical Fiction*'. Should the book be assigned a children (2926) category?

④ $T2_{10177}$: The book is '*Historical Fiction*'. Should the book be assigned an adult (10177) category?

⑤ $T2_{17437}$: The book is '*Historical Fiction*'. Should the book be assigned a teens (17437) category?

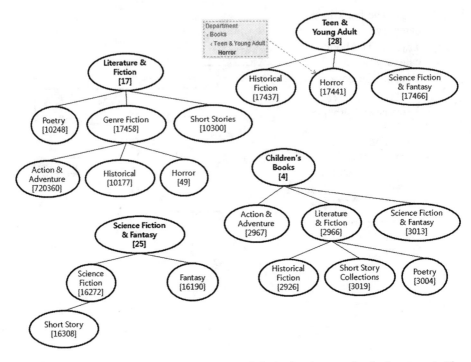

Fig. 6. Partial book category hierarchy selected from the Amazon book department. The numbers inside the square brackets indicate Browse Node ID given by Amazon.

Table 1. Conceptual category (*CC*), category assignment task ID, Browse Node ID of *CRW*, number of books and book reviews used to generate *CRW*.

Conceptual category (*CC*)	Category assignment task ID	Browse Node ID of *CRW*	# of books	# of reviews
Action & Adventure	$T1_{2967}$	2967	95	119
	$T1_{720360}$	720360	233	312
Historical Fiction	$T2_{2926}$	2926	474	606
	$T2_{10177}$	10177	687	993
	$T2_{17437}$	17437	300	390
Horror	$T3_{49}$	49	136	192
	$T3_{17441}$	17441	299	386
Poetry	$T4_{3004}$	3004	312	388
	$T4_{10248}$	10248	109	138
Science Fiction & Fantasy	$T5_{25}$	25	443	639
	$T5_{3013}$	3013	64	86
	$T5_{17466}$	17466	319	415
Short Story	$T6_{3019}$	3019	312	387
	$T6_{10300}$	10300	285	360
	$T6_{16308}$	16308	80	120

⑥ $T3_{49}$: The book is '*Horror*'. Should the book be assigned an adult (49) category?

⑦ $T3_{17441}$: The book is '*Horror*'. Should the book be assigned a teens (17441) category?

⑧ $T4_{3004}$: The book is '*Poetry*'. Should the book be assigned a children (3004) category?

⑨ $T4_{10248}$: The book is '*Poetry*'. Should the book be assigned an adult (10248) category?

⑩ $T5_{25}$: The book is '*Science Fiction & Fantasy*'. Should the book be assigned an adult (25) category?

⑪ $T5_{3013}$: The book is '*Science Fiction & Fantasy*'. Should the book be assigned a children (3013) category?

⑫ $T5_{17466}$: The book is '*Science Fiction & Fantasy*'. Should the book be assigned a teen (17466) category?

⑬ $T6_{3019}$: The book is '*Short Story*'. Should the book be assigned a children (3019) category?

⑭ $T6_{10300}$: The book is '*Short Story*'. Should the book be assigned an adult (10300) category?

⑮ $T6_{16308}$: The book is '*Short Story*'. Should the book be assigned a science fiction (16308) category?

The category assignment tasks were evaluated using precision, recall, and F-measure. Each measure is defined as follows:

$$precision = \frac{|category\ relevant\ reviews \cap category\ assigned\ reviews|}{|category\ assigned\ reviews|}$$

$$recall = \frac{|category\ relevant\ reviews \cap category\ assigned\ reviews|}{|category\ relevant\ reviews|}$$

$$F\text{-measure} = 2 \cdot \frac{precision \cdot recall}{precision + recall}$$

Table 2 shows the category assignment performance of individual reviews. We compared the performance of exact word match between the *NRW* and *RW*, and the cosine similarity between the *NRW* and *RW* as defined in Sect. 3.3 formula (3). The exact word match only counted the number of matching words between the *NRW* and *RW*. The number of matching review words among different *CRW* included in the same *CC* were compared, and the book category linked to *CRW* with the greatest number of matching words were selected and assigned. For example, in the case of task $T1_{2967}$, the matching words were counted for (a) CRW_{2967} and the new book review and (b) CRW_{720360} and the new book review; if the number of matching words of (a) was greater, then the '*Children (2967)*' category was assigned to the new review, and this in turn assigned the category to the book linked to the new review. Similarly, in the case of the cosine similarity, i.e., *Sim (NRW, RW_i)*, the normalized frequency of the *RW* and *NRW* were compared for different *CRW* within the same *CC*, and the *CRW* with greater similarity was selected and the book category linked to the

Table 2. Performance of book category assignment to individual reviews.

Task	Exact word match			Cosine Similarity ($Sim (NRW, RW_i)$)			# of reviews (# relevant)
	Precision	Recall	F-measure	Precision	Recall	F-measure	
$T1_{2967}$	49.2 %	60.8 %	54.3 %	70.8 %	62.5 %	66.4 %	3468 (904)
$T1_{720360}$	76.1 %	95.2 %	84.6 %	86.5 %	92.4 %	89.3 %	3468 (2564)
$T2_{2926}$	42.5 %	64.5 %	51.2 %	62.7 %	69.8 %	66.0 %	15135 (4602)
$T2_{10177}$	59.3 %	88.7 %	71.1 %	79.4 %	81.2 %	80.3 %	15135 (7397)
$T2_{17437}$	29.2 %	58.2 %	38.9 %	52.4 %	53.5 %	52.9 %	15135 (3182)
$T3_{49}$	41.6 %	94.7 %	57.8 %	60.6 %	90.2 %	72.5 %	4630 (1532)
$T3_{17441}$	75.2 %	65.6 %	70.1 %	92.4 %	73.9 %	82.2 %	4630 (3159)
$T4_{3004}$	82.1 %	83.0 %	82.5 %	92.4 %	86.7 %	89.5 %	4070 (2887)
$T4_{10248}$	48.8 %	92.6 %	64.0 %	68.5 %	87.9 %	77.0 %	4070 (1187)
$T5_{25}$	64.5 %	81.8 %	72.2 %	84.4 %	80.1 %	82.2 %	9230 (5351)
$T5_{3013}$	19.7 %	35.9 %	25.4 %	42.6 %	39.6 %	41.1 %	9230 (646)
$T5_{17466}$	47.4 %	80.1 %	59.5 %	69.8 %	76.9 %	73.2 %	9230 (3387)
$T6_{3019}$	60.9 %	80.9 %	69.5 %	80.1 %	85.1 %	82.5 %	6803 (2984)
$T6_{10300}$	53.3 %	81.2 %	64.3 %	79.5 %	74.2 %	76.8 %	6803 (3011)
$T6_{16308}$	31.3 %	66.2 %	42.5 %	58.1 %	65.8 %	61.7 %	6803 (914)
AVG	52.1 %	75.3 %	**60.5 %**	72.0 %	74.7 %	**72.9 %**	

CRW was selected and assigned to a book review. Note that the book review can be assigned multiple book categories.

We see in Table 2 that cosine similarity performs better than the exact word match in every assignment tasks in terms of precision and F-measure. However, in terms of recall, exact word match performed slightly better than the cosine similarity on average; only in six tasks, $T1_{2967}$, $T2_{2926}$, $T3_{17441}$, $T4_{3004}$, $T5_{3013}$, and $T6_{3019}$, the cosine similarity performed better than the exact word match. All in all, the cosine similarity performed better than the exact word match; the average F-measures of the fifteen book category assignment tasks were 60.5 % and 72.9 % for exact word match and cosine similarity, respectively.

We also performed book category assignments to individual books by conducting majority voting of the assigned categories of the individual reviews; we used the category assignment results in Table 2. This experiment applies to the dynamic updating of the book categories. Each book contained one or more book reviews with the assigned book category, and if the proportion of the assigned book category was greater than 50 %, then that category was automatically assigned to the book. Table 3 shows the result of majority voting of grouped reviews' category assignment results to books. The average F-measure of was 73.6 %. Note that the performances of book category assignment to book reviews (Table 2) and to books (Table 3) do not differ greatly. This implies that even a small number of book reviews can be exploited in assigning book categories to a book to a certain degree. Finally, CRW_{50} and CRW_{100} in Table 3 indicate the F-measure using the CRW constructed with fifty and a hundred book reviews, respectively. We see that the category assignment performance generally increases as more book reviews are incorporated in the construction of CRW.

Table 3. Performance of book category assignment to individual books.

Task	Voting (cosine similarity)					# of books (# relevant)
	Precision	Recall	F-measure	CRW_{50}	CRW_{100}	
$T1_{2967}$	83.9 %	58.3 %	68.8 %	56.9 %	73.2 %	722 (206)
$T1_{720360}$	85.1 %	95.9 %	90.2 %	66.0 %	89.7 %	722 (516)
$T2_{2926}$	69.3 %	69.5 %	69.4 %	55.5 %	54.4 %	3295 (1185)
$T2_{10177}$	79.2 %	84.9 %	82.0 %	31.2 %	72.5 %	3295 (1383)
$T2_{17437}$	58.9 %	46.9 %	52.2 %	39.9 %	47.6 %	3295 (736)
$T3_{49}$	59.3 %	91.7 %	72.0 %	64.6 %	72.1 %	1035 (313)
$T3_{17441}$	97.1 %	73.5 %	83.7 %	88.0 %	75.4 %	1035 (742)
$T4_{3004}$	92.7 %	86.9 %	89.7 %	90.5 %	80.9 %	1174 (818)
$T4_{10248}$	72.1 %	85.8 %	78.3 %	79.0 %	71.6 %	1174 (358)
$T5_{25}$	81.8 %	82.8 %	82.3 %	78.9 %	75.4 %	1815 (918)
$T5_{3013}$	52.1 %	24.7 %	33.5 %	32.6 %	38.6 %	1815 (150)
$T5_{17466}$	77.0 %	76.9 %	77.0 %	63.0 %	69.6 %	1815 (772)
$T6_{3019}$	85.3 %	86.6 %	86.0 %	77.8 %	76.8 %	1752 (845)
$T6_{10300}$	83.7 %	72.4 %	77.7 %	59.4 %	69.1 %	1752 (768)
$T6_{16308}$	57.1 %	65.7 %	61.1 %	39.8 %	49.0 %	1752 (166)
AVG	75.6 %	73.5 %	**73.6 %**	61.5 %	67.7 %	

5.2   Discussion

The automatic assignment of book categories using book reviews is enabled under the assumption that readers' book reviews contain defining word features across different book categories; consequently, we started out with the assumption that category-specific review words (*CRW*) contain distinguishing words that semantically relate to a specific book category. We aimed to extract such distinguishing words by sorting the *CRW* in the descending order of normalized frequency, and removing those words shared by the different *CRW* within the same conceptual category (*CC*). Moreover, when conducting the evaluation experiment, we freshly collected book reviews with a word size greater than fifty words in order to increase the presence of the distinguishing words that match the words in *CRW*. This is similar for humans, because if the book review consisted of a single sentence, "I strongly recommend this book," even humans could not guess the relevant book category.

The effect of the book review size involved in constructing *CRW* must be studied more carefully. A naive assumption could be that the more reviews there are, the better the *CRW* may be. Examining the actual data tells a different story. The *CRW* with Browse Node ID 49 ($T3_{49}$) and 17466 ($T5_{17466}$) in Table 1 were generated with 192 and 415 reviews respectively, one double the size of the other, but the category assignment performance was similar with 72.5 % and 73.2 % (Table 2), respectively. Note that the overall/relevant review size used in the experiment was different (Table 2. $T3_{49}$: 4630/1532, $T5_{17466}$: 9230/3387), but the overall/relevant review ratio of the two tasks were similar. Looking at this evidence, we might hypothesize that the effectiveness of *CRW* may not increase after a certain number of reviews; or the effectiveness will vary according to different book categories or conceptual categories. These questions should be further investigated.

The automatic book category assignment we have presented can be viewed as a superclass detection problem from a different angle. By determining the correct leaf node of the book category hierarchy (for example '*Horror*' [17441] in Fig. 6), we can infer the correct superclass ('*Teen & Young Adult*' [28] in Fig. 6). Moreover, subclass detection is also possible by restricting the conceptual category to a single book category, and constructing *CRW* for all children nodes of the (single-book-category) conceptual category.

One of the concerns regarding our method is the minimum number of book reviews required for the system to work. If too many reviews of a single book are needed to assign a category to that book, then the system may not be useful since the book may be already popular at that stage (as proven by the large number of book reviews); consequently, additional category assignment will not be useful so much. Moreover, collecting a large number of reviews for each book is time consuming. We dispel such concerns by constructing *CRW* using all available books within a given category. That is, *CRW* is not restricted to the book reviews of a specific book. By incorporating all book reviews at the book category level and not at the individual book level, we can easily obtain abundant book reviews required for constructing effective *CRW*. Nevertheless, we performed an experiment to see how the differences in the number of book reviews used in *CRW* affect the book category assignment. Table 3 displays the F-measure of the book category assignment using fifty and a hundred reviews (CRW_{50} and CRW_{100}). We see that a hundred reviews can be useful to a certain degree with 67.7 % category assignment performance.

Another concern is that category assignment is impossible if no book reviews are provided. While this is true, it is often the case that a new book accompanies some form of review such as editorial review from the publisher or the media. Our system can exploit these reviews initially and then go on to collect readers' reviews.

Lastly, we presented a dynamic updating of book category by incorporating majority voting of individual reviews' assigned book categories. As more book reviews are added, more category instances will be accumulated for each book. The key to dynamic updating of book category is to limit the incorrectly assigned book category to fewer than 50 % of the all assigned instances.

6   Conclusion and Further Study

We proposed concept-based book ontology which stands at the center of our prototype book category assignment system which automatically and dynamically assigns relevant book categories to books based on reader-provided book reviews. The proposed book ontology defines the relationship between the existing book category hierarchy, conceptual category (*CC*), and category-specific review words (*CRW*) in order to extract semantically related category-specific review words. The cosine similarity between a book review's review words and the *CRW* are calculated and compared for different *CRW* included in the same *CC* to determine the relevant book categories. We tested the effectiveness of our prototype system through an automatic book category assignment experiment and confirmed the F-measure to be 73.6 %. Since our system's performance still has approximately 30 % error rate, we plan to improve the performance by testing other similarity measures such as Naïve Bayes and SVM in the future. Furthermore, we plan to apply our approach to other products and customer reviews to verify whether our approach can be generalized.

Acknowledgments. This research was supported by the Korean Ministry of Science, ICT and Future Planning (MSIP) under the "IT Consilience Creative Program" supervised by the National IT Industry Promotion Agency (NIPA) of Republic of Korea (NIPA-2013-H0203-13-1002).

References

1. Cantador, I., Castells, P.: Multilayered semantic social network modeling by ontology-based user profiles clustering: application to collaborative filtering. In: Staab, S., Svátek, V. (eds.) EKAW 2006. LNCS (LNAI), vol. 4248, pp. 334–349. Springer, Heidelberg (2006)
2. Celma, O., Cano, P.: From hits to niches? Or how popular artists can bias music recommendation and discovery. In: Proceedings of the Netflix-KDD Workshop (2008)
3. Cho, H., Ishida, T., Takasaki, T., Oyama, S.: Assisting pictogram selection with semantic interpretation. In: Bechhofer, S., Hauswirth, M., Hoffmann, J., Koubarakis, M. (eds.) ESWC 2008. LNCS, vol. 5021, pp. 65–79. Springer, Heidelberg (2008)
4. Ekstrand, M.D., Riedl, J.T., Konstan, J.A.: Collaborative filtering recommender systems. Found. Trends Human–Comput. Interact. **4**(2), 81–173 (2010)
5. Haav, H.M., Lubi, T.L.: A survey of concept-based information retrieval tools on the web. In: Proceedings of the 5th East-European Conference ADBIS, vol. 2, pp. 29–41 (2001)
6. Lee, C.-S., Wang, M.-H., Hagrea, H.: A type-2 fuzzy ontology and its application to personal diabetic-diet recommendation. IEEE Trans. Fuzzy Syst. **18**(2), 374–395 (2010)
7. Lee, J.H., Waterman, N.M.: Understanding user requirements for music information services. In: Proceedings of the ISMIR, pp. 253–258 (2012)
8. Linden, G., Smith, B., York, J.: Amazon.com recommendations item-to-item collaborative filtering. IEEE Internet Computing, pp. 76–80 (Jan–Feb 2003)
9. Mobasher, B., Jin, X., Zhou, Y.: Semantically enhanced collaborative filtering on the web. In: Berendt, B., Hotho, A., Mladenič, D., van Someren, M., Spiliopoulou, M., Stumme, G. (eds.) EWMF 2003. LNCS (LNAI), vol. 3209, pp. 57–76. Springer, Heidelberg (2004)
10. Schafer, J.B., Frankowski, D., Herlocker, J., Sen, S.: Collaborative filtering recommender systems. In: Brusilovsky, P., Kobsa, A., Nejdl, W. (eds.) Adaptive Web 2007. LNCS, vol. 4321, pp. 291–324. Springer, Heidelberg (2007)
11. Sohn, M., Kim, H., Lee, H.J.: Personalized recommendation framework based on CBR and CSP using ontology in a ubiquitous computing environment. Comput. Syst. Sci. Eng. **27**(6) (2012)
12. Voogdt, M.L.D., Leman, M., Baets, B.D., Meyer, H.D., Martens, J.P.: How potential users of music search and retrieval systems describe the semantic quality of music. J. Am. Soc. Inf. Sci. Technol. **59**(5), 695–707 (2008)
13. Whitman, B., Lawrence, S.: Inferring description and similarity for music from community metadata. In: Proceedings of the International Computer Music Conference, pp. 591–598 (2002)

An Automatic Instance Expansion Framework for Mapping Instances to Linked Data Resources

Natthawut Kertkeidkachorn[1]([⊠]), Ryutaro Ichise[2],
Atiwong Suchato[1], and Proadpran Punyabukkana[1]

[1] Department of Computer Engineering, Faculty of Engineering,
Chulalongkorn University, Bangkok, Thailand
natthawut.k@student.chula.ac.th,
{atiwong.s,proadpran.p}@chula.ac.th
[2] National Institute of Informatics, Tokyo, Japan
ichise@nii.ac.jp

Abstract. Linked Data is an utterly valuable component for semantic technologies because it can be used for accessing and distributing knowledge from one data source to other data sources via structured links. Therefore, mapping instances to Linked Data resources plays a key role for consuming knowledge in Linked Data resources so that we can understand instances more precisely. Since an instance, which can be aligned to Linked Data resources, is enriched its information by other instances, the instance then is full of information, which perfectly describes itself. Nevertheless, mapping instances to Linked Data resources is still challenged due to the heterogeneity problem and the multiple data source problem as well. Most techniques focus on mapping instances between two specific data sources and deal with the heterogeneity problem. Mapping instances particularly relying on two specific data sources is not enough because it will miss an opportunity to map instances to other sources. We therefore present the Instance Expansion Framework, which automatically discover and map instances more than two specific data sources in Linked Data resources. The framework consists of three components: Candidate Selector, Instance Matching and Candidate Expander. Experiments show that the Candidate Expander component is significantly important for mapping instances to Linked Data resources.

Keywords: Instance expansion · Instance matching · Linked data · Linking open data and similarity metric

1   Introduction

Generally, Linked Data [1] is proposed to provide a simple method of publishing and connecting the structured data on the web. It is aimed to construct Linked Data resources, called the Web of Data, by interlinking between different data sources. Building the linked data resources gives many advantages because it can be use for accessing and consuming knowledge from one data source to other data

W. Kim et al. (Eds.): JIST 2013, LNCS 8388, pp. 380–395, 2014.
DOI: 10.1007/978-3-319-06826-8_28, © Springer International Publishing Switzerland 2014

sources via structured links. For example when we look up for an information about something on the source A, the result can be enriched by the information in source B, which have explicitly links to the source A. For constructing Linked Data resources, there are still high barriers between different data sources such as the data representation, the data integration and so on. Therefore, some criteria are needed to be considered in order to overcome the barriers. The criteria are that the data must be published under the Resource Description Framework (RDF) [2] and use Uniform Resource Identifier (URI) to represent things. In the RDF, the data is published in triples (subject, predicate, object).

Recently, there is an ongoing effort to construct Linked Data resources called the Linking Open Data (LOD) cloud project [3]. The LOD cloud is a collection of billion RDF triples. It contains over 31 billion RDF triples from 295 data sources and is categorized into seven domains: cross-domain, geographic, media, life, sciences, government, user-generated content and publication. As we can notice from the LOD diagram[1], each data source has linked to the other sources by various link properties. The most important link property is owl:sameAs [4], which is one property of the Web Ontology Language (OWL). The owl:SameAs property is designed to identify that the couple of things or instances, which are linked by owl:SameAs, refers to the same real-world object. Consequently, if we can establish such property between the instance pairs, it means that we can explore and exploit other information relating to an instance via another instance in another different data source. Therefore, the interesting problem in this study is to discover and establish owl:sameAs links between any instances toward instances in the LOD cloud.

For establishing owl:SameAs property, an instance matching system is a practical solution so many instance matching systems are proposed [5–13]. Most studies have focused on mapping instances between two given-specific data sources. They are highly efficient to match instance pairs. Nevertheless, the techniques still require end users to be aware of which data sources containing matching pairs. In case of the LOD cloud, we deal with many data sources. We therefore cannot know that which data source contains an instance corresponding to our instances. Although it is possible to compare one data source by one until we found the correspondence, it is not convenience. According to the growing of LOD cloud, we have to compare a million or even more instance pairs in order to discover the just one of matching pairs in the LOD cloud. Note that, when the LOD cloud project started in 2007, there are only twelve data sources while currently there are at least 295 data sources as of 2011.

In this paper, we therefore present a novel framework for automatic discovering and mapping instances to the LOD cloud. The basic idea is that given source instances, we firstly select one data source in the LOD cloud in order to generate the candidate pairs between source instance and that source by using a set of keywords. Then, we perform the instance matching by employing a statistical machine learning technique based on similarity vectors. Our approach also utilizes the useful information, which obtains from selecting the candidate

---

[1] http://lod-cloud.net/

pairs, to compute the similarity feature vectors. After we verify that candidate pairs are correctly matched, then, we establish owl:sameAs property between instance pairs. If there are some instances which cannot be correctly matched with instances in initial data source, we can assume that these instances are missing from this data source. In order to discover those missing instances, we expand the data source from the initial data source to another data source and then we match the instances again until we can match the remained instances. In order to satisfy our basic idea, three components, which are candidate selector, instance matching and candidate expander, are included in our framework. For testing our framework, Instance Matching at Ontology Alignment Evaluation Initiative 2012 (IM@OAEI 2012) data set is chosen. We show that our approach significantly improves mapping instances toward LOD cloud.

The remainder of the paper is organized as follows. In the following section, some related works are reviewed and discussed. Section 3 gives problem definition of mapping instances toward LOD cloud. The details of our approach are described in Sect. 4. Experiments are conducted in the Sect. 5 and results also show in the same section. Eventually, in the Sect. 6, we conclude the study and discuss some future directions.

2   Related Work

An instance matching problem is widely attended by many researchers. Based upon our literature reviews, proposed instance matching systems can roughly be divided into two categories. One is a Domain-Dependent system and the other is a Domain-Independent system.

In the Domain-Dependent system, there are many proposed systems: Silk [5], AgreementMaker [6] and Zhishi.Links [7]. Silk is a framework for discovering and manipulating owl:sameAs links between different data sources. Silk provide three components for interlinking links. The first component is a discovery engine computing links between different data sources. The second component is intended for tuning and evaluating the correct of links. The third component is designed to deal with changing of data sources. AgreementMaker is a system for both ontology matching and instance matching. For instance matching, AgreementMaker conducts three phrases. Firstly, in the candidate generation phrase, candidate instances are selected by considering the similarity of label. In disambiguation phrase, similarity vectors between instance pairs are computed by many techniques and then similarity vectors between instance pairs are used to identify matching of instances in the matching phrase. Zhishi.Links is an improved version of Silk. It utilizes some weighting schema to enhance the matching results. Although the systems in this category show the interesting performance, they still need a human effort to drive some background information.

In the Domain-Independent system, a system is not required any background information. Many studies are presented [8–13]. ObjectCoref [8] is a self-learning system, which iteratively finds discriminative property-value pairs for identifying matching of instances. SERIMI [9] utilizes the predicates, which higher entropy, in order to match instances without the prior knowledge. Nguyen et al. proposed the instance matching based on a learning-based approach [10]. They build a binary classifier to predict that instance pairs are matched correctly. SLINT [11] and SLINT+ [12], which is an extended version of SLINT, are proposed. Both systems select and align important predicates between source and target to generate instance pairs, then, instance pairs are verified whether they are matched correctly. Rong et al. presented an instance matching approach based on similarity metrics [13]. They train a binary classifier based on similarity metrics to classify that instance pairs are matched. Even though the approaches can predict instance matching without a domain or background information, they still perform instance matching between two given-specific data sources.

Since the LOD cloud consists of many data sources, instance matching systems based on two given-specific data sources cannot efficiently deal with it. Furthermore, we will miss an opportunity to map instances to the LOD cloud if match pairs do not exist in the data sources. We therefore introduce our framework, which is Domain-Independent system that can expand the data source from one data source to another data source via utilizing owl:sameAs links, in order to discover and map instances toward the LOD cloud efficiently.

3   Problem Definition

To clarify the research problem, we describe mapping instances toward LOD cloud. An instance or individual, which usually refers to a real-world object, is a basic component of ontology. Generally, it is described by property-value descriptions, which mention about the object. Some property-value descriptions can be literal information, which directly describe the instance, or some property-value descriptions may refer to other instances. Given the instance p, we call those referred instances of p as "adjacent instances" denoted by $Adj(p)$.

Given a pair of instance x and y, x and y are matched denoted by $x \equiv y$ if they describe the same object. In order to find the relation between instance pairs, a classifier C is trained. The classifier C is designed to decide an instance pair, is correctly matched. Since we define the instance matching problem as a binary classification problem, the results of classifier C can be 1 or -1 where 1 represent that the instance pair is a match and -1 represent that the instance pair is a non-match.

In LOD cloud, it is not similar to match instances between two specific data sources because there are many data sources. We therefore define the instance matching problem to LOD cloud as follows. Let $D_1, D_2, D_3, ..., D_n \in LOD$ by D_i represent the data source i in LOD and given a new data source A, the aim of mapping instances toward LOD cloud is to compute set $\omega = \{(x, y) | x \equiv y, x \in A, \exists i (y \in D_i)\}$

4   Approach

In this section, we present our framework for automatic mapping instances to LOD. The framework is shown in Fig. 1. There are three components as follows:

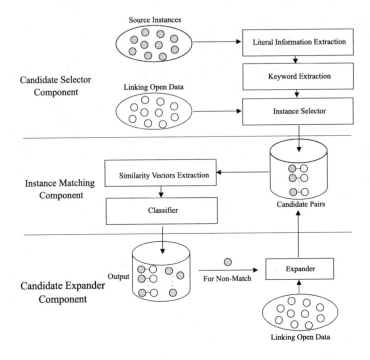

Fig. 1. The diagram of our proposed framework

(1) Candidate Selector, (2) Instance Matching and (3) Candidate Expander. The Candidate Selector aims to construct a set of candidate pairs. A candidate pair is a pair of instances, which is likely to match, between a source instance and an instance in LOD. After we acquire the set of candidate pairs, each candidate pair is conveyed to the Instance Matching in order to verify that it is a correct match. If the candidate pair is a match, the owl:sameAs link is established between instances. Otherwise, the candidate pair is unpaired. From the Instance Matching, outputs are candidate pairs, which already matched, and source instances, which still cannot map to instances in LOD. A non-matched instance from source is passed to the Candidate Expander in order to re-generate new candidate pairs. The detail of each component is given in the following section.

4.1   Candidate Selector

The Candidate Selector consists of three modules: Literal Information Extraction, Keyword Extraction and Instance Selector as shown in Fig. 1. The Literal information Extraction module intends to extract information, which can sufficiently represent an instance. After the information of instance is extracted, the information is passed thorough to the Keyword Extraction module. The Keyword Extraction module is to select cardinal words, which can capably represent an identity of instance, and then construct the set of keywords for an instance.

When we obtain the set of keywords, the Instance Selector module utilizes the set of keywords as representing of instance in order to select instances in the LOD, which are likely to match, and pair them as candidate pairs. The following subsections explain the detail of each module.

Literal Information Extraction. In the Literal Information Extraction module, most studies extract the literal information, which relate to an instance by considering two common properties, rdfs:label and rdfs:comment. Those two properties are widely included in most instances, especially instances in the LOD. Nonetheless, instances are defined dependently by many publishers that caused a heterogeneity problem. Extracting information limiting to rdfs:label property and rdfs:comment property might not be sufficient because some publishers might omit those two properties. If those two properties are absent, we do not get any information about the instance. We therefore select all properties, which are literal, and treat them as string. We define string into two types. One is a short string (l_s). l_s is a string, which contain at most one sentence or one phrase. Most l_s can specify instances well because a short string usually is likely to be a name of object, e.g. a label of instance, a symbol that represent object or a concise description of instance. Therefore, all information in l_s is quite essential to represent instances. The other string is long string (l_l). l_l generally give a finer detail, which describe the instance, than l_s. In case that l_s cannot disambiguate between instances, l_l can efficiently support to differentiate instances. However, we do not consider all information in l_l as l_s because l_l may contain information whether directly related to instance or not. Therefore, in the Literal Information Extraction module, we extract literal information from an instance as l_s and l_l respectively.

Keyword Extraction. After we obtain literal information from an instance, a set of keywords relating to an instance is selected and constructed by the Keyword Extraction module. The process of selecting a set of keywords depends on type of literal information. In case of l_s, it contains crucial information so omitting some words in l_s may lack ability to represent an identity of instance correctly. Consequently, we utilize the whole word in l_s as keywords by the following patterns. There are four patterns for selecting keywords. In the first pattern, we use a whole sentence or a whole phrase to be a keyword. In other patterns, we use an N-gram technique to generate the rest of three patterns by setting $n = 1, 2, 3$ respectively. For l_l, we cannot utilize the same strategy as l_s because l_l contains a lot of words and sentences. If we use the same strategies, we will obtain a huge set of keywords. It is unacceptable since if we acquire a lot of keywords that mean in the Instance Selector module, we will obtain a lot of candidate pairs. Consequently, the computational time is not feasible. To avoid the computational time problem, we carefully select the words, which highly reflect instances by using a Name Entity Recognition (NER) technique. Since l_l contains a long description of text, it therefore comprises of many name entities such as person, location and organization name, which highly probably represent an instance. If we select name entities in order to represent an instance, we can limit relative keywords to a small set. Therefore, we use a NER system[2] to retrieve related name entities from l_1 and handle them as keywords. After that we combine keywords, which obtain from l_s and l_1, to be a set of keywords.

Instance Selector. The LOD cloud consists of many data sources and many instances as well. If we perform instance matching for every instance, it causes a lot of computational cost and time consuming. In order to limit possible candidate pairs, we install the Instance Selector module into the Candidate Selector component. The Instance Selector module is to select instances, which relate to instance in source instances, from the LOD cloud by using the set of keywords. Before collecting the candidate in the LOD cloud, we need to specify the primary source for searching instances in the LOD cloud. We intend to look up instances at the DBpedia[3] [14] first because DBpedia is widely attended by many researchers. Hence its data is revised frequently and quality of data is quite correct. Furthermore, many techniques are created to support working with DBpedia. In addition, DBpedia is also a common hub of the LOD cloud because it has a lot of connections to other data sources since it contain a lot of owl:sameAs links. As a result, it is necessary for the Candidate Expander component. We therefore select DBpedia as a primary data source for collecting related instances in the LOD cloud.

To collect related instances from DBpedia, we input the set of keywords, which is obtained from an instance by the Keyword Extraction module, into the Instance Selector module. When the Instance Selector module receives the set of keywords, each of keyword is used to query the DBpedia's instances via the DBpedia Lookup Service[4] interface. Note that, the DBpedia Lookup Service is the service, which can look up the unique URI of instances in DBpedia by using a keyword. For example, if we input the keyword "tokyo", we will obtain the URI "http://dbpedia.org/resource/Tokyo", which is an URI of Tokyo instance. Thanks to this service we can collect URI of instances, which related to the keyword, easily. As a result we acquire candidate pairs, which prompt to verify similarities for instance matching.

4.2   Instance Matching

As we can see in Fig. 1., there are two modules in the Instance Matching: the Similarity Vector Extraction and Classifier. After we acquire candidate pairs from the Candidate Selector, the Similarity Vector Extraction module computes similarity vectors between candidate pairs. Various similarity metrics are applied in order to compute similarity vectors. The similarity vectors then are used to determine that a candidate pair is a correct match or not by the Classifier

---

[2] http://nlp.stanford.edu/software/CRF-NER.shtml#Extensions
[3] http://dbpedia.org/
[4] http://wiki.dbpedia.org/lookup/

module. The Classifier module establishes owl:sameAs link if the candidate pair is a correct match. The detail of each module is given as follows.

Similarity Vector Extraction. The Similarity Vector Extraction module is designed to extract similarity vectors for each candidate pair by computing similarities between each pair. The literal information from each candidate pair is extracted and the Similarity Vector Extraction module then computes similarity vectors between candidate pairs. Before we can compute similarity vectors, we need to consider that which similarity metrics are properly suitable for computing similarity between instances.

Generally, mapping instance toward the LOD cloud is challenged since data sources are heterogeneity and some data is distorted or ambiguous. Limiting similarity metrics to some metrics might not be sufficient enough. We therefore need to conduct an experiment to find suitable metrics. Firstly, the basic similarity metrics, which are frequently used in [13,15,16], are taken into account. There are 6 similarity metrics: TF-IDF (Term Frequency and Inverse Document Frequency) Cosine similarity, IDF Cosine similarity, TopIDF similarity, Edit Distance similarity, Count similarity as well as Jaro-Winkler similarity. Please refer to study in [13,15,16] for further information about similarity metrics. We also introduce the 2 novel similarity metrics regarding the relevance of source instance and instance in LOD.

Our two similarity metrics take advantages from the Candidate Selector. From the Keyword Extraction module, we can get the set of keywords, which related to instances. Consequently the set of keyword can provide implicitly information. Instances, which share the same keyword, are likely to match because they probably describe the same thing. We therefore capture similarity vectors by using CommonKeyword similarity metric as in Eq. 1 where K_S is a set of keywords of source instance and K_{LOD} is a set of keywords of instance in LOD.

$$\text{CommonKeyword}(K_S, K_{LOD}) = \frac{|K_S \cap K_{LOD}|}{|K_S \cup K_{LOD}|} \tag{1}$$

The other similarity metric uses information obtained from the Instance Selector module. In the Instance Selector module, we select a set of URIs of instance from the LOD cloud. The selecting process may select the URI more than once. The URI, which was selected frequently, has a high chance to match because it shows that information between instances is corresponded. We therefore assume that how many times the URI was selected, how likely they are matched. We compute the similarity metric via Eq. 2 where URI_i is how many times the URI of instance i is selected and I is a set of URIs in the LOD, which selected by the Instance Selector module.

$$\text{URIHits}(URI_i) = \frac{n(URI_i)}{\sum_{i \in I} n(URI_i)} \tag{2}$$

After defining similarity metrics, we perform similarity vector extraction process by the following steps. Firstly we extract literal information from instances. The procedure is similar to the Literal Information Extraction module. After we obtain literal information, we then compute similarity vectors.

Classifier. To identify that each candidate pair is correctly matched, we use a machine learning approach. There are many machine learning techniques such as neural network, decision tree and support vector machine. In this study, we choose the Support Vector Machine (SVM) based approach. The SVM is capable to predict that which class is candidate pairs belonging to. In our study, there are two classes: matched class and non-matched class. If the prediction result is positive or 1, it means candidate pair are correctly match so it belong to matched class. Otherwise, they are non-matched class.

In our study, the major contribution is to show that how we can automatically map source instances to Linked Data resources. Although we limit instance matching component to the basic approach, it should be considered we can improve it by replacing with powerful approaches such as AgreementMaker [6], ObjectCoref [8], SERIMI [9] and SLINT+ [12].

4.3   Candidate Expander

The Expander module is designed to expand instances to other data sources in the LOD cloud in order to pair new candidate pairs, which are more likely to match, and find missing instances in some data sources. The Expander module is an agent, which can traverse through toward the LOD cloud by using owl:sameAs properties, in order to find new candidate pairs. We assume that adjacent instances hold useful information for the instance. If we cannot map the instance to one data source, we can consider adjacent instances of the instance and we can find other instances, which relate to the instance. If adjacent instances have explicitly links to the LOD cloud, we can traverse through the LOD cloud by adjacent instances and we can discover instances, which are likely to match. We show the example of instance expanding in Fig. 2. The instance "Baron Papanoida" does not contain in DBpedia. Considering adjacent instances, we can find the adjacent instances of "Baron Papanoida", which are "Star war iii Revenge of the Sith" and "George Lucas". By owl:sameAs property, we can link to DBpedia and then to Freebase. By utilizing owl:sameAs and adjacent instances, we can reach to the instance in Freebase, which is likely to match to the instance "Baron Papanoida". Therefore, it is possible to find other instances in other data sources in the LOD cloud by utilizing adjacent instances.

By utilizing adjacent instances, we can gradually map instances to the LOD cloud step by step. However, we need some heuristic functions to limit the expanding space because if we do not limit the expanding space, the expanding space will grow up rapidly due to the characteristic of adjacent instances in the LOD cloud. To limit the expanding space, we assume that the expanded instance should share some keywords, which are similar to the source instance. If the expanded instance does not share some similar keywords, the instance

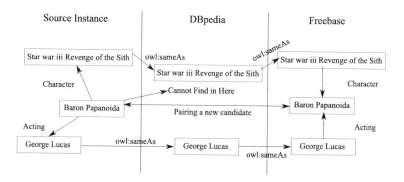

Fig. 2. The example of expanding instance from one source to any sources

is pruned. Based upon our preliminary experiment, we can reduce the expanding space from approximately more than 360,000 instances per instance to 45 instances per instance.

The algorithm of Candidate Expander module is described in Algorithm 1. The idea of algorithm is that we have a queue (Q) of instances. We attempt to add adjacent instances, which share the keywords to Q. If the adjacent instance has an owl:sameAs property to other instances, the other instances are also added to Q. The instances in Q are explored until there is no any instances in Q. When the instance and the adjacent instance in the other source, which share the same keywords, is paired to be a new candidate pair, the new candidate pair is added to a set of candidate pairs (C). As a result we can obtain new candidate pairs. Note that, For each instance $r \in I_S$, we extract the set of keywords by function *ExtractKeywords*. The adjacent instances are obtained from function *Adj*. The function *sameAsLink* is to return the instance which links by an owl:sameAs property.

5 Experimental Evaluation

In this section, the detail of experiments is described. Experimental setup and Experiments and Results are involved. Experimental setup is to explain the configuration for experiments, especially the data set and the evaluation method. In Experiments and Results section, the experiments are conducted. The details of Experimental setup and Experiments and Results are given in the following section.

5.1 Experimental Setup

The aim of experiment is to investigate the efficient of the framework for instance matching toward LOD cloud. The source instances are selected from Ontology

---

Algorithm 1. Expander Module for Generating New Candidate Pairs

---

Input: I_s (Set of instance which still do not match to any instances)
Output: C (Set of Candidate pair)
1: $C \leftarrow \emptyset$
2: $Q \leftarrow \emptyset$ # Q is a queue of instances, which will explore and expand
3: **foreach** $r \in I_s$ **do**
4:    $Q \leftarrow r$
5:    $K \leftarrow ExtractKeywords(r)$ #Extract a set of Keywords from instance r
6:    **while** $Q \neq \emptyset$ **do**
7:       $p \leftarrow Q.Dequeue()$ #Get the first instance of Q
8:       $N \leftarrow Adj(p)$ #Get the adjacent instances of instance p ans store in N
9:       **foreach** $e \in N$ **do**
10:        **if** $e \in K$ **then**
11:           $Q \leftarrow e$ #If an adjacent instance e share the same keyword with
             instance r then add to Q for exploring in the next round
12:           $C \leftarrow (r, e)$ #Pair instances between instance r and instance e
13:        **end if**
14:     **end foreach**
15:     $Q \leftarrow sameAsLink(p)$ #Get other instances, which were linked to the
          instacne p by owl:sameAs for expanding instances and store in Q
16:   **end while**
17: **end foreach**
18: **return** C

Alignment Evaluation Initiative 2012 (IM@OAEI 2012). All instances are manually matched and aligned to instances in LOD cloud. Before alignment processing, we need to consider that which data source in LOD cloud we aim to do alignment because there are many sources. As we mentioned in the Instance Selector section, we initialize the first source in the LOD cloud as DBpedia. After we setup the first data source, we manually align instance from IM@OAEI2012 to it. However there are still some instances missing from DBpedia. The coverage in DBpedia is only 90.36 % for IM@OAEI2012 data set. As we expected, only one data source is not sufficient to align instances.

In order to find instances in other source, we need to determine which data source we should traverse to find instances next. If we do not consider the second source, we will suffer from huge search space since there are many data sources connecting to DBpedia. We therefore use Freebase as the second data source because instances in IM@OAEI2012 data set are closely related to Freebase. Therefore, we also manually do alignment for each instance from IM@OAEI2012 to Freebase as well. By manual alignment, we can find the rest of instances, which are missing from DBpedia. As a result, we acquire alignment pairs, which match instance pairs between the IM@OAEI2012 data set and the LOD cloud.

To evaluate our framework, we use three standard evaluation metrics: precision, recall and F-measure. The precision aims to measure the correctness of system while the recall is to measure the coverage of system. Usually if the precision is high, the recall will low, conversely, if the recall is high, the precision will low. Therefore, we also report the F-measure, which is interpreted result from the precision and the recall.

5.2   Experiments and Results

In this section, we evaluate our framework. The evaluations are divided into three experiments. Experiment 1 aims to evaluate correctness of selecting candidate pairs. In the experiment, Candidate Selector component and Candidate Expander component are involved. The purpose of Experiment 2 is to find the appropriate similarity vectors for Instance Matching component. Experiment 3 is designed to evaluate the whole framework and show the importance of Candidate Expander component. The details of each experiment are as follows.

Experiment 1. We evaluate the Candidate Selector together with the Candidate Expander since their purpose is similar. They aim to select candidate pairs between a source instance and a LOD instance. However, the Candidate Selector is to select candidate pairs in a general case while the Candidate Expander selects candidate pairs, which are missing. Therefore, the Candidate Expander is performed after matching some instances by the Candidate Selector and the Instance Matching. In Experiment 1, we manually verify candidate pairs in order to avoid the effect of the Instance Matching component. To evaluate the Candidate Selector and the Candidate Expander, we therefore conduct an experiment to find the coverage of selecting correct candidate pairs. The Candidate Selector and the Candidate Expander are evaluated. We use lookup phrase in extended version of AgreementMaker in study [17], which is used only label to retrieve candidate pairs for matching instance in LOD as a Baseline system. We also conduct the experiment by changing the first source from DBpedia to Freebase in order to compare the effect of data sources since instances from IM@OAEI 2012 are closely related to Freebase. We evaluate the experimental result by focusing on three evaluation types: selecting, non-selecting and missing. Selecting means that we can generate existing pairs between source instances and instance in the first source of LOD. Non-Selecting is that we cannot successfully generate existing pairs. Missing is that there are no existing matched pairs. The results of experiment are reported in Table 1.

The results show that although Candidate Selector with DBpedia and Candidate Selector with Freebase generate a number of average candidates more than the Baseline system [17], Both of them still give better performance in selecting the candidate than the Baseline system. In addition, both of Candidate Selector with DBpedia and Candidate Selector with Freebase can reduce the non-selecting candidate pairs 4.41 % and 1.1 % respectively. Considering the Candidate Expanders result, we can also notice that Candidate Expander yields the best performance due to decreasing of non-selecting candidate pairs and missing pairs. Consequently, we can conclude that the set of keywords, which obtain from Name Entity Recognition, really helps to generate candidate pairs and the Candidate Expander can find missing instances effectively. Furthermore,

Table 1. The result of generating candidate pairs

Type	Selecting (%)	Non-selecting (%)	Missing (%)	Average candidate
Baseline [17]	84.30	6.06	9.64	**13.81**
Candidate Selector with DBpedia	88.71	1.65	9.64	82.41
Candidate Selector with Freebase	95.04	4.96	-	754.79
Candidate Expander	**97.80**	**0.00**	**2.20**	92.53

although our data set is closely related to Freebase, the results of Candidate Expander still give the better result for selecting correct candidate pairs than Candidate Selector with Freebase. It can conclude that if we initial any data source we can reach the best of selecting candidate pairs by using the Candidate Expander. This experiment therefore shows that the set of keywords of instance is essential for selecting candidate pairs and the Candidate Expander is necessary to increase the coverage for mapping instances toward the LOD cloud.

Experiment 2. In this experiment, we aim to evaluate the Classifier module in the Instance Matching. The Classifier module in Instance Matching is highly importance because it use for verifying that each candidate pair is correctly matched. If the performance of Classifier module is not good enough, the performance of framework is also not satisfying. For the Classifier module, similarity vectors representing the similarities between candidate pairs numerously affect to the performance of the classifier module because if the similarity vectors well represent between candidate pairs, the SVM model, which trained by those similarity vectors, will verify candidate pairs correctly. We therefore conduct the experiment on Instance Matching to find the suitable similarity vector set, which yield the best performance.

There are many studies [13,15,16], which proposed similarity vectors for instance matching. Rong et al. [13] presented the similarity vectors for instance matching. The similarity metric, which Rong used are TF-IDF cosine similarity, IDF cosine similarity, TopIDF similarity, edit distance and count similarity. RiMON [16] is a multi-strategy ontology matching framework. For instance matching, RiMON utilize edit distance between name of instances and TF-IDF cosine similarity between information of instances as similarity vectors. In the record linkage, there is a study proposed similarity vectors for name-matching [15]. The study show that the TF-IDF with Jaro-Winkler similarity metric give the interesting performance. We therefore use TF-IDF between information of instance with Jaro-Winkler between name of instance as similarity vectors and called "TF-IDF_Jaro-Winkler". We combine the similarity vectors of three above approaches [13,15,16] in order to construct new similarity vectors and we referred this set as "Combined set". We improve Combined Set by adding our two

Table 2. The results of similarity vector set

Approach	Precision	Recall	F-Measure
Rong et al. [13]	**0.97333**	0.81111	0.88485
TF-IDF_Jaro-Winkler [15]	0.91358	0.81319	0.86047
RiMON [16]	0.92500	0.81319	0.86550
Combined set	0.96104	0.82222	0.88623
Combined set with 2 features	0.94118	**0.88889**	**0.91429**

similarity vectors, which presented in similarity vector extraction section and call "Combined with 2 features".

The similarity vectors are used to train a classifier model as we describe in the Classifier section. We randomly select 25 % of the data set for the experiment and use 10-fold cross-validation to evaluate each similarity vector set. The results of each similarity vector set are reported in Table 2.

The results show that the highest precision is 0.97333. It acquires from the similarity vectors, which proposed by Rong et al. [13]. When we combine three approaches [13, 15, 16], the recall of Combined set similarity vectors outperforms other three baselines while the precision slightly decreased from similarity vectors proposed by Rong et al. [13]. However, considering the F-measure between Combined set and Rong proposed set, we can see that the Combined set yield the better F-measure. For Combined set with 2 features, the recall and the F-measure of Combined set with 2 features yields the highest result. It shows that the similarity vectors, which compute from information of the Candidate Selector, are useful to discover the candidate pairs because recall is evidently increased. The experiment shows that Combined set with 2 features outperforms the other similarity set. We therefore utilize this similarity vectors for other experiments.

Experiment 3. In the last experiment, we conduct an experiment to evaluate the whole framework. In order to evaluate our idea and show importance of the Candidate Expander, we use the framework without the Candidate Expander to be a baseline and compare it with our framework. The 10-fold cross-validation technique uses to evaluate the results. The results are shown in Table 3.

The results show that the Candidate Expander immensely enhances the recall of framework while it slightly drops the precision. Therefore, the F-measure of framework with the Candidate Expander substantially outperforms the

Table 3. The results of system with and without the Candidate Expander

Candidate Expander	Precision	Recall	F-Measure
Without	**0.94314**	0.77260	0.84940
With	0.93805	**0.85027**	**0.89201**

F-measure of Framework without the Candidate Expander. We also conduct the significant testing in order to test the statistical difference between frameworks. We use paired t-test among precision recall and F-measure of each fold. The significant testing showed that there is a statistical significant difference between the F-measure and the recall of our framework with the Candidate Expander and without the Candidate Expander. The p-value of recall and F-measure are less than 0.001 while the p-value of precision is 0.53. It means that the Candidate Expander improves recall and F-measure of framework significantly while it slightly decreases the precision without significance. Therefore, the Candidate Expander is essential for mapping instances toward the LOD cloud since it showed the contribution for improving the overall framework.

6   Conclusion and Future Work

In this study we proposed a framework for mapping instances to Linked Data resources. Our framework consists of three components: Candidate Selector, Instance Matching and Candidate Expander. The Candidate Selector is designed to select candidate pairs between a source instance and a LOD instance. The Instance Matching is to verify that candidate pairs correctly match. The Candidate Expander aims to find other candidate pair in different data sources. From our experiment, we showed that the Candidate Expander is necessary for discovering other new candidate pairs because the coverage of selecting correct candidates is increased. Furthermore, we also showed that the information obtaining from the Candidate Selector component really help the Instance Matching component to discover candidate pairs more correctly. Eventually, we showed that our framework with the Candidate Expander yield better performance than the framework without the Candidate Expander. This also confirms the importance of Candidate Expander component.

For our future work, although we have already shown the contribution of Candidate Expander, we still conducted the experiments on two data sources in the LOD cloud. We therefore plan to conduct the experiments more than two data sources in order to study deeply in the detail of Candidate Expander. In addition, for the Candidate Expander, we limit the expanding space by the simple heuristic function, which only consider common keywords. There is still a lot of room for investigating an appropriate heuristic function for instance expansion in the LOD cloud.

References

1. Berners-Lee, T.: Linked data - design issues (2006). http://www.w3.org/DesignIssues/LinkedData.html
2. Klyne, G., Carroll, J.J.: Resource Description Framework (RDF): Concepts and abstract syntax, W3C Recommendation (2004). http://www.w3.org/TR/rdf-concepts/
3. Bizer, C., Heath, T., Berners-Lee, T.: Linked data - the story so far. Int. J. Semant. Web Inf. Syst. 4(2), 1–22 (2009)

 4. Bechhofer, S., Harmelen, F., Hendler, J., Horrocks, I., McGuinness, D.L., Patel-Schneider, P.F., Andrea Stein, L.: OWL web ontology language reference, W3C Recommendation (2004). http://www.w3.org/TR/owl-ref/

 5. Volz, J., Bizer, C., Gaedke, M., Kobilarov, G.: Discovering and maintaining links on the web of data. In: Bernstein, A., Karger, D.R., Heath, T., Feigenbaum, L., Maynard, D., Motta, E., Thirunarayan, K. (eds.) ISWC 2009. LNCS, vol. 5823, pp. 650–665. Springer, Heidelberg (2009)

 6. Euzenat, J., Ferrara, W., Hage, A., Hollink, L., Meilicke, C., Nikolov, A., Scharffe, F., Shvaiko, P., Stuckenschmidt, H., Zamazal, O., Trojahn, C.: Final results of the ontology alignment evaluation initiative 2011. In: Proceedings of the 6th Workshop on Ontology Matching, pp. 85–113 (2011)

 7. Niu, X., Rong, S., Zhang, Y., Wang, H.: Zhishi.links results for OAEI 2011. In: Proceedings of the 6th Workshop on Ontology Matching, pp. 220–227 (2011)

 8. Hu, W., Chen, J., Qu, Y.: A self-training approach for resolving object coreference on the semantic web. In: Proceedings of the 20th International Conference on World Wide Web, pp. 87–96. ACM (2011)

 9. Araujo, S., Tran, D., de Vries, A., Hidders, J., Schwabe, D.: SERIMI: Class-based disambiguation for effective instance matching over heterogeneous web data. In: The 15th Workshop on Web and Database Proc., pp. 19–25 (2012)

10. Nguyen, K., Ichise, R., Le, B.: Learning approach for domain-independent linked data instance matching. In: Proceedings of the 2nd Workshop on Mining Data Semantics, no. 7 (2012)

11. Nguyen, K., Ichise, R., Le, B.: SLINT: a schema-independent linked data interlinking system. In: Proceedings of the 7th Workshop on Ontology Matching, pp. 1–12 (2012)

12. Nguyen, K., Ichise, R., Le, B.: Interlinking linked data sources using a domain-independent system. In: Proceedings of the 2nd Joint International Semantic Technology Conference, pp. 113–128 (2012)

13. Rong, S., Niu, X., Xiang, E.W., Wang, H., Yang, Q., Yu, Y.: A machine learning approach for instance matching based on similarity metrics. In: Cudré-Mauroux, P., et al. (eds.) ISWC 2012, Part I. LNCS, vol. 7649, pp. 460–475. Springer, Heidelberg (2012)

14. Auer, S., Bizer, C., Kobilarov, G., Lehmann, J., Cyganiak, R., Ives, Z.G.: DBpedia: a nucleus for a web of open data. In: Aberer, K., et al. (eds.) ASWC 2007 and ISWC 2007. LNCS, vol. 4825, pp. 722–735. Springer, Heidelberg (2007)

15. Cohen, W., Ravikumar, P., Fienberg, S.: A comparison of string metrics for matching names and records. In: Proceedings of the Workshop on Data Cleaning and Object Consolidation (2003)

16. Li, J., Tang, J., Li, Y., Luo, Q.: RiMON: a dynamic multistrategy ontology alignment framework. IEEE Trans. Knowl. Data Eng. 21(8), 1218–1232 (2009)

17. Caimi, F.: Ontology and instance matching for the linked open data cloud. Master Thesis of University of Illinois at Chicago (2012)

Learning and Discovery

An Automatic sameAs Link Discovery from Wikipedia

Kosuke Kagawa, Susumu Tamagawa, and Takahira Yamaguchi[✉]

Keio University, 3-14-1 Hiyoshi, Kohoku-ku, Yokohama-shi,
Kanagawa 223-8522, Japan
yamaguti@ae.keio.ac.jp

Abstract. Spelling variants of words or word sense ambiguity takes many costs in such processes as Data Integration, Information Searching, data preprocessing for Data Mining, and so on. It is useful to construct relations between a word or phrases and a representative name of the entity to meet these demands. To reduce the costs, this paper discusses how to automatically discover "sameAs" and "meaningOf" links from Japanese Wikipedia. In order to do so, we gathered relevant features such as IDF, string similarity, number of hypernym, and so on. We have identified the link-based score on salient features based on SVM results with 960,000 anchor link pairs. These case studies show us that our link discovery method goes well with more than 70 % precision/recall rate.

Keywords: Spelling variants · Disambiguation · Synonym · sameAs link · Wikipedia · Ontology

1 Introduction

Spelling variants of words or word sense ambiguity takes many costs in such processes as Data Integration, Information Searching, data preprocessing for Data Mining, and so on. It is useful to construct relations between a word or phrases and a representative name of the entity to meet these demands. As Wikipedia already becomes the hub resource which many resources refer to, it is useful to construct a word or phrases-to-Wikipedia's URI link. It seems easy to extract "sameAs" link from Wikipedia's redirect link or extract "meaningOf" link from Wikipedia's disambiguation page, considering the purpose of them. However, since they are too noisy, they need to be cleansed for extraction.

In this paper, we propose a method for extracting "sameAs" and "meaningOf" links using Wikipedia's anchor link sets, redirect link sets, and disambiguation page. We construct a large-scale and wide-coverage word or phrases-to-URI network.

We have used just machine learning after manually labeled positive/negative data. We have extracted more than 960,000 links between words and Wikipedia URIs. The result and experiments are based on Japanese Wikipedia, but our method can be applied for another language's Wikipedia.

In the remaining of this paper, Sect. 2 discusses related work on synonym discovery, disambiguation, semantics of owl:sameAs. Section 3 describes our

W. Kim et al. (Eds.): JIST 2013, LNCS 8388, pp. 399–413, 2014.
DOI: 10.1007/978-3-319-06826-8_29, © Springer International Publishing Switzerland 2014

background idea, definition for "sameAs" and "meaningOf" links, overview of Wikipedia's resource focusing on links with statistical information. Section 4 shows us feature modeling for machine learning, the environment of extraction, and how experimental results is going.

2   Related Work

Here we discuss how related work is going in the field of the semantics such as Linked Open Data, Ontology, word sense or spelling variants disambiguation.

Christian Bøhn et al. [1] has proposed the method for constructing synonym dictionaries from Wikipedia using its anchor link sets. The goal of the method is to expand the query for information search. In the paper, their criteria of "synonym" are different from ours. Their goal is to expand vocabularies for information search query, on the other hand, our goal is for Data Integration. What they do causes the improvement for query expansion, however, when we conduct the Data Integration by their criteria, there can be mismatch. For example, in their criteria, a pair of "United Nations" and "UN" is defined as "synonym", but if there is "UN" as a meaning of "Urea Nitrogen" in the database, the data integration fails. "UN" is one of the representative names of the "United Nations". It is true "United Nations" is used as "UN", but "UN" itself is not always used as "United Nations". It depends on the context. In addition, their method is based on the heuristic rules, so it will be less robust especially when applied to Wikipedia in multiple languages.

T. Michishita et al. [2] has proposed the method for word sense disambiguation using Wikipedia's disambiguation pages. They create word sense annotated corpus focusing on Wikipedia's disambiguation pages. They conduct machine learning for word sense disambiguation by using Wikipedia's resource and record high precision/ recall rate.

Harry Halpin et al. [3] discusses owl:sameAs link. They claim that there are potential misuse concerning owl:sameAs link. They categorize the situation when people incorrectly put the owl:sameAs link such as "Identical but referentially opaque", "Identity as claims", "Matching", "Similar" and "Related".

3   Extraction for sameAs Link from Wikipedia

In this section, we describe our basic idea for usefulness and necessity of "sameAs" link, the definition of "sameAs" and "meaningOf" link itself, overview of Wikipedia's resources, and then method for "sameAs" and "meagningOf" link extraction from Wikipedia.

3.1   Basic Idea

Here we discuss the costs in the process of computing and describe the basic idea for it. Usually, humans indicate or describes a meaning with one or more written form. On the other hand, there is more than one meaning with one written form. Usually, humans can disambiguate a meaning considering the situation. On the other hand, the machine cannot recognize which meaning does the written form indicate (Fig. 1). So, it needs many costs for humans to match one written form to one meaning when they compute beforehand. If there is a structure where each of written form links to one meaning as a same relation before we compute (Fig. 2), it reduces costs for us. In Wikipedia, since a URI represents one meaning without ambiguity (we mention it later in Sect. 3.3), we focus on how to annotate the "sameAs" link between one written form to one Wikipedia's URI.

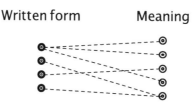

Fig. 1. Written form and meaning

Fig. 2. Our goal (Example)

Our target for "word" or "written form" mentioned above is so-called word or phrases that consist of more than one tokens, in general use such as "President", or "The President of United States".

When we annotate the "sameAs" link, our method mainly focuses on the cases when the different written forms indicate the same meaning and the cases when the almost the same spellings indicate different meanings (Table 1). In a situation like A or D, we can discover the "sameAs" link in relatively easy way such as using regular expression, converting uppercase/lowercase or some heuristic rules. On the other hand, in a situation like B, there can be decrease of recall score often when the entity has several types of written form. In a situation like C, there can be decrease of precision score often when the word itself has ambiguity. Our method overcomes these difficulties.

3.2 The Definition for sameAs

What is "sameAs"? The OWL defines "two URI references actually refer to the same thing: the individuals have the same identity [4]." For example, "World War II" and

Table 1. Our main target

	Spelling Matching	Spelling NOT Matching
Meaning Matching	A	B (Main focus)
Meaning NOT Matching	C (Main focus)	D

Table 2. Anchor link

Property type	Percentage
sameAs	0.401
meaningOf	0.184
Hypernym/Hyponym	0.089
Other relations	0.326

"Second World War" actually refer to the same thing(=meaning). Another example, "World War II" and "WWII" also refer to the same thing. However, how about the following case?

- "PM" and "Prime Minister"
- "PM" and "Project Manager"
- "Washington" and "George Washington"
- "Washington" and " Washington D.C."

Depending on context, "PM" can refer to "Prime Minister", or "Project Manager".

What is the difference between the "WWII" case and the "PM" case? Generally speaking, there is nothing except the meaning for "World War II" associated with "WWII". On the other hand, the word "PM" itself has more than one meaning. "Washington" indicates the person in the context of history, for example, on the other hand, it indicates the state in the context of geography.

The "sameAs" means there is no ambiguity between two words; the relation should be independent of the context. We should consider another issue. How can we recognize between the case where a word has only one meaning, and the case where a word has more than one meaning? In an extreme case, if someone defines the "WWII" for another meaning, can we annotate "sameAs" for "World War II"? We assume we can distinguish them apart on the basis of common sense. That is why we utilize the Wikipedia's resource that is commonly shared by many editors' knowledge. Namely, we decide that the "sameAs" property in our method limits within Wikipedia's resource (closed world). We do not refer to outside of Wikipedia's resource in this paper. For example, even if there is another meaning of "WWII" outside of the Wikipedia's resource, we discard the entity about it, as far as Wikipedia has only one entity of "WWII". The evaluation and the discovery of "sameAs" link are conducted by this criterion in this paper.

We also would like to mention that although example above consists of abbreviated word and unabbreviated word, we do not draw a line between them. It is true that abbreviated words tend to have several meanings, but there are also the exceptions such as "WWII".

In our method, we annotate the "sameAs" link between a written form that is unambiguous and Wikipedia's URI, on the other hand, we annotate the "meaningOf" link between a written form that is ambiguous and Wikipedia's URI. The "meaningOf" link is our original one. It is annotated when a word is one of the representative written form of Wikipedia's URI(=meaning). Figure 3 shows the example.

In the example above, since the "Washington" is an ambiguous word as we mentioned before and it is a representative word of "Washington, D.C." or "George Washington", we annotate "meaningOf" label between them. However, since we do not always represent "Washington" as "President of the United States", we annotate "other relation" label between them. On the other hand, since the "Second World War" and the "WWII" are unambiguous word, as well as they are same meaning of "World War II", we annotate "sameAs" label between them. As in Fig. 3, the "meaningOf" link tends to put when a written form has more than one Wikipedia's URI destination, and the "sameAs" link tends to put when a Wikipedia's URI is linked by more than one written form. We utilize these relations for one of the features in machine-learning.

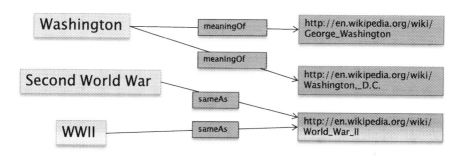

Fig. 3. Example of "sameAs" and "meaningOf" link

3.3 Wikipedia

In Wikipedia, editors are recommended to name a title of article without ambiguity. For instance, a title such as "Washington" is not recommended. In Wikipedia, the word "Washington" is not used for title but "Washington (State)", "Washington, D.C.", "George Washington" or "Washington University", with these unique identifiers. In addition, there is no article which is entitled with the same title as another article. Based on these rules of Wikipedia's guideline, we assume the Wikipedia's title does not have any ambiguities.

In an article of Wikipedia, there are many anchor link texts in the contents. Wikipedia editors put an anchor link in the format like this;

[[Colony of Virginia | Colonial Virginia]]

In an article, words are written in many ways. In this case, readers look at the word "Colonial Virginia". When readers click the anchor link, the page jumps to an article entitled with "Colony of Virginia" which is the unique name in Wikipedia. "[[ ]]" is an anchor link tag."|" is called piped tag which divides inside the tag space for two words. The left one is the Wikipedia title as a link destination. Since the right one is a word or phrases in the article. In this paper, we call the left one "Wikipedia URI" and the right one "written form". The Wikipedia URI usually seems to be entitled as a word or phrases. In addition, there is a sentential title such as "Everybody's Got Something To Hide Except Me And My Monkey" song by the Beatles. We also call sentential titles as Wikipedia URIs. Since editors put a word or phrase inside the anchor link tag, there is no need to consider the length of the phrase or some grammatical issues.

There are several relations between the written form and the Wikipedia URIs. The relation is sometimes the exactly same relation (so-called "sameAs") or an instance of the other word (so-called "type") or the other some relations.

The same Wikipedia URI has several types of written forms depending on the context. For instance, the "World War II" URI could be sometimes described as written forms of "WW2", "WW II", "Second World War" or "World War 2". On the other hand, the same written forms could refer to several types of Wikipedia URIs. For instance, the word "Washington" sometimes refers to "Washington (State)" in an article such as "Presidential Election campaign in 2008". So, the relation's feature between written forms and Wikipedia URIs is many-to-many. We would like to mention that it does not mean we discuss the page-to-page links in this paper. We discuss the pairs given in the anchor link format.

Wikipedia's Redirect Links are usually the alternative title from one written form to a Wikipedia URI. For instance, "President Obama" redirects to "Barack Obama" in any case. So, the relation between written forms and Wikipedia URIs is many-to-one in the Redirect Links, on the other hand, the relation is many-to-many in usual anchor links.

Wikipedia's Disambiguation pages are created for ambiguous word. The page has more than two Wikipedia URIs with simple explanations to distinguish one entity from another. For example, the ambiguous word such as "USA" is entitled with "USA_(disambiguation)". This resource is strongly effective to distinguish whether a word is ambiguous one or not.

3.4   Link Statistics in Wikipedia

We trimmed the HTML tags or other specific Wiki tags and then extracted anchor link pairs which are composed of a Wikipedia URI and a written form. As we mentioned before, we just target the pairs given in the anchor link format, we discard the information such as where the pairs appear, or which kind of context near the pairs. As a result, we extracted more than 960,000 unique pairs including redirect link sets from Wikipedia.

We randomly extracted 2,000 pairs from anchor link sets and then we manually surveyed the component of these and categorized the relations between a written form and a Wikipedia URI. We mainly focused on the popular properties such as rdf:type or owl:sameAs. Since our method focuses on ambiguous words and unambiguous Wikipedia URIs, we propose "meaningOf" property when we categorize as we mentioned in Sect. 3.2. The statistical results are below. In Table 3, "meaningOf" property is excluded because each written form in redirect link always links to only one destination.

Table 3. Redirect link

Property type	Percentage
sameAs	0.671
meaningOf	-
Hypernym/Hyponym	0.157
Other relations	0.172

Interestingly you can easily find the fact that even redirect link does not always represent the "sameAs" link. The percentage of "sameAs" relation is not so high, as in Table 3. Usually, a redirect link pair represents the "sameAs" relation, firstly because the written form of a redirect link pair always links to only one Wikipedia's URI regardless of context, secondly because Wikipedia's redirect link pairs are prepared for the purpose of linking between the same meanings to guide readers. However, the Wikipedia's redirect link pair is also prepared for the situation such as "Usually, when most people look at or think of the characters of 'UK', they intuitively associate it with United Kingdom, regardless of context". There is another entity represented by "UK" such as the "UK (a band team)". In addition, sometimes editors put an anchor link or redirect link between not so popular word or phrases and Wikipedia URI. For instance, "I Just Don't Understand" which is one of the songs by "the Beatles" links to the "Live at the BBC (The Beatles album)", just because the song is yet to be created as a title of Wikipedia's article. The song should have the relation like "recorded album" with "Live at the BBC (The Beatles album)". In a case like this, the reliability of redirect link as the "sameAs" tag is doubtful (Figs. 4, 5, 8).

Table 4 shows the statistics in disambiguation pages. The Number of pages means the number of disambiguation pages in Japanese Wikipedia. Total Number of items means the total number of listed items in disambiguation pages (see also Fig. 6). We also would like to mention that there can be some decrease of recall, because we have collected the listed items by scraping rules which is heuristically generated, so there can be a little bit mistaken items. As in Fig. 6, the listed items in a disambiguation page of Wikipedia normally begin with meaning, but sometimes they put the meaning at the end of the sentence. It is no doubt that disambiguation pages are created for the purpose of disambiguating the ambiguous word. Following our definition in Sect. 3.2, we can put "meaningOf" property on relations between a title of disambiguation page to each listed item, such as "USA" to "United States of America", "USA" to "Union of South Africa", "USA" to "United Space Alliance" and so on. However, there are also exceptions such as hyponym, hypernym relations. For example, "Yoshinoya" that is one of the restaurants links to "a restaurant". We manually surveyed the percentages of "meaningOf" relations among 1,000 pairs randomly extracted from Disambiguation pages.

Washington was born into the provincial gentry of [[Colony of Virginia|Colonial Virginia]] his wealthy became personally and professionally attached to the powerful [[William Fairfax]], who promoted his car [[French and Indian War]]. Chosen by the [[Second Continental Congress]] in 1775 to be commander-in-chi

Fig. 4. Wiki format

written form   Wikipedia_URI   written form   Wikipedia_URI

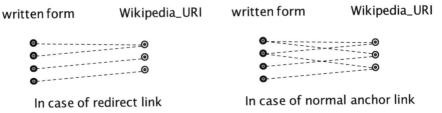

In case of redirect link   In case of normal anchor link

Fig. 5. Redirect links and normal anchor links

Fig. 6. Disambiguation page

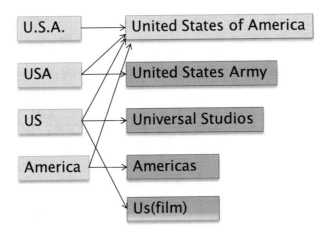

Fig. 7. Candidate written forms and Wikipedia URI (Example)

In order to clarify the number of "meaningOf" relations properly from our method, we acquire the "meaningOf" pairs from disambiguation pages beforehand on the assumption that the title of disambiguation page is trustful. In the Sect. 4, we show the properly-extracted number of "meaningOf" relations from anchor link pairs (without using disambiguation pages).

Hillary Rodham Clinton

From Wikipedia, the free encyclopedia

Hillary Diane Rodham Clinton /ˈhɪləri daɪæn ˈrɒdəm ˈklɪntən/; born October 26, 1947) is an American politician and diplomat who was the previously a United States Senator for New York from 2001 to 2009. As the wife of President Bill Clinton, she was also the First Lady of the presidential nomination.

Fig. 8. Candidate written forms and Wikipedia URI (Example)

Table 4. Disambiguation page

Number of pages	31,077
Total number of items	203,418
Average number of meanings with an ambiguous word	6.55
MeaningOf relations (among random 1,000 pairs)	794
Expected number of meaningOf relations	Approximately 150,000

4   Experimental Results and Discussion

4.1   Feature Modeling

Our method focuses not on the classifier selection or algorithm, but on the feature modeling in this paper. The criteria for the feature modeling are from the views of Information Search, Ontology, Link Analysis, Natural Language Processing, and Wikipedia Mining. The reason or validity for defining these features is also described below. Following the definition in Sect. 3.2, we try to put "sameAs", "meaningOf", or "Other relations" on the link pairs, namely we try to conduct 3 values classification.

1. Number of link pairs

    This feature is the frequency of given anchor link pairs which are written in all of the Wikipedia articles. We roughly assume when this score is higher, the probability of Positive gets higher. For instance, a pair of "World War II" and "Second World War" frequently appears in a text, on the other hand, a pair of "World War II" and "British Army during the Second World War" rarely appears. However, there is also probability that not so popular word can be labeled Negative because the number of pair is small.

2. Number of Candidate URIs and Number of Candidate written forms

    The former one is the number of Wikipedia URIs which are linked by the given written forms. The latter one is the number of written forms which link to the given Wikipedia URI. Figure 7 shows how to calculate these scores using an example.

    In this case, when given "U.S.A" and "Wikipedia URI: Unites States of America" pair, "U.S.A.", "USA", "US" and "America" link to " Wikipedia URI: United States of America", so the Number of candidate written forms is 4, on the other hand, "U.S.A." just links to " Wikipedia URI: United States of America", so the Number of URIs is 1.We assume these numbers represents how strong the pair is connected.

3. Appearance in the First passage

Sometimes, a written form itself is written in the Wikipedia URI's article as an alias. In our research, 22 % of alias among the "sameAs" links appears in the first passage of Wikipedia URI's article. This feature calculates whether written forms appear in the destination article (i.e. Wikipedia URI's article).

4. IDF score

This feature calculates how specific or popular a written form itself is(i.e. to eliminate words used as a pronoun). In (1), "D" is the total number of Wikipedia articles, and "d" is the total number of Wikipedia articles which contains the written forms in the text of article.

$$idf(w, D) = \log|D|/|d| \tag{1}$$

5. String Similarity score

This score is calculated by Levenshtein Distance and length of the words as (2).

$$sim(w1, w2) = LevenshteinDistance(w1, w2) * 2/\{len(w1) + len(w2)\} \tag{2}$$

6. Number of Hypernym

Most of Wikipedia URIs have its hypernym which appears in the lead sentence of the article. We extracted hypernyms by using the heuristic method like pattern matching or grammatical parsing beforehand. Most of Wikipedia articles have lead sentence in the first passage where the title of article is defined by simple explanation. For example, the first passage of "Hillary Rodham Clinton" starts with "Hillary Diane Rodham Clinton is an American politician and diplomat who was the 67th United States Secretary of State from 2009 to 2013, serving under President Barack Obama.". We collected the relations of Hillary Rodham Clinton-American politician and Hillary Rodham Clinton-diplomat from the passage by using the pattern matching rules.

The pattern matching examples are like these.

"[Title name] is a [hypernym]"
"[Title name] is a kind of [hypernym]"
"[Title name] refers to [hypernym]"
"… also known frequently as the [hypernym] …"

This method is partly based on [5, 6]. We altered the method for Japanese Wikipedia.

Table 5 shows in detail. We extracted hypernym by using the heuristic method such as pattern matching or parsing rules beforehand. We use the number of hypernym as a feature score which the Wikipedia URI has. It is based on the assumption that the number of hypernym which a word has is one of an indicator which represents the abstraction level and its popularity.

Table 5. Statistics about hypernym sets

Number of relations	Distinct hypernym	Precision	Recall
713,371	97,790	0.929	0.887

7. Number of Hyponym
   Like the number of hypernym above, we also use the number of hyponyms which Wikipedia's URI has.
8. Specific Mark-up
   () indicator and # tags are specifically used in Wikipedia. The former one is a unique indicator, as we mentioned before. The latter one is used like a hash-tag in HTML (i.e. it refers not to the title of Wikipedia itself, but to a written form in the article).

4.2   Proposed Method

Figure 9 describes the overview of our proposed method. Firstly we extract anchor link and redirect link pairs(i.e. written form and Wikipedia URI) from normal articles of Wikipedia. In this paper, we consider articles that are not created for particular use (such as listed pages or Wikipedia's guideline pages) as normal articles. Then we annotate "sameAs", "meaningOf", or other relations label on randomly extracted 2,000 link pairs. By using the annotated data, we conduct machine learning with feature sets described in the previous section. Finally, we automatically annotate three types of label(i.e. "sameAs", "meaningOf", or other relation) on anchor link and redirect link pairs with built model of classifier. On the other hand, we extract

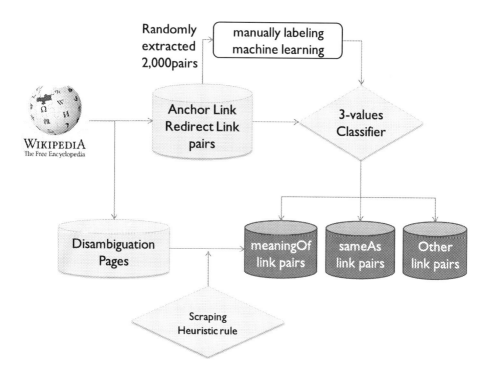

Fig. 9. System overview

disambiguation pages to find "meaningOf" link pairs by using scraping or heuristic rules, as we mentioned in Sect. 3.4.

4.3   Experimental Results

We used the 2013-May-13 version dump data of Japanese Wikipedia. First, we trimmed the HTML tags or other specific Wiki tags and then extracted anchor link pairs from all each article of Wikipedia. As a result, we extracted 960,000 distinct anchor link pairs. Next, we calculated each feature and then conducted the machine learning with C-SVC [7] with RBF kernel and given parameters. We used 2,000 training datasets from 960,000 anchor link pair sets for C-SVC. Then we manually annotated positive or negative label. After we built the classifier by C-SVC, we extracted 392,102 "sameAs" and "meaningOf" links from 960,000 anchor link pairs by using it.

The criteria of evaluation come from the definition of Sect. 3.2. Table 6 shows the results. The precision and the recall are calculated by randomly extracted 2,000 pairs from 960,000 distinct anchor link pairs. Note that the 2,000 pairs used for evaluation are different from training datasets.

Table 6. sameAs and meaningOf link discovery performance by C-SVC

Number of relations	Precision	Recall
392,102	0.71	0.77

Table 7 shows examples of the extraction(Note that in English Wikipedia there is disambiguation page entitled with "Goal", but in Japanese Wikipedia there is no disambiguation page entitled with "Goal"). In Table 8, we compare our result with Japanese Wikipedia ontology[8] concerning the number of "sameAs" and "meaningOf" links.

Table 7. Successful cases

Written forms	Wikipedia URI	Label
Play station2	Play Station 2	sameAs
Goal	Goal Inc.	meaningOf
Chinatsu Ito	Yuuho Mita	sameAs
米国(=United States)	アメリカ合衆国	sameAs
諭吉	福澤諭吉	meaningOf

Table 8. Extension for Japanese Wikipedia Ontology

Wikipedia's resource	sameAs link	meaningOf link
Info boxes (Japanese Wikipedia Ontology)	91,613	
Disambiguation Page	-	Approximately 150,000
Our method	266,529	125,573

5   Discussion

As shown in Table 7, the results show the method is effective regardless of string-similarity. As the first row shows, the string-similar pair tends to be annotated as "sameAs", on the other hand, the second row shows even the string-similar pair can be annotated as "meaningOf". "Goal" represents several meanings. Obviously, the string-similarity score well functions in the first case. In the second case, document frequency or hypernym/hyponym feature probably functions by calculating the abstract level or popularity of a word itself. The person's name is in the third row. Both of them actually indicate the same person. The spelling between the written form and the Wikipedia URI is totally different. In this case, the word itself appears in the first passage as an alias. In worst case, the spelling is totally different and the alias does not appear in the first passage. However, we observed the link features support to annotate "sameAs" label like the third row. These examples demonstrate the classifier of our method is composed of well-balanced feature sets.

As shown in Table 8, In Wikipedia, some article has its alias or another name inside the infobox. The method for extracting "sameAs" link in Japanese Wikipedia Ontology is based on infobox-scraping. Also in DBpedia[9], there are several types of vocabularies which are strongly related to Wikipedia URIs as instances, such as name, nickname, alias, and so on. However, since their method strongly depends on Wikipedia's infobox, the coverage of vocabularies is limited to the number of Wikipedia infoboxes. The percentage of articles with infobox is about 26 % of all the Wikipedia articles. So, their vocabularies are not sufficient from the point of Wikipedia's vocabulary-richness.

The percentage of each feature's contribution is described below as cumulative sum graph (Fig. 10) and (Table 9) based on the information gain ratio.

Most contributing feature is the link feature (i.e. Number of anchor link pairs, Number of URIs, and Number of Candidate written forms). We suppose this is caused strongly by our definition of "sameAs". However, it is not easy to formulate these three relations. When the number of anchor link is high, the situation provides us just

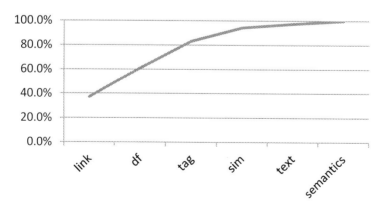

Fig. 10.   Feature performance (cumulative sum)

Table 9. Each feature's contribution (ratio)

Feature	link	df	tag	sim	text	semantics
Contribution	37.0 %	23.8 %	22.4 %	11.1 %	3.2 %	2.6 %

the information that the pair (i.e. written forms and URI) is popular, or the written forms are often used as the representative names of the URI. It does not provide us anymore, but this score contributes to excluding noisy pairs. When the number of URIs is high, the written forms tend to be an ambiguous word such as "Goal" in Table 7. It contributes to excluding the ambiguous word. When the number of written forms is high, this situation contributes to acquiring several synonyms. To sum up, the result testifies these three link features contribute to excluding noisy pairs, acquiring unambiguous word, and excluding ambiguous word. If these three features are not combined, the precision/recall score drops by 4–5 %.

Secondly, the document frequency feature is also quite effective. Probably, the feature functions for excluding the pairs in which written forms are used as a pronoun or popular and ambiguous word such as "Goal" in Table 7. It is quite natural because such common words account for about one-third of not "sameAs" link pairs, as shown in Table 2.

Mark-up tag feature is effective simply because the guideline of naming a title of Wikipedia recommends using it when the title itself has ambiguity.

The string-similarity score is not so high percentage of the method. It must be because our dataset is Japanese version. Japanese has several types of characters (hiragana, katakana, kanji, and alphabet), so the similarity score itself does not highly contribute to the classifier. In English or other language which uses just one kind of character, we are strongly sure the string similarity score well functions compared with Japanese.

The appearance in the first passage is not so effective. Even if 22 % of alias exists in the first passage of the article, our method cannot sufficiently acquire them. In the future work, we need to try to adopt the pattern-matching method. We can also improve the contribution of the features of semantics (i.e. the number of hypernym/hyponym). In English version of Wikipedia, we can utilize the resource of YAGO [10] where the depth of categories is defined instead of "the number of hypernym".

6   Conclusion

Our method can extract many of the "sameAs" links from Wikipedia without dependency on the linguistically specific feature, thesaurus, and other costing resources. We are sure these links reduce the costs

However, no one can evaluate the true meaning of recall score of the "sameAs" link. There must be another "Barack Obama" entity. However, we are sure of the general usefulness of Wikipedia's large-scale, wide-coverage and high-updated vocabularies.

We have applied our method for Japanese Wikipedia in this paper for the purpose of our resource extension for Japanese Wikipedia Ontology [2], however, the method can also be applied for English or other languages. Statistics and evaluation are also based on the Japanese Wikipedia in this paper, so we should evaluate English version Wikipedia in the near future.

In the future work, we should investigate into how many costs the "sameAs" and "meaningOf" network is going to reduce, by conducting the Data Integration or searching the cross-domain documents. Moreover, we should annotate not only "sameAs" link and "meaningOf" link, but also "rdf:type" or other label.

References

1. Bøhn, C., Nørvåg, K.: Extracting named entities and synonyms from wikipedia. In: Advanced Information Networking and Applications (AINA), pp.1300–1307 (2010)
2. www-nishio.ist.osaka-u.ac.jp/Thesis/master/2009/michishita/thesis.pdf
3. Halpin, H., Hayes, P.J., McCusker, J.P., McGuinness, D.L., Thompson, H.S.: When owl:sameAs isn't the same: an analysis of identity in linked data. In: Patel-Schneider, P.F., Pan, Y., Hitzler, P., Mika, P., Zhang, L., Pan, J.Z., Horrocks, I., Glimm, B. (eds.) ISWC 2010, Part I. LNCS, vol. 6496, pp. 305–320. Springer, Heidelberg (2010)
4. Bechhofer, S., van Harmelen, F., Hendler, J., Horrocks, I., McGuinness, D.L., Patel-Schneider, P.F., Stein, L.A.: OWL Web Ontology Language Reference (2004)
5. Yamada, I., Torisawa, K., Kazama, J., Kuroda, K., Murata, M., DeSaeger, S., Bond, F., Sumida, A., Hashimoto, C.: Hyponymy relation acquisition based on distributional similarity and hierarchical structure of wikipedia. Inf. Process. Soc. Jpn. **52**, 3435–3447 (2011)
6. Hearst, M. A.: Automatic acquisition of hyponyms from large text corpora. In: 14th International Conference on Computational Linguistics, pp.539–545 (1992)
7. http://www.csie.ntu.edu.tw/~cjlin/libsvm/
8. Tamagawa, S., Sakurai, S., Tejima, T., Morita, T., Izumi, N., Yamaguchi, T.: Learning a large scale of ontology from japanese wikipedia. In: Web Intelligence and Intelligent Agent Technology (WI-IAT), pp.279–286 (2010)
9. Auer, S., Bizer, C., Kobilarov, G., Lehmann, J., Cyganiak, R., Ives, Z.G.: DBpedia: a nucleus for a web of open data. In: Aberer, K., et al. (eds.) ASWC 2007 and ISWC 2007. LNCS, vol. 4825, pp. 722–735. Springer, Heidelberg (2007)
10. Hoffart, J., Suchanek, F., Berberich, K., Weikum, G.: YAGO2: a spatially and temporally enhanced knowledge base from wikipedia. Research Report MPI-I-2010-5-007, Max-Planck-Institut für Informatik (2010)

Concept Learning Algorithm for Semantic Web Based on the Automatically Searched Refinement Condition

Dongkyu Jeon and Wooju Kim[✉]

Department of Information and Industrial Engineering,
University of Yonsei, Seoul, Korea
jdkclub85@gmail.com, wkim@yonsei.ac.kr

Abstract. Today, the web is the huge data repository which contains excessively growing with uncountable size of data. From the view point of data, Semantic Web is the advanced version of World Wide Web, which aims machine understandable web based on the structured data. For the advent of Semantic Web, its data has been rapidly increased with various areas. In this paper, we proposed novel decision tree algorithm, which called Semantic Decision Tree, to learning the covered knowledge beyond the Semantic Web based ontology. For this purpose, we newly defined six different refinements based on the description logic constructors. Refinements are replaced the features of traditional decision tree algorithms, and these refinements are automatically searched by our proposed decision tree algorithm based on the structure information of ontology. Additional information from the ontology is also used to enhance the quality of decision tree results. Finally, we test our algorithm by solving the famous rule induction problems, and we can get perfect answers with useful decision tree results. In addition, we expect that our proposed algorithm has strong advantage to learn decision tree algorithm on complex and huge size of ontology.

Keywords: Semantic Web · Ontology · Decision tree · Concept learning

1 Introduction

Today, the web is the huge data repository which contains excessively growing with uncountable size of data. From the view point of data, the Semantic Web [2], which called 'Web of data', is an one of approaches that converting the existing web data(not or semi-structured) to structured knowledge. The Semantic Web provides a common framework that allows data to be shared and reused across application, enterprise, and community boundaries [23]. There are diverse approaches on the increasing of Semantic Web data to achieve the advent of Semantic Web. For the most representative examples of the growth of Semantic Web data are Semantic Web search engine and the sharing of linked data. One of the Semantic Web document search engine Sindice [20] has been crawled more than 700 million RDF base document, and another one, Swoogle [21] has been gathered four millions of Semantic Web documents and more than one billions of triples. The Linking Open Data [14] project to

W. Kim et al. (Eds.): JIST 2013, LNCS 8388, pp. 414–428, 2014.
DOI: 10.1007/978-3-319-06826-8_30, © Springer International Publishing Switzerland 2014

expand of the web on the basis of large amount of RDF Link data collected from different sources. Since 2007, the LOD project started with only twelve data sets, but as of September 2011, it is increased to 295 data sets that consist of more than 316 billion of RDF triples, and out-going links are more than 503 million. As we can see the all above circumstances, the Semantic Web data is increased rapidly in various categories. In the context of these data growth, knowledge discovery from the Semantic Web is increasingly important in Semantic Web application area. Logical reasoning is generally used to derive implicit information by deductive inference. However, reasoning has own limitations. It is hard to work on the large size of data and uncertain information, and also it must require ontological knowledge [24]. On the other hand, data mining algorithm is a definitely useful solution to achieve successful knowledge discovery application [24]. Data mining algorithms have been applied in huge size of data to extract implicit knowledge in it. Classification, one of the representative data mining areas, is also widely used for data learning. References [11, 13, 15] said, the important factor of data mining is depends on the quality of data. Especially, in classification area, choosing the closely related feature with target class of classifier and gathering the exact value of feature are important points of successful classification [15, 17]. For this reason, generally, the set of right features are selected by experts of classification domain.

However, in the case of the Semantic Web, it is too large and complex to understand the character of various data sets linked each other. This means that it is difficult to find proper features of classification by human selection based on the perfect understand of complex relationship between domain ontologies and its composed data. In addition to feature selection problem, because of the characteristic of the Semantic Web ontology, classification needs modified methods to fully use the every information represented on the ontology. To solve all these issues, we developed a novel classification algorithm based on the decision tree that can be applied to Semantic Web based ontology. Our proposed decision tree algorithm builds the top-down induction decision tree [18] with automatically defined features based on the structured information of ontology. The newly defined features of decision tree are constructed based on the description logics representation system.

The rest of this paper is organized as follows: In Sect. 2, we discuss the brief of Semantic Web technologies and the related works about linked data based classification algorithms. Description logic constructor based refinements are demonstrates in Sect. 3, and the detail process of learning procedure are addressed in Sect. 4. The implemented system and its test experiment result are discussed in Sect. 5. Lastly, we end with conclusions of the paper and future works in Sect. 6.

2   Related Work

The Semantic Web is the extension of the World Wide Web which enables people to share content beyond the boundaries of applications and web sites through the meaning of web resources. It is based on ontology that is a formal, explicit specification of a shared conceptualization to provide a shared vocabulary [22]. Extensible Markup Language (XML) [6], Resource Description Framework (RDF) [3, 7],

RDF-Schema (RDFS) [3], Web Ontology Language (OWL) [5] are standard languages suggested by W3C to exchange and use information on the Semantic Web. The Ontology that used in the Semantic Web can be constructed by these languages. Since the Semantic Web data can represents relationship between objects, and its properties with structured form, general decision tree algorithm for single table data is not possible to apply directly to Semantic Web data.

Related works about developing the decision tree algorithm for Semantic Web ontology are separated into two categories. The first category is the methodology about doing decision tree on the multi relational data which has similar structure of sematic web ontology and the second one is about doing concept learning on the Description Logic(DLs). References [1] and [4] suggest data refine method to apply the decision tree algorithm on the network structured data such as relational database and social network data. Both researches are making their own data structure to describe the variable of decision tree. Reference [1] developed the 'Selection Graph' which is composed of node and edges from database schema and the conditions to use the graph as an attribute of decision tree. In this paper, decision tree is constructed by searching the set of instances which satisfies a selection graph and its complement set. Reference [4] also suggest similar data structure 'QGraph' to extract local network from whole social network. The 'QGraph' consist of nodes, edges and conditions based on the aggregation functions, and it can query many objects and variables of network data at once. Based on result of 'QGraph', Relational Probability Tree (RPT) algorithm is used to learning decision tree. These researches are suggested partial solution about multi-value issue and learning decision tree on the network data which is also occurred in Semantic Web based ontology. However, both approaches are focused on the change of data structure into attribute-value framework to apply existing decision tree algorithm. Therefore, they are not enough to represent all information of ontology, and also not give a solution of choosing proper variables of decision tree.

There are two related works about concept learning in DLs. Reference [12] suggests concept learning method in description logics which called 'DL-Learner' by using refinement operators. DL-learner searches the best refinement operator which explains as many as instances from description logics with the lack of well-structured knowledge. So, in this paper, refinement operators are generated based on the combination of randomly selected description logic constructors. This research aims to find best refinement operator which can define the concept of existing instances without using the structure information of ontology. Reference [16] develops the concept learning algorithm for description logic representations based on the FOIL algorithm [19] which is developed for inductive logic programming. They use the same refinement operators from DL-learner, and also generate refinements based on the randomly selected description logic constructors. In addition, they suggested the methodology which can handle the unlabeled individuals from ontology.

These researches are performed based on the data written by OWL, because the Semantic Web language 'OWL' can represent $SHOIN(D)$ [10] of DLs. However, both algorithms of researches are based on the randomly generated refinements, and they are not giving the tree base decision rules but only finding general rules that cover all examples of data. Therefore, we suggested our novel algorithm to induce classification rules according to the decision tree algorithm with automatically generated DLs based refinements.

3 Description Logic Based Refinement

In this section, we demonstrate the definition of DLs constructor based refinement which represents the split condition of decision tree. Since the purpose of split condition of our suggest algorithm is to find instances of Semantic Web ontology that satisfy the meaning of given condition, the refinement must express the precise and informative condition by using the structural information of ontology. Therefore, the scope of DLs expression is $SHOIN(D)$ which can describe the information equally with OWL DL [9] to fully cover the expression power of Semantic Web. By using the constructors of $SHOIN(D)$, we defines six different kinds of refinement operators. The schema of OWL based Semantic Web ontology $O = \{C, O.P, D.P\}$ is structured by using the set of classes $C = \{c_1, c_2, c_3, \cdots, c_n\}$, object properties $O.P = \{op_1, op_2, \cdots, op_k\}$ and datatype properties $D.P = \{dp_1, dp_2, \cdots, dp_m\}$. These ontology schema components are matched with atomic concept, atomic role and concrete domain respectively. Basically, the expression of DLs is built on the atomic concept constructors and atomic role constructors [8]. Thus, refinement operators are defined as three types of categories according to DLs constructors: atomic concept based refinement, atomic role based refinement and concrete domain based refinement.

3.1 Atomic Concept Based Refinement

The atomic concept based refinements are defined based on the concept represented as an *owl:Class* type resource. This type refinement can separate instances according to its 'is-a' relationship. The Concept Constructor Refinement, Conjunction Constructor Refinement and Disjunction Constructor Refinement are belonging to atomic concept based refinement.

- **Concept Constructor Refinement:** concept constructor refinement classifies the set of instances that its class type is given concept c_i of the refinement. eg: 'c_i' or negation form : '$\neg c_i$'
- **Concept Conjunction Constructor Refinement:** concept conjunction constructor refinement represents the intersection of two different concept constructor refinements. This refinement classifies the set of instances whose type is both concept c_i and c_j. eg: The demonstration is '$c_i \wedge c_j$' or its negation form '$\neg(c_i \wedge c_j)$'
- **Concept Disjunction Constructor Refinement:** concept disjunction constructor refinement represents the union of two different concept constructor refinements. This refinement classifies the set of instances whose type is concept c_i or c_j. eg: '$c_i \vee c_j$' or negation form '$\neg(c_i \vee c_j)$'

3.2 Atomic Role Based Refinement

The atomic role constructor describes the specific meaning of property relation between concepts. In Semantic Web ontology, role constructor relationship between concepts is represented by the object property which connects two classes. In OWL DLs representation, object property type resource represents the role constructor. Especially, object property can express the direction of property in OWL DL expressions by using the domain class and range class declaration of object property. The cardinality restriction refinement and qualification refinement are belonging to atomic role based refinement.

- **Cardinality Restriction Refinement:** cardinality restriction refinement is based on the at least restriction constructor of DLs. Thus, it classifies the set of instances which have values of object property op_i and satisfy the restriction that the number of property value is more than a specific standard number n. eg: '$n > op_i$' or negation form '$n \leq op_i$'
- **Qualification Refinement:** qualification refinement classifies the set of instances which have the value of given object property OP_i, and the class type of value is given concept c_k. Since the instance can have more than one values of object property, quantifier is needed to clarify the quantification of refinement. In case of qualification refinement, existential quantifier specifies that at least one value of property must satisfy the given concept, and universal quantifier specifies that all of values of object property must satisfy the given concept. eg: '$\exists, \forall\ op_i.c_k$' or negation form '$\neg\exists, \forall\ op_i.c_k$'

According to the direction of object property, inverse direction can also be expressed in atomic role based refinement. Through the inverse directed role based refinement, we can classify the set of instances which are being the value of object property. Inverse direction is marked as an upper hyphen of property. eg: '$\exists, \forall\ op_i^-.c_k$' or '$n > op_i^-$'

3.3   Concrete Domain Based Refinement

The concrete domain based refinement has similar representation with attribute-value based split condition which is generally used in traditional decision tree algorithm. This refinement describes the split condition through the restriction of literal values. Each literal value is represented by datatype property of OWL DL.

- **Domain Restriction Refinement:** There are three kinds of data types of literal values in domain restriction refinement; *integer*, *Boolean* and *string*. The *integer* type domain restriction refinement classifies the set of instances which have the value greater than standard. The other remain types can classify instances that those values of the refinement are equals to standard.

4   Concept Learning Algorithm

In this section, we present the detail procedure of concept learning for classifying given example instances and inducing the set of representation rules to retrieve rule matched instances from ontology. The start of learning procedure is defining the training examples which are composed of positive marked example instances and negative marked example instances. After that, candidate refinements are searched to split decision tree. Candidate refinements generating procedure requires the specific concept which determines the starting point of searching candidate refinements from ontology structure. Therefore, at the beginning of candidate refinement searching, all *rdf:type* values of example instances are selected as a concept constructor refinement is defined from all *rdf:type* classes of given example instances. Among the set of concept constructor refinements, the best refinement is selected as an optimal classifiers to split or expand the root node of tree. Let us assume the example case that the set of instance examples $I = \{+ I_1, + I_2, + I3, -I_4, -I_5\}$ have only one *rdf:type* class c_1 as in Fig. 1(a). In this case, root node is expanded into new child nodes with all example instances by concept constructor refinement c_1. Since new child node still needs to be further classified, c_1 is defined as the start class of candidate searching. Let's consider another example case that if I_1, I_2 and I_4 are c_1 type, and I_3 and I_5 is c_2 type as in Fig. 1(b). If the optimal concept constructor refinement is based on the c_1, then root node is split into two child nodes n_1 and n_2, that n_1 satisfies c1 and n_2 is not. In this case, n_1 will be further classified though new candidate refinements from the start class c_1, and n_2 is further classified based on the candidate refinements from the start class c_2. After selecting the start class, all types of candidate refinements are defined based on the structure information of ontology related to the start class.

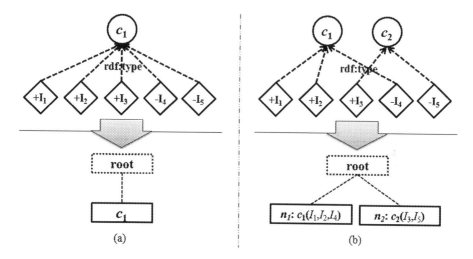

Fig. 1. Example of the start of learning procedure

4.1   Generation of Refinement

Candidate refinement searching of atomic concept based refinement is based on the hierarchically related classes to the start class. Therefore, atomic concept based refinement is defined in accordance with the subclass of the start class. Each subclass of the start class can be a condition of concept constructor refinement, and example instances of the current node are split by its *rdf:type*. For example, let us assume that the start class is c_1 in Fig. 2, and then its subclasses, c_2, c_3 and c_4 are being possible candidate concept constructor refinements. Concept conjunction constructor refinement is defined based on the intersection relationship between subclasses of the start

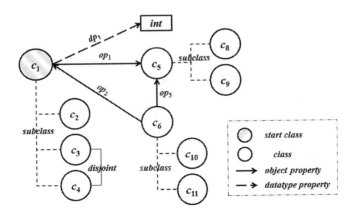

Fig. 2. Example ontology schema

class. Thus, the start class must have more than one subclass, and selected two classes for intersection condition are not restricted by disjoint restriction. All combination pairs of subclasses can be a condition of possible candidate refinements, and example instances of the current node which have both paired subclasses as a value of *rdf:type* are classified by this refinement. For example, from the start class c_1, combination set of its subclasses c_2 and c_3, c_2 and c_4 are being the candidate concept conjunction constructor refinement, and the combination of c_3 and c_4 is excluded by their disjoint restriction. Concept disjunction constructor refinement is also defined based on the combination pair of subclasses of the start class. Following the previous example, all combination pairs of subclasses, c_2 and c_3, c_2 and c_4 and c_3 and c_4, are being the candidate concept disjunction refinement of the start class c_1. Summarization of generated atomic concept based refinement based on the example ontology of Fig. 2 is reported in Table 1.

Candidate refinement searching of atomic role based refinement is based on the relationship between the start class and its connected classes through the object property. Therefore, the base of atomic role based refinement is the set of object properties that its domain class or range class is the starting class of the present node. In addition, if the starting class has super classes then object properties whose domain or range class is the super class are also being the base of atomic role based refinement. This is because subclass inherits the property relation of its super class in Semantic Web environments. The generation of Cardinality Restriction Refinement depends on how many values of the base object property does the present node's example instances have. The standard number of refinement condition is basically zero and the median value among the all observation numbers of the base object property's values. Qualification refinement is defined based on the class type of instances which are values of the base object property. Therefore, the set of classes, connected to the starting class by the base object property, are the condition of refinement. This refinement requires additional restrictions to describe the quantification of condition satisfied values. Thus, existential quantifier and universal quantifier are applied on the object property of refinement. Let us assume that current node

Table 1. Example refinements

Atomic Concept Based Refinement		Atomic Role Based Refinement	
Concept Constructor Refinement	c_2	Cardinality Restriction Refinement	$n > op_1$
	c_3		$n > op_2$
	c_4		$\exists\, op_1.(c_8)$
Concept Conjunction Refinement	$c_2 \wedge c_3$		$\exists\, op_1.(c_9)$
	$c_2 \wedge c_4$		$\forall\, op_1.(c_8)$
Concept Disjunction Refinement	$c_2 \vee c_3$		$\forall\, op_1.(c_9)$
	$c_2 \vee c_4$	Qualification Refinement	$\exists\, op_2^{-}.(c_8)$
	$c_3 \vee c_4$		$\exists\, op_2^{-}.(c_9)$
Atomic Role Based Refinement			$\forall\, op_2^{-}.(c_8)$
Domain Restriction Refinement	$dp_1.(> n)$		$\forall\, op_2^{-}.(c_9)$

has three example instances $\{I_1, I_2, I_3\}$, and the starting class is c_1 of Fig. 2. According to the ontology schema of Fig. 2, possible base object properties for atomic role based refinement are op_1 and op_2. For Cardinality Restriction Refinement, the standard number for restriction is needed to generate refinement for present node. In this example, since the maximum number of observation values of op_1 is two and op_2 is one, standard numbers for the base property op_1 and op_2 are zero and one (Fig. 3). Thus, there are four different Cardinality Restriction Refinement is defined as in Table 1. Qualification Refinement requires the condition class to define meaning. The connected class of op_1 is c_5 and its subclasses are c_8 and c_9, and connected class of op_2 is c_6 and its subclasses are c_{10} and c_{11}. For each combination of properties and related subclasses, qualification refinements of present node are generated.

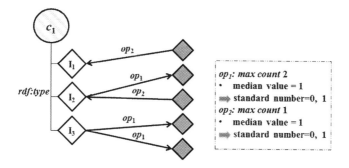

Fig. 3. Example case of cardinality restriction refinement

Domain Restriction Refinement is generated according to a datatype property whose domain class is a center class is exists. The range datatype of datatype property should be one of the 'Boolean', 'integer', or enumerated type of 'string'. If the type of range is *Boolean* or *string* then condition of refinement will follow *datatype value* DL constructor, and if range is *integer*, condition will be defined as a *datatype at least*.

4.2   Automatically Expanded Generation of Refinement

While traditional decision tree algorithm is performed on the fixed attributes and values, our proposed decision tree algorithm is possible to induce decision tree with the automatically generated refinements. In other words, graph data based refinements can be defined by not only directly connected values with target instances but also other values which are consecutively linked through the object properties. For example, let us assume that the decision tree is induced for classifying whether the person has a hereditary or not, and its base graph data is family tree ontology. Possible candidate refinements are based on the information such as target instance's father, target instance's father of father (grandfather's information) or his mother (great-grandmother's information), and these are automatically defined according to the link of ontology. This is the important feature of our proposed algorithm, because it removes the need of predefined attributes, and automatically enlarges the information required to generate decision tree.

The searching space of candidate refinement generating procedure is expanded according to the object property which is the base of atomic role based refinement. When the atomic role based refinement is selected as the best refinement R_B of present node, new starting classes, which are linked through the base object property of the R_B, for candidate refinement searching are added to the child node that not only satisfies RB but also still needs to be classified (not a leaf node). The child node, which is not satisfied the R_B, is no needed to expand its candidate refinement searching space, because it is not clear that whether target instances belongs to this child node have relation with the object property of R_B or not. This expansion rule can reduce the search space of each node's candidate refinement searching procedure.

The search space of candidate refinements for child node is expanded along with the object property of the best refinement. Thus, in addition to parent node's candidate refinements, additional candidate refinements based on the new starting classes are added to the child node's candidate refinements. Since newly searched refinements based on the new starting class are not directly related to target instances, the object property must be noted with quantifiers to clarify the meaning of refinement. For example, as shown in Fig. 4(a), let us assume that the best refinement of node n_1 is '0 $> op_1$', and n_2 is satisfied with it, then the searching space of candidate refinement generation is expanded along with the object property a'. Therefore, new candidate refinements according to class c_5 is added to n_2 in addition to the set of candidate refinements of parent node n_1 (Fig. 2). If the best refinement of n_2 is '$\exists op_1.(0 > op_3)$' then the searching space for candidate refinements of n4 is also expanded to class c_6 (Fig. 2). Examples of candidate refinements of each node are reported in Fig. 4(b).

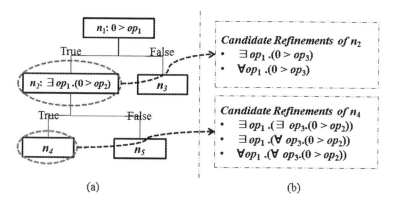

Fig. 4. (a) Example decision tree (b) Example candidate refinements of *true* nodes

4.3   Growing Decision Tree

Since the proposed learning algorithm follows divide and conquer method, decision tree starts with the root node which contains all training examples. The induced decision tree is grown by dividing or expanding the root node repeatedly until the purity of all expanded leaf node reaches zero. When the purity of node is not zero, algorithm searches every possible candidate refinements for the present node, and it is split into two child nodes based on the searched candidate refinements, one is satisfying the condition of refinement, the other is not. Due to the characteristic of graph data, we do only binary splitting the node according to whether the instance examples of the node are satisfying the condition of refinement or not. Not like fixed data for other decision tree algorithm, in Semantic Web, the refinement can be largely varied along with the structure of ontology and the instance can be also satisfied multi refinements at once.

Among the all candidate refinements for present node, the best one for decomposing node is selected based on a statistic measure. We use the *information gain* measures which is also used in ID3 [18] to choose the best refinement. The entropy of node N is calculated by function (1), where p_+ is proportion of positive examples and p_- is proportion of negative examples belongs to node N. Information gain $G(N, R)$ is the expected reduction in entropy through the split nodes by refinement R. N_{R+} is the subset of N_p which satisfies refinement R, and N_{R-} is remain subset of N_p. $E(N_p)$ is entropy of parent node, and each entropy of subset $E(N_{R+})$ and $E(N_{R-})$ is weighted by its proportion of instances.

$$E(N) = -p_+ \log_2 p_+ - p_- \log_2 p_- \tag{1}$$

$$G(N_P, R) = E(N_P) - (p_{R+} \cdot E(N_{R+}) + p_{R-} \cdot E(N_{R-})) \tag{2}$$

5 Implementation and Performance Evaluation

We implemented the prototype system for decision tree algorithm as we proposed in this paper. Our system is developed by using Jena [11] for processing the ontology, and provides interface for users to support setting the target instances for learning and browsing the result decision tree and its rules. The result of decision tree can be displayed as a tree form or induced set of rules. Each edge of tree is notated by its satisfied refinements. Figure 5 illustrates the architecture of the prototype system. In order to test the prototype system, we did a representative learning problem which is called FORTE experiment.

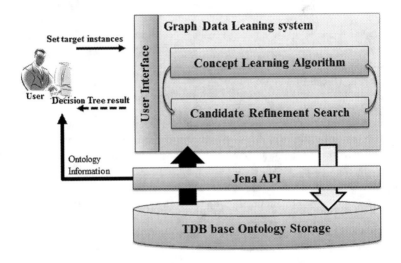

Fig. 5. The architecture of prototype system

5.1 Algorithm Test: FORTE Problem

FORTE family data set describes the simple family trees of 86 persons. In the research of [12] they found a generic rule of Uncle concept based on this FORTE family data, and its result rule is Fig. 6(b). This rule explains the general definition of uncle and part of maternal uncle exactly. Following this experiments, we also try to find the decision tree of classifying the uncle of FORTE. Figure 6(a) is the ontology schema of FORTE; there are 86 person type instances (46 man and 40 woman) and only 23 man are uncle. The result of decision tree learning is in Fig. 7. As we can see, the upper part of decision tree contains the both rules of Fig. 7. The dotted rectangle assigns the definition of uncle, and the bold lined rectangle assigns the maternal uncle.

5.2 Performance Experiment

In order to evaluate the performance of proposed algorithm, we applied our result prototype system to several classification problems and concept retrieval problems

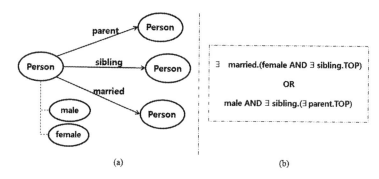

Fig. 6. (a) FORTE ontology schema (b) Answer rule of uncle relationship

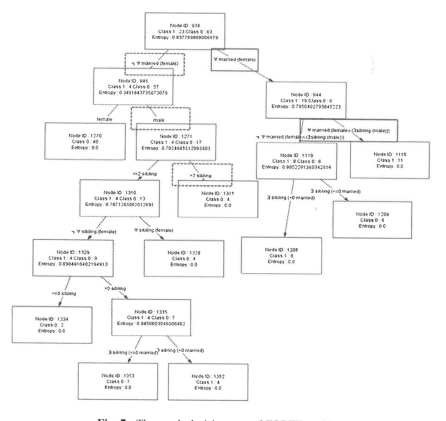

Fig. 7. The result decision tree of FORTE problem

based on the ontologies designed to describe specific domains for Semantic Web. The performance is assessed in accordance with efficiency of learning process during the candidate searching procedure.

While other concept learning algorithm searches candidate refinements in stochastic way, our proposed algorithm incrementally expands the searching space of generating candidate refinements according to the structural information of ontology and importance of ontology resources. To verify the efficiency of our algorithm, we compared the number of candidate refinements which directly indicates the traversing size of candidate searching space with the result of *ocel* algorithm of [12] which is based on the stochastic searching method. Five different classifying problems are applied to both algorithm, and solving results of both algorithms completely coincide with each other. The important details of ontologies that used for problem is shown in Table 2.

Table 2. Description of test ontologies

Ontology	Number of class	Number of property	Number of instances	Number of example instances
Trains	10	5	50	10
Forte	5	3	86	86
Poker1	2	6	294	49
Poker2	2	6	330	55
Moral	44	0	202	202

Each problem gives positive labeled instances and negative labeled instances among individuals of ontology. We count the number of candidate refinements until the algorithm completes classifying given example instances. From the result of all experiment, we can sure that our algorithm requires much smaller size of candidate refinements to complete given problems. These differences are getting higher in accordance with the number of properties of ontology. Therefore, we expect that our algorithm has strong advantages on the ontology which have complex structure with various links. The comparison of result is reported in Table 3.

Table 3. The result of experiment

	Number of candidate refinements	
	Our algorithm	*ocel* algorithm (DL-Learner)
Trains	28	286
Forte	107	2159
Poker1	15	724
Poker2	55	11139
Moral1	43	194

6   Conclusion

In this paper, we proposed the concept learning algorithm for Semantic Web based ontology. Based on the procedure of decision tree algorithm, our proposed algorithm can induce the decision tree according to concepts and roles of Semantic Web based node split conditions. For this purpose, we define the six different refinements which can represent the features of ontology based on the Description Logic constructors. The expression power of our proposed refinement is much more complex and rich than the condition of traditional decision tree. Moreover, candidate refinements for each tree node are generated automatically in accordance with the structure information of Semantic Web ontology. User just needs to set the target instances that user wanted to learn concepts. Our proposed algorithm is much more efficient in searching and testing the candidate refinements than other stochastic generation based learning algorithms.

One of future works of our research is to avoid local optimize solution by developing the parallel learning environment for this algorithm. Lastly, we will try to do performance test in accuracy perspectives.

References

1. Knobbe, A.J., Siebes, A., Van Der Wallen, D.: Multi-relational decision tree induction. In: Żytkow, J.M., Rauch, J. (eds.) PKDD 1999. LNCS (LNAI), vol. 1704, pp. 378–383. Springer, Heidelberg (1999)
2. Berners-Lee, T.: The Semantic Web. Sci. Am. **501** (2001)
3. Dan, B., Guha, R.V.: RDF vocabulary description language 1.0: RDF schema. W3C Recommendation (2004)
4. David, J., Jennifer, N.: Data mining in networks. In: Papers of the Symposium on Dynamic Social Network Modeling and Analysis, National Academy of Sciences. National Academy Press, Washington DC (2002)
5. Deborah, L.M., Frank, V.H.: OWL web ontology language overview. W3C Recommendation (2004)
6. Stefan, D., Serge, M., Frank, V.H., Dieter, F., Michel, K., Jeen, B., Michael, E., Ian, H.: The Semantic Web: the roles of XML and RDF. IEEE Internet Comput. **15**(3), 63–74 (2000)
7. Frank, M., Eric, M.: RDF primer. W3C Recommendation (2004)
8. Franz, B., Ulrike, S.: An overview of tableau algorithms for description logics. Stud. Logica **69**, 5–40 (2001)
9. Horrocks, I., Patel-Schneider, P.F., van Harmelen, F.: From SHIQ and RDF to OWL: the making of a web ontology language. J. Web Semant. **1**(1), 7–26 (2003)
10. Horrocks, I., Patel-Schneider, P.F.: Reducing OWL entailment to description logic satisfiability. In: Fensel, D., Sycara, K., Mylopoulos, J. (eds.) ISWC 2003. LNCS, vol. 2870, pp. 17–29. Springer, Heidelberg (2003)
11. Jena. http://jena.sourceforge.net/
12. Jens, L., Pascal, H.: Concept learning in description logics using refinement operators. Mach. Learn. **78**(1–2), 203–250 (2010)
13. Hermiz, K.: Critical success factors for data mining projects, information management magazine. http://www.informationmanagement.com/issues/19990201/164-1.html (2010). Accessed 05 Apr 2010
14. Linking open data project. http://linkeddata.org/
15. Dash, M., Liu, H.: Feature selection methods for classification. Intell. Data Anal. **1**(3), 131–156 (1997)

16. Fanizzi, N., d'Amato, C., Esposito, F.: Induction of concepts in web ontologies through terminological decision trees. In: Balcázar, J.L., Bonchi, F., Gionis, A., Sebag, M. (eds.) ECML PKDD 2010, Part I. LNCS, vol. 6321, pp. 442–457. Springer, Heidelberg (2010)
17. Petra, P.: Improving the accuracy of decision tree induction by feature pre-selection. Appl. Artif. Intell. **15**(8), 747–760 (2001)
18. Quinlan, J.R.: Induction of decision trees. Mach. Learn. **1**, 81–106 (1986)
19. Georgopoulos, L.J., Emmett, S.: The yorkshire enterprise fellowship scheme. In: Howlett, R.J. (ed.) Innovation through knowledge transfer. SIST, vol. 5, pp. 239–247. Springer, Heidelberg (2010)
20. Sindice. http://sindice.com
21. Swoogle. http://swoogle.umbc.edu/
22. Thomas, G.: A translation approach to portable ontology specifications. Knowl. Acquis. **5**, 199–220 (1993)
23. World Wide Web Consortium (W3C).: W3C Semantic Web Activity (2011)
24. Achi, R., Uta, L., Volker, T., Claudia, A., Nicola, F.: Mining the Semantic Web. Data Min. Knowl. Disc. **24**(3), 613–662 (2012)

Reasoning Driven Configuration of Linked Data Content Management Systems

Stuart Taylor, Nophadol Jekjantuk[✉], Chris Mellish, and Jeff Z. Pan

dot.rural Digital Economy Hub, University of Aberdeen, Aberdeen AB24 5UA, UK
n.jekjantuk@abdn.ac.uk
http://www.dotrural.ac.uk

Abstract. The web of data has continued to expand thanks to the principles of Linked Data, increasing its impact on the web both in its depth and range of data sources. However tools allowing ordinary web users to contribute to this web of data are still lacking. In this paper we propose Linked Data CMS, an approach allowing existing web content management system (CMS) software to be configured to display a web site based on a group of ontology classes, by making use of a *configuration* to map ontological entities to CMS entities. We have implemented a prototype of our Linked Data CMS approach using the popular Drupal CMS. This approach provides the tools for semantic web application developers to rapidly develop an entire website based on linked data, while allowing ordinary web users to contribute directly to the web of data using familiar CMS tools.

Keywords: Linked data · Ontologies · Content management systems

1 Introduction

There are many mature development platforms that are well suited to traditional data-based applications and which could benefit from integration with Linked Data [1]. In particular, nowadays web content is frequently authored by a user using some form of Content Management System (CMS), where the content may be structured or relatively unstructured (e.g. wikis and blogs). Integrating Linked Data into the CMS paradigm could not only allow users unfamiliar with the complexities of semantic web technologies to consume Linked Data [2], but also to produce it in a familiar way based on structured data already held in a CMS. This would greatly benefit the web in terms of reuse and aggregation of data held within CMSs.

In this paper, we introduce the notion of a *Linked Data Content Management System* (Linked Data CMS). A Linked Data CMS performs similar operations to a traditional CMS but whereas a traditional CMS uses a data model of content types stored in some relational database back end, a Linked Data CMS deals with performing CRUD operations on linked data held in a triple store, in the context of a content management system. High quality Linked Data is provided

W. Kim et al. (Eds.): JIST 2013, LNCS 8388, pp. 429–444, 2014.
DOI: 10.1007/978-3-319-06826-8_31, © Springer International Publishing Switzerland 2014

in ways compatible with W3C standards such as RDF [3] and SPARQL. We make the additional assumption that the Linked Data to be managed is described by a (possibly very lightweight, but possibly also arbitrarily complex) ontology described in OWL, which is underpinned by description logics [4]. To facilitate the Linked Data CMS we then define a mapping between certain structures of the OWL classes and entities into a traditional CMS. We call this mapping *Class Based Browsing* because the ontology classes are the central entities in our mapping. Because complete and correct retrieval of information from an ontology is more complex than simple database lookup, we also introduce a process of *Reasoning Driven Configuration*, which configures the class based browsing and Linked Data CMS based on the source ontological schema. In reasoning driven configuration, an ontology reasoner, such as the highly efficient reasoner TrOWL [5], is used to automate the process of configuring the CMS.

In the paper, Sect. 2 briefly surveys Linked Data, ontologies and CMS, as well as the previous work that has attempted to integrate them. In Sect. 3 we present a formalisation of the entities that are maintained in a traditional CMS. Section 4 describes the mapping from ontology entities to CMS entities and the abstract configuration process that creates a Linked Data CMS from an ontology. Section 5 describes how the information used in the mapping can itself be represented in an ontology and the configuration achieved by standard (meta-) ontology reasoning. Finally in Sect. 6 we discuss the implementation of a prototype of our class based browsing and reasoning driven configuration approach using the Drupal CMS, and in Sect. 7 we present a proof of concept with a case study for the CURIOS project, which deals with managing cultural heritage linked data using the Drupal Linked Data CMS.

2   Background

2.1   Linked Data

Linked Data was coined by Tim Berners-Lee [6], describing a set of conventions for publishing RDF [7] datasets on the web. The main ideas are to use HTTP URIs to refer to things, allowing these URIs to be dereferenced in order to discover further related information, in W3C standard formats such as RDF and OWL [8]. These datasets should also refer to further datasets where necessary so that other related resources can be found. This has led to a vast increase in the number of RDF *Linked Data* datasets being published on the web [9]. In 2011, the Linked Data cloud totalled approximately 31.6 billion RDF triples [10].

2.2   Ontologies

High quality Linked Open Data (data which would be given four "stars" by Berners-Lee [6]) is made available using W3C standards such as RDF. These formats use vocabularies to describe the entities in the world. Descriptions of the vocabularies are themselves in RDF, ideally should themselves be open and can be expressed in various ways, the most general ways being the different versions of the OWL Web Ontology Language [11]. Vocabulary descriptions expressed in OWL are called *ontologies*. Since an OWL description can include not only a description of a vocabulary but also the uses of that vocabulary to state facts about individual things, in general one can think of any high quality Linked Data as being specified simply by an OWL ontology.

In this paper, we assume the use of OWL DL ontologies, which are based on ideas from the Description Logic (DL) [12] family of knowledge representation languages. An OWL DL ontology $\mathcal{O}$ consists of a set of classes $\mathbf{C}$, object properties $\mathbf{OP}$, datatype properties $\mathbf{DP}$, individuals $\mathbf{I}$, and the axioms describing the relationships between them.

The axioms in the OWL DL ontology are divided into the TBox $\mathcal{T}$ (terminological box; or *schema*) and the ABox $\mathcal{A}$ (assertional box; or *data*). TBox axioms make statements about how classes and properties relate to each other, e.g.,

$$\text{Person} \sqcap \text{PostgraduateDegree} \sqsubseteq \bot$$

("nothing can be both a Person and a PostgraduateDegree"). On the other hand, ABox axioms make statements about class and property membership, e.g.,

$$\text{Person}(\text{Jek}), \; knows(\text{Stuart}, \text{Jek})$$

("Jek is a Person and Stuart knows Jek"). In the following we consider an OWL DL ontology as a tuple $\mathcal{O} = \langle \mathbf{C}, \mathbf{OP}, \mathbf{DP}, \mathbf{I}, \mathcal{A}, \mathcal{T} \rangle$; we use these names to refer to the sets of ontology entities throughout the remainder of the paper.

2.3   Content Management Systems

In this paper we focus solely on Web Content Management Systems. A Web Content Management System is a system that allows website maintainers and contributors to manage the content of a website via a central graphical user interface, without relying on any significant knowledge of the underlying technologies involved in running the website.

In addition to managing page content, a CMS can typically model various *content types*. These content types define specialisations of a web page, e.g., a content type for a blog post will require some additional attributes such as author and a topic or category. Most modern CMSs allow the user to define custom content types, which can consist of the page structure, page attributes, layout and style.

2.4   Semantic Web Content Management

There has been much interest in integrating CMS and Semantic Web technologies. Semantic MediaWiki (SMW) [13] is a semantic extension of MediaWiki[1],

---

[1] MediaWiki is a popular wiki-engine used throughout the web, most notably it is used by http://www.wikipedia.org/.

which allows users to semantically annotate wiki mark-up in the content of wiki articles, using OWL classes and properties. This approach allows the content of the wiki to be queried using SPARQL and reused by other semantic web tools, however it does not support presenting existing Linked Data within the wiki, and therefore is limited to managing the Linked Data (OWL assertions) generated only from its own wiki articles.

The BBC's Dynamic Semantic Publishing (DSP) architecture [14] was designed to allow BBC journalists to annotate news articles[2] with metadata from a domain ontology. DSP both consumes and produces Linked Data, however the approach is geared towards producing articles with annotations to a fixed ontology, or one that is managed outside of the CMS.

Semantic Drupal[3] [15] is of particular interest, since it provides a platform to combine Drupal with the Semantic Web. The Drupal Semantic Web Group[4] focus on the integration between Drupal and Semantic Web technologies such as RDF, RDFa and SPARQL, and have successfully produced a number of Drupal extensions addressing these areas. Semantic Drupal allows site administrators to export data from *nodes* as RDF. In this approach, *content types* and *fields* are given mappings to specified classes and properties. In general however this approach does not allow management of existing Linked Data with the Drupal CMS, since it is intended only to export Drupal data to Linked Data and import existing Linked Data to Drupal's database.

Some of the limitations of consuming Linked Data in Content Management Systems are overcome by Clark [16] with the SPARQL Views Drupal module. This approach essentially adds a SPARQL query builder to the powerful Views module[5]. This module allows site administrators to specify a view of a set of fields via a GUI. This specification is then used to generate the appropriate SPARQL query to instantiate the view and therefore not relying on any SPARQL knowledge by the user. The strength of SPARQL Views is in its utilisation of the Views module's powerful GUI building capabilities, allowing users to building complex GUIs without requiring any web programming knowledge. However, this approach is intended to be read only, i.e., views only display the results from SPARQL queries, but do not allow the corresponding SPARQL endpoint to be updated. Another limitation of this approach, is that when the number of required SPARQL Views resource types increases, e.g., when building views for a large set of OWL classes (where each class would typically be specified as an individual resource type) the task of initially building and maintaining the set resource types and associated views becomes a serious burden for the user. For example, an ontology with 10 classes, each with 10 properties would correspond

---

[2] DSP was first used to build the BBC World Cup 2010 website http://news.bbc.co.uk/sport1/hi/football/world_cup_2010/default.stm.

[3] A series of guides for configuring Drupal 7 as "Semantic Drupal" can be found at http://semantic-drupal.com/.

[4] The Drupal Semantic Web Group can be found at http://groups.drupal.org/semantic-web.

[5] Drupal Views module: http://drupal.org/project/views

to 10 resource types, 100 fields with 100 RDF mappings and at least one single view for each resource type.

The approaches presented in this section go a long way to integrating the advantages of modern Content Management Systems with Linked Data. However so far no approach for general Linked Data content management exists. In this paper we propose an extension to existing content management systems which allows the full management of ontology instances data; i.e., create, retrieve, update, delete. We propose to automate the process of configuring the CMS based on the target ontology schema.

3   Formalisation of CMS Entities

Here we propose a straightforward way of formally describing the entities of a traditional content management system. The idea is to capture the features of the CMS that can be used to represent ontology entities in our Linked Data CMS.

Definition 1. *A **CMS** is a tuple $S = \langle T, F, R, P \rangle$ where T is set of page templates, which define the structure of particular types of page in the CMS (i.e., sets of fields and relationships); F is a set of fields, which are used to display literal values; R is a set of relationships, which define links to another pages in the CMS; P is a set of page instances, which use a page template to display some specific information.*

Definition 2. *A field is a single $f = \langle f_{lab} \rangle \in F$, where f_{lab} is a human readable label for the field.*

Definition 3. *A relationship is a pair $r = \langle r_{lab}, r_T \rangle \in R$, where r_{lab} is the relationship label; $r_T \in T$ is a template for the pages for the objects of the relationship.*

Definition 4. *A page template for a CMS $\langle T, F, R, P \rangle$ is a tuple*

$$t = \langle t_{lab}, t_F, t_R \rangle \in T$$

where t_{lab} is the page template label; t_F is a set of fields, such that $t_F \subseteq F$; t_R is a set of relationships, such that $t_R \subseteq R$.

A page template consists of a set of fields, and a set of relationships. It is used as a template for creating page instances in the CMS (i.e., a page instance has the same fields and relationships as its page template). Page templates also contain the information on how to render page instances, however we don't include the display concerns this paper. A page instance is an instantiated page template, for a particular set of field and relationship values:

Definition 5. *A page instance for a CMS $\langle T, F, R, P \rangle$ is a tuple*

$$p = \langle p_{lab}, p_T, p_F, p_R \rangle \in P, \quad where$$

- p_{lab} is the page title;
- p_T is a page template $t = \langle t_{lab}, t_F, t_R \rangle \in T$;
- p_F is a set of pairs $\langle f_i, f_{val} \rangle$ where $f_i \in t_F$ and f_{val} is a literal value (the field value);
- p_R is a set of pairs $\langle r_i, r_{val} \rangle$, where $r_i \in t_R$, $r_{val} \in P$ and the page template of r_{val} is the same as the template specified in r_i.

3.1 CMS Entity Example

We can now show how the CMS entities defined in the previous section could be used to describe a CMS about people. This example is for a CMS $\mathcal{S}$ containing pages about a person *May*, with two children *Pam* and *Chris*. It can be described using our CMS entities as follows:

$$\mathcal{S} = \langle \{person\}, \{name, dob\}, \{parentOf\}, \{may, pam, chris\} \rangle;$$

where *person*, *name*, *dob*, *parentOf*, *may*, *pam*, *chris* are entities defined below.

The page template *person* is defined using the fields and relationships in $\mathcal{S}$, and the page instance *may* would be populated with the data about May:

$$person = \langle \text{``Person Page''}, \{name, dob\}, \{parentOf\} \rangle;$$
$$may = \langle \text{``May's Page''}, person, may_F, may_R \rangle;$$
$$may_F = \{\langle name, \text{``May''}\rangle, \langle dob, \text{``1972-01-05''}\rangle\};$$
$$may_R = \{\langle parentOf, pam \rangle, \langle parentOf, chris \rangle\};$$

where *pam* and *chris* are page instances for Pam and Chris respectively, which are defined using the *person* page template in a similar manner as for *may*. The fields and relationships are defined as:

$$name = \langle \text{``Full Name''} \rangle;$$
$$dob = \langle \text{``Date of birth''} \rangle;$$
$$parentOf = \langle \text{``Children''}, person \rangle.$$

4 Linked Data Content Management

4.1 Class Based Browsing

Our approach focuses on browsing ontological entities in a CMS, based on ontology classes. The idea is that a particular set of classes from an ontology's class hierarchy is selected to provide *views* of their instances in the CMS. The Linked Data CMS configuration (which we refer to as *the configuration*) is a set of these classes, along with some metadata to allow the Linked Data CMS to configure the CMS entities in $\mathcal{S}$ described in the previous section. We call the combination of these classes and associated metadata *browsing classes*.

Definition 6. *A Linked Data CMS configuration for an ontology $\mathcal{O}$ with components $\langle C, OP, DP, I, \mathcal{A}, \mathcal{T} \rangle$ is a pair $\mathcal{C} = \langle B, lab \rangle$, where B is a set of browsing classes and $lab : (C \cup OP \cup DP \cup I) \to String$ is a function that maps ontology entities to human-readable labels.*

Definition 7. *A browsing class in a configuration $\langle B, lab \rangle$ for the ontology $\mathcal{O} = \langle C, OP, DP, I, \mathcal{A}, \mathcal{T} \rangle$ is a tuple*

$$b = \langle b_C, b_{DP}, b_{OP}, b_\omega \rangle \in B, \quad where$$

- b_C *is the base OWL class* $\in C$;
- b_{DP} *is a set of datatype properties* $\subseteq DP$, *such that* $\forall p : p \in b_{DP} : \mathcal{O} \models b_C \sqsubseteq$ domain(p);
- b_{OP} *is a set of object properties* $\subseteq OP$, *such that* $\forall p : p \in b_{OP} : \mathcal{O} \models b_C \sqsubseteq$ domain(p);
- b_ω *is a function that assigns to each object property* $p \in b_{OP}$ *a class* $c \in C$, *such that* $\mathcal{O} \models$ range$(p) \sqsubseteq c$ *and* $\exists b' \in B : c = b'_C$.

A browsing class contains an ontology class (called the *base class*) and some associated metadata which is used as a view of a set of individuals in the CMS. This includes the datatype and object properties that have been chosen for presentation for instances of the base class. The Linked Data CMS mapping uses the set of browsing classes to create the set of page templates in the CMS that correspond to the chosen base classes.

4.2   Linked Data CMS Mapping

Given an ontology $\mathcal{O}$ and a configuration $\mathcal{C} = \langle B, lab \rangle$ for $\mathcal{O}$, the algorithm in Fig. 1 can be used to construct a CMS. Because the definition of a Linked Data CMS mapping makes use of the notion of entailment ("$\models$") from $\mathcal{O}$, ontology reasoning is required to validate a potential configuration (test that all the conditions are satisfied). We address the issue of validation in Sect. 5.2 below.

The Linked Data CMS mapping can be summarised as follows. First a field is created for each datatype property in any b_{DP}. A template t can indirectly contain other templates, via t_R, and so templates and relationships are created in two phases. The first phase creates the outer structure of each template and stores these structures in $Temp[]$, which associates an ontology class with the template that will represent it. The second phase fills in the links between templates, via the relationships, updating the template structures in place. Where a relationship is required to hold a template, a pointer to the relevant template is retrieved from $Temp[]$. Finally the page instances are created, in a similar two phases, since a page instance p can indirectly contain another page instance, via p_R. Here $Page[]$ is used to keep a mapping between ontology individuals and the page instances that will be used for them.

The CMS configuration process implemented in this algorithm is *reasoning-driven* because the page instances depend on what is entailed by the ontology, and in general a reasoner is needed to establish this.

Inputs: An ontology $\mathcal{O}$ and a configuration $\mathcal{C} = \langle B, lab \rangle$ for $\mathcal{O}$
Output: A CMS $\langle T, F, R, P \rangle$

$Temp[]$: A hash table from concepts to templates, initially empty
$Page[]$: A hash table from individuals to pages, initially empty
r: a relationship
T, F, R, P :- sets of templates, fields, relationships and pages

$F := \{ \langle lab(p) \rangle : b \in B, p \in b_{DP} \};$
$R := \phi;$
$T := \phi;$
// Initialise template structures
For each $b \in B$
       $Temp[b_C] := \langle lab(b_C), \{ \langle lab(p) \rangle : p \in b_{DP} \}, \phi \rangle;$
// Finalise templates and relationships
For each $b \in B$
    For each $p \in b_{OP}$
        For each $b' \in B$
            If $b'_C = b_\omega(p)$ then
                $r := \langle lab(p), Temp[b'_C] \rangle;$
                $R := R \cup \{r\};$
                $Temp[b_C]_R := Temp[b_C]_R \cup \{r\};$
        $T := T \cup \{Temp[b_C]\};$
// Initialise page strructures
$P := \phi;$
For each $b \in B$
    For each a such that $\mathcal{O} \models b_C(a)$
        $Page[a] := \langle lab(a), Temp[b_C],$
                $\{ \langle lab(p), v \rangle : p \in b_{DP}, \mathcal{O} \models p(a, v) \}, \phi \rangle;$
// Connect pages
For each $b \in B$
    For each a such that $\mathcal{O} \models b_C(a)$
        For each $p \in b_{OP}$
            For each v such that $\mathcal{O} \models p(a, v)$
                $r := \langle lab(p), Temp[b_\omega(p)] \rangle;$
                $Page[a]_R := Page[a]_R \cup \{ \langle r, Page[v] \rangle \};$
        $P := P \cup \{Page[a]\};$
Return $\langle T, F, R, P \rangle;$

Fig. 1. Linked Data CMS Mapping Algorithm

The Linked Data CMS mapping also allows page instances to be used as *update views*, i.e. entities in the CMS can be mapped to entities in the ontology by looking at the browsing classes. Finally, we can define the Linked Data CMS.

Definition 8. *A Linked Data CMS is the combination of an ontology $\mathcal{O}$, a configuration $\mathcal{C} = \langle B, lab \rangle$ and the resulting CMS $\mathcal{S} = \langle T, F, R, P \rangle$ once the Linked Data CMS mapping has been applied.*

4.3   Class Based Browsing Example

We now revisit the CMS entity example presented in Sect. 3.1 and show how this CMS structure can be created from an example ontology $\mathcal{O}$ using our Linked Data CMS approach.

The ontology $\mathcal{O}$ consists of:

$$Person(may); \; name(may, \text{``May''}); \; dob(may, \text{``1972-01-05''});$$
$$Person(pam); \; name(pam, \text{``Pam''}); \; parentOf(may, pam);$$
$$Person(chris); \; name(chris, \text{``Chris''}); \; parentOf(may, chris);$$
$$rdfs\,label(may, may); \; rdfs\,label(pam, pam); \; rdfs\,label(chris, chris).$$

We have omitted the rdfs:label axioms for classes and properties, however we assume they have been defined in the same way as for the individuals.

We can define a configuration $\mathcal{C} = \langle B, lab \rangle$ as follows.

$$B = \{ \langle Person, \{name, dob\}, \{parentOf\}, \{ \langle parentOf, Person \rangle \} \rangle \}.$$
$$lab(x) = l : \mathcal{O} \models rdfs\,label(x, l).$$

Although these labels are not the same as those used in Sect. 3.1, $lab(x)$ could easily be defined to produce labels based on the type of x, e.g., if x is a class, then $lab(x) = concat(l, \text{``Page''}) : \mathcal{O} \models rdfs\,label(x, l)$; where $concat$ is the string concatenation function. However, for ease of presentation we use the simpler definition of $lab(x)$ in this example.

Now the Linked Data CMS mapping is applied to $\mathcal{C}$ to produce a CMS $\mathcal{S}$ as follows. First the basic CMS structure and fields are created:

$$\mathcal{S} = \langle T, \{name, dob\}, R, P \rangle;$$
$$name = \langle name, \text{``name''} \rangle; \quad dob = \langle dob, \text{``dob''} \rangle;$$

A page template is then created for each browsing class in B and the relationships are created to connect these together:

$$person = \langle \text{``Person''}, \{name, dob\}, \{parentOf\} \rangle;$$
$$T = \{person\};$$
$$parentOf = \langle \text{``parentOf''}, person \rangle;$$
$$R = \{parentOf\}$$

Then for each new page template, a page instance is created for each individual in the ontology that has a corresponding browsing class:

$$P = \{may, pam, chris\};$$
$$may = \langle \text{``may''}, person, may_F, may_R \rangle;$$
$$pam = \langle \text{``pam''}, person, \{\langle name, \text{``Pam''}\rangle\}, \emptyset \rangle;$$
$$chris = \langle \text{``chris''}, person, \{\langle name, \text{``Chris''}\rangle\}, \emptyset \rangle;$$
$$may_F = \{\langle name, \text{``May''}\rangle, \langle dob, \text{``1972-01-05''}\rangle\};$$
$$may_R = \{\langle parentOf, pam\rangle, \langle parentOf, chris\rangle\}.$$

In this example we have created the CMS entities in $\mathcal{S}$ to reproduce the CMS structure described in Sect. 3.1, based on the ontology $\mathcal{O}$ and configuration $\mathcal{C}$. This example illustrates that given an ontology and a fairly simple configuration, a significant CMS can be created in a straight forward manner. The ontology in this example was small (since we only used three individuals) however using our approach we can increase the number of page instances at no extra cost to the user, since the complexity of defining a configuration is only related to the number of classes chosen for display.

5   Representing the Browsing Classes

In the previous section, we described the configuration, the central part of the Linked Data CMS, as a formal entity. This entity is referred to by the mapping algorithm which, using ontology reasoning as needed, constructs the CMS from the information in the ontology. In this section, we propose the use of a meta ontology to represent the configuration. This has the advantage that validation of the configuration can be done by standard meta ontology reasoning.

OWL 2 supports a very basic, but decidable, approach to metamodeling called "punning", which means that the sets of names for classes, properties and individuals do not have to be disjoint. Punning allows one to, for instance, treat :Eagle as separately denoting both a class of Birds and an individual of the class :Species. This can be quite useful for some sorts of modelling. However, there is no logical relation between classes, individuals or properties with the same name. The semantics of punning is based on contextual semantics [17]. There are some restrictions on punning - the same identifier (IRI) cannot be use to denote both a class and a datatype property, also datatype properties and object properties cannot have the same name.

5.1   The Linked Data CMS Using a Meta Ontology

A Linked Data CMS consists of a configuration meta-ontology[6] O_{config} which is applied to a domain ontology O_{domain} in order to map entities in the domain

---

[6] The meta-ontology schema can be found at: http://www.abdn.ac.uk/~csc363/ldcms/ldcms.owl.

ontology to entities in a CMS, such as Drupal. O_{config} has as individuals a set of classes **C**, object properties **OP** and datatype properties **DP** from O_{domain}; and a set of browsing classes **B**. Each browsing class $b \in B$ is an ontology individual in the meta-ontology describing how a particular base class **bc** (OWL class in the domain ontology) should be rendered by the CMS.

The following example shows part of O_{config} for a configuration with a single browsing class in Turtle syntax:

```
: PersonPage  rdf:type  cms:BrowsingClass  ;
  rdfs:label  'Person Page'  ;
  cms:baseClass  domain:Person  ;
  cms:relationship  [
     rdf:type  cms:Relationship  ;
     rdfs:label  'Child Of'  ;
     cms:target  domain:Person  ;
     cms:property  domain:childOf ]  ;
  cms:relationship  [
     rdf:type  cms:Relationship  ;
     rdfs:label  'Lived At'  ;
     cms:target  domain:Residence  ;
     cms:property  domain:livedAt ]  ;
  cms:field  [
     rdf:type  cms:Field  ;
     rdfs:label  'Full Name'  ;
     cms:property  domain:name ]  .
```

5.2   Validating the Linked Data CMS

Given the definition of browsing class in Sect. 4.1, a configuration such as the above must satisfy a number of constraints. For instance, each relationship target must be a base class in the configuration and the base class should include the range of the relationship property. In the above, we need, for instance, to have:

$$O_{domain} \models range(livedAt) \sqsubseteq Residence$$

A browsing class can be validated using the following SPARQL query where **bc** is the browsing class to be validated. If the ASK query returns false, then the browsing class is invalid.

```
# Validate relationship against domain ontology.
prefix cms:<http://www.abdn.ac.uk/~csc363/ldcms/ldcms.owl#>
ASK {
  # Query configuration meta-ontology.
  SERVICE <http://sparql.example.org:3030/config/query> {
   <bc> a cms:BrowsingClass  ;
     cms:relationship  [
       cms:property ?P  ;
```

```
      cms:target ?C ].
  [] a cms:BrowsingClass ;
    cms:baseClass ?C .
}
# Query domain ontology.
SERVICE <http://sparql.example.org:3030/dataset/query> {
  ?P rdfs:range [ rdfs:subClassOf ?C ].
}
}
```

In this query a SPARQL 1.1 Federated Query [18] is used to query O_{config} hosted at the SPARQL endpoint http://sparql.example.org:3030/config/query and O_{domain} at the SPARQL endpoint http://sparql.example.org:3030/dataset/query. The query against O_{config} makes use of the entailments of O_{domain} when validating the relationship target in specified in the browsing class <bc>.

6   Drupal Linked Data CMS

We have implemented our Linked Data CMS approach using the Drupal platform. Drupal 7 was chosen due to having RDF support built into the core system and many contributed modules. Our Linked Data CMS approach is implemented as a Drupal module which builds on a number of existing semantic web based modules. In this section we provide an overview on how our approach has been implemented within Drupal.

In our Drupal implementation we allow the user to specify the configuration in much the same manner as described in Sect. 4.1. The configuration is used to create entities in Drupal using the Views, Panels and SPARQL Views modules to represent the page templates. At runtime the page instances are generated on the fly by the SPARQL Views module (see Sect. 2.4). We configure Drupal with three main types of page template: (i) the *listings view*, which allows users to browse and search ontology individuals; (ii) the *details view*, which allows users to view all the fields and relationships for a particular individual (Fig. 2); (iii) the *update view*, which allows users to create, update and delete individuals in the ontology.

The SPARQL Views module allows users to map Drupal fields to RDF predicates and then build a view based on those fields. The field mappings are grouped into SPARQL Views Resource Types, each intended to represent a particular type of RDF resource. In the background SPARQL Views generates a SPARQL query based on the RDF mapping and Views specification. The view can be configured by users selecting the appropriate *fields*, *filters* (usually based on URL parameters), *relationships* to other SPARQL Views resources and display configuration.

Whereas maintaining a set of SPARQL Views by hand can be very complex (Sect. 2.4), in essence our Linked Data CMS automatically configures a set of SPARQL Views Resources and Views based on an OWL ontology and user specification of how that ontology should be represented in Drupal; along

Fig. 2. Drupal Linked Data CMS: Details View

with an additional set of pages allowing those resources to be maintained (create/update/delete). The browsing classes identified by the user, along with their associated datatype and object properties are mapped to resource types, fields and relationships respectively. A default SPARQL Views view is then created for each browsing class (listings and details views), this view can then be modified by the user using the standard Views interface.

Our Drupal module uses the configuration to create an entire Drupal site using SPARQL Views, based on the ontology and configuration specified by the user. This approach also centralises the maintenance of the structure of the CMS w.r.t. the ontology, e.g., if a new browsing class is required, the user can update the configuration and then execute the Linked Data CMS mapping algorithm to create the required Drupal entities. Additionally our approach can handle changes to the schema of the ontology. For example if a change in the ontology occurs, such as a domain/range, additional classes or a change of URIs, then the configuration can be used to synchronise CMS with the ontology schema.

Drupal site administrators can also maintain the Drupal Linked Data CMS generated by the configuration in the same way as a regular Drupal site. Specifically once the configuration mapping has been applied, site administrators are free to change the labels, fields, RDF mappings, page layouts and so on using the administration interface provided by the various Drupal modules.

7   Proof of Concept: Cultural Heritage Linked Data

We have used our Linked Data CMS approach to build a system that manages a repository of cultural heritage linked data. The Hebridean Connections cultural repository is a repository about people, their relationships, their occupations, the places they have lived, events they have been involved in and even the historical artefacts they have interacted with. The repository was originally a database of information collected by several historical societies in the Western Isles of Scotland. It has been recreated as an OWL ontology by the CURIOS project [19] at dot.rural in the University of Aberdeen[7].

7.1   Cultural Heritage Ontology

As an OWL ontology, the repository consists of approximately 32,000 individuals, 520 classes, 250 object properties, 55 datatype properties. The ontology schema uses an OWL 2 RL [20] level of OWL expressivity, where most of the expressive power is used to express relationships between object properties.

This ontology makes use of domain, range, functional, reflexive, symmetric and transitive property axioms, while having a relatively simple set of atomic classes. Using our reasoning driven configuration, the Linked Data CMS has been configured with 16 browsing classes, with 106 datatype and object properties from the ontology. The browsing classes contain a total of 211 field assignments (datatype properties) and 118 relationship assignments (object properties)[8]. The expressivity of the Hebridean Connections fits within the OWL 2 RL profile, allowing for efficient runtime reasoning performance.

The browsing classes used for the ontology have been selected at a fairly high level close to the top of the classification hierarchy. Some examples of the browsing classes are: Person, Occupation, Residence, Business; each class subsumes a number of more specific classes which we have also used to configure search filters in the CMS. The browsing classes and their sub-classes appear in the ranges of the object properties used as relationships in the Linked Data CMS.

In this case study we use the Fuseki SPARQL server [21], since it can be integrated with Jena API reasoners and supports SPARQL 1.1 [18]. We use SPARQL 1.1 for all communication with the endpoint (query and update services). This allows the selection of any SPARQL 1.1 compatible endpoint for the system.

8   Conclusion

In this paper, we have presented a general approach to automatically setting up a standard web content management system to manage semantic web data based on an OWL ontology and a user specification of the views of the ontology to be presented (the *configuration*). The entities generated by the mapping algorithm

---

[7] CURIOS Project Homepage: http://www.dotrural.ac.uk/curios/.

[8] An *assignment* is an occurrence of a field or relationship being used in a browsing class.

(Sect. 4.2) rely on ontology entailments [4], which are used to automatically infer additional relationships between pages in the CMS. The validation of the user specification and the generation of the CMS are both *reasoning-driven*, in that they make essential use of ontology reasoning [2].

The approach relies on using a particular method of managing ontology data in the content management system, called class based browsing. Using our approach the gap between linked data and popular the CMS tools currently in use on the web is greatly reduced, allowing ordinary web users to contribute directly to the semantic web using familiar CMS tools. We showed how our Linked Data CMS approach can be implemented in the Drupal CMS by building on top of some popular Drupal modules. An additional advantage of this implementation is that once the class based browsing has been configured in Drupal, users can still use all of the standard Drupal tools to customise the Linked Data CMS. Finally, we presented an instantiation of the Linked Data CMS with cultural heritage data. As for future work, we plan to investigate how to consider other user requirements [22] of ontology enabled software [23], when setting up content management systems.

Acknowledgement. This work is partially supported by the RCUK dot.rural Digital Economic Hub and the EU K-Drive (286348) projects.

References

1. Hogan, A., Pan, J.Z., Polleres, A., Ren, Y.: Scalable OWL 2 reasoning for linked data. In: Polleres, A., d'Amato, C., Arenas, M., Handschuh, S., Kroner, P., Ossowski, S., Patel-Schneider, P. (eds.) Reasoning Web 2011. LNCS, vol. 6848, pp. 250–325. Springer, Heidelberg (2011)
2. Pan, J.Z., Thomas, E., Ren, Y., Taylor, S.: Tractable fuzzy and crisp reasoning in ontology applications. IEEE Comput. Int. Mag. **7**(2), 45–53 (2012)
3. Heino, N., Pan., J.Z.: RDFS Reasoning on massively parallel hardware. In: Proceedings of the 11th International Semantic Web Conference (ISWC2012) (2012)
4. Pan, J.Z.: Description logics: reasoning support for the semantic web. Ph.D. thesis, School of Computer Science, The University of Manchester (2004)
5. Thomas, E., Pan, J.Z., Ren, Y.: TrOWL: tractable OWL 2 reasoning infrastructure. In: Proceedings of the Extended Semantic Web Conference (ESWC2010) (2010)
6. Berners-Lee, T.: Linked-data design issues. W3C design issue document. http://www.w3.org/DesignIssue/LinkedData.html, June 2009
7. Manola, F., Miller, E.: RDF Primer. W3C Recommendation. http://www.w3.org/TR/rdf-primer/ (2004)
8. Hitzler, P., Krötzsch, M., Parsia, B., Patel-Schneider, P.F., Rudolph, S. (eds.): OWL 2 Web Ontology Language: Primer. W3C Recommendation. http://www.w3.org/TR/owl2-primer/ (2009)
9. Bizer, C., Heath, T., Berners-Lee, T.: Linked data - the story so far. Int. J. Semant. Web Inf. Syst. **5**(3), 1–22 (2009)
10. Cyganiak, R.: The linking open data cloud diagram. http://richard.cyganiak.de/2007/10/lod/ (2011). Accessed 16 Oct 2012

11. Motik, B., Patel-Schneider, P.F., Grau, B.C.: OWL 2 Web Ontology Language: Direct Semantics. W3C Recommendation. http://www.w3.org/TR/owl2-direct-semantics/ (2009)
12. Baader, F., Horrocks, I., Sattler, U.: Description logics for the semantic web. KI - Künstliche Intelligenz **16**(4), 57–59 (2002)
13. Krötzsch, M., Vrandečić, D., Völkel, M.: Semantic MediaWiki. In: Cruz, I., et al. (eds.) ISWC 2006. LNCS, vol. 4273, pp. 935–942. Springer, Heidelberg (2006)
14. Rayfield, J.: Dynamic semantic publishing. In: Maass, W., Kowatsch, T. (eds.) Semantic Technologies in Content Management Systems, pp. 49–64. Springer, Heidelberg (2012)
15. Corlosquet, S., Delbru, R., Clark, T., Polleres, A., Decker, S.: Produce and Consume Linked Data with Drupal!. In: Bernstein, A., Karger, D.R., Heath, T., Feigenbaum, L., Maynard, D., Motta, E., Thirunarayan, K. (eds.) ISWC 2009. LNCS, vol. 5823, pp. 763–778. Springer, Heidelberg (2009)
16. Clark, L.: SPARQL views: a visual SPARQL query builder for Drupal. In: Polleres, A., Chen, H. (eds.) ISWC Posters&Demos. CEUR Workshop Proceedings, vol. 658. http://CEUR-WS.org (2010)
17. Motik, B.: On the properties of metamodeling in OWL. J. Logic Comput. **17**(4), 617–637 (2007)
18. SPARQL: SPARQL 1.1 overview, W3C Working Draft. http://www.w3.org/TR/sparql11-overview/ (2012)
19. Mellish, C., Wallace, C., Tait, E., Hunter, C., Macleod, M.: Can digital technologies increase engagement with community history? In: Digital Engagement 2011 (2011)
20. Motik, B., Grau, B.C., Horrocks, I., Wu, Z., Fokoue, A., Lutz, C.: OWL 2 Web Ontology Language - Profiles. Technical report, W3C (2009)
21. Seaborne, A.: Fuseki: serving RDF data over HTTP. http://jena.apache.org/documentation/serving_data/ (2011). Accessed 27 Oct 2012
22. Siegemund, K., Zhao, Y., Pan, J.Z., Assmann., U.: Measure software requirement specifications by ontology reasoning. In: Proceedings of the 8th International Workshop on Semantic Web Enabled, Software Engineering (SWESE2012) (2012)
23. Pan, J.Z., Staab, S., Amann, U., Ebert, J., Zhao, Y.: Ontology-Driven Software Development. Springer, Heidelberg (2013). (ISBN: 978-3-642-31225-0)

A Comparison of Unsupervised Taxonomical Relationship Induction Approaches for Building Ontology in RDF Resources

Nansu Zong, Sungin Lee, and Hong-Gee Kim[✉]

Biomedical Knowledge Engineering Lab, School of Dentistry,
Seoul National University, Seoul, Korea
{zongnansu1982,sunginlee,hgkim}@snu.ac.kr

Abstract. Automatically generated ontology can describe the relationship of meta-data in Linked Data or other RDF resources generated from programs, and advances the utility of the data sets. Hierarchical document clustering methods used to generate concept hierarchies from retrieved documents or social tags can be used for constructing taxonomy or ontology for Linked Data and RDF documents. This paper introduces a framework for building an ontology using the hierarchical document clustering methods and compares the performance of three classic algorithms that are UPGMA, Subsumption, and EXT for building the ontology. The experiment shows EXT is the best algorithm to build the ontology for RDF resources and demonstrates that the quality of the ontology generated can be affected by the number of concepts that are used to represent the entities and to formalize the classes in the ontology.

Keywords: Taxonomical relationship induction · Hierarchy generation · Ontology generation · RDF · Linked data

1 Introduction

The popularity of RDF resources has grown gradually since 2001. For example, Linked Data increased in the number to 32 billion RDF triples by 2011[1]. The growing needs of the RDF resources push organizations to publish their own RDF format data, and they create RDF data by transforming their legacy data, such as relational database (RDB) or Web pages, by using transformation programs [5,11,18,30]. However, the use of the RDF data sets automatically generated from the programs has decreased due to lack of an ontology describing the relationship resident in the meta-data. For example, Linked Life Data[2] realizes mappings at both instance and predicate levels, but not at the class level because of lack of an ontology or concept hierarchy[3].

---

[1] http://events.linkeddata.org/ldow2012/

[2] http://linkedlifedata.com/

[3] http://linkedlifedata.com/sources

W. Kim et al. (Eds.): JIST 2013, LNCS 8388, pp. 445–459, 2014.
DOI: 10.1007/978-3-319-06826-8_32, © Springer International Publishing Switzerland 2014

There are three ways for solving the problem: (1) generate an ontology from the data sources before publishing RDF data [1,26,35,36]; (2) map entities to existing ontologies during RDF data generation [3,28]; and (3) generate an ontology directly from RDF resources or Linked Data sets [24,40], a task similar to generating a concept hierarchy from retrieved documents or social tags in Information Retrieval (IR) known as hierarchical document clustering [31,33]. Linked Data sets and RDF resources store all the triples for entities as RDF documents [4]. This makes hierarchical document clustering methods potentially applicable to ontology generation in the Semantic Web.

In this paper, we introduce a framework for generating ontology from Linked Data or other RDF resources, based on hierarchical document clustering algorithms. We adapted the three most popular methods known as UPGMA [20], Subsumption [31], and EXT [19], to generate ontology in our framework. We evaluate the three algorithms with a preliminary experiment and the experiment shows that EXT is the best algorithm to build the ontology for RDF resources. We also notice that the quality of the ontology generated can be affected by the number of concepts that are used to represent the entities and to formalize the classes in the ontology, and suggest to improve the quality by changing the parameters of the algorithms to adjust the number of the layers in the generated ontology.

The rest of the paper is organized as follows: Sect. 2 introduces the related works; Sect. 3 introduces the framework designed for ontology generation; Sect. 4 systematically describes the methods using for the framework; Sect. 5 presents our experiment results; and finally, we provide conclusions in Sect. 6.

2   Related Works

The ontology generation or the concept hierarchy generation is considered as one branch of Knowledge Discovering and can be used for describing the relationship of meta-data [16,23]. Previous several studies used clustering and classification algorithms to build the taxonomical relationships based on inter-correlation attributes for databases [13,17]. Later, as the popularity of the Semantic Web grew, more studies used other machine learning approaches to build ontology, specially taxonomical relationship [7,32] to help transform databases into RDF data or to boost the application of the Semantic Web.

The taxonomical relationships or the concept hierarchy extracted from text data has been used to represent the search results in a hierarchy. The most traditional methods for building the hierarchy are based on hierarchical clustering algorithms known as agglomerative UPGMA and bisecting k-means [20]. The bisecting k-means is considered a better solution than UPGMA in performance [34]. Probability models are used to build the hierarchy and are reported as better algorithms than traditional ones in performance. The Subsumption [31], which is a kind of co-occurrence relationship of two concepts extracted from documents, is used to organize the concept hierarchy for retrieved documents. The subsumption is a kind of probability method to calculate the probability of is_A relationship of two concepts and is considered as one of the most classical methods for concept hierarchy generation. Studies, such as [9,33], improved the subsumption-based approaches for different usages. Other studies are inspired to advance the precision of the subsumption-based method using a probability model. The DSP [22] computes the importance of every concept by topicality and predictiveness. The most important concepts are put to the top level of the hierarchy using the greedy approximation of the Dominating Set Problem (DSP). DisCover [21] maximizes hierarchy coverage while maintaining distinctiveness of concepts to identify the concepts in the hierarchy and receives better precision than DSP. FIHC [14] measures the cohesiveness of a cluster by using the frequent item sets that are calculated by the Global support and the Cluster support. FIHC is reported as better than agglomerative UPGMA and bisecting k-means [34]. Other famous studies that apply Formal Concept Analysis (FCA) to build a hierarchy are introduced in [8,12]. These studies build the concept-feature lattices and remove the features of the formalized graph to make a human readable hierarchy. Reference [25] uses Self-Organizing Map (SOM) to build the hierarchy by three different feature extraction approaches.

The concept hierarchy and the taxonomical relationships are also studied for social tags. Reference [33] uses Subsumption to induce the hierarchy from flicker tags. Reference [37] builds the hierarchy by using heuristic rules and deep syntactic analysis. EXT [19] induces a similarity graph to build the hierarchical taxonomy. An extensible greedy algorithm is used to place concepts that are the center of the similarity graph into the hierarchy. Reference [19] replaces the extensible greedy algorithm in [19] with a Directed Acyclic Graph (DAC) allocation algorithm, which allows the classes to maintain multiple super classes in the hierarchy.

In this research, we chose three most representative algorithms from the existing studies: a classic hierarchical clustering algorithm known as UPGMA [20,34], a popular probabilistic algorithm called Subsumption [9,31], and the most latest approach EXT [19].

3   Framework of Building Taxonomical Relationship of Entities

We separated the procedure of taxonomical relationship generation into four parts as shown in Fig. 1: Data Preparation, Pre-processing, Taxonomical Relationship Induction and Post-processing.

Data Preparation is designed to collect every piece of information about an entity to form an RDF document. We partitioned the RDF data graphs into small graphs called RDF documents, each of which contains the description of an entity. The triples or the quads are extracted to form a star-shaped document [4]. For example, an RDF document named "Acetylsalicylic acid" formalized by Data Preparation contains all the triples of the entity "Drugbank:drugs/DB00945" as the subject.

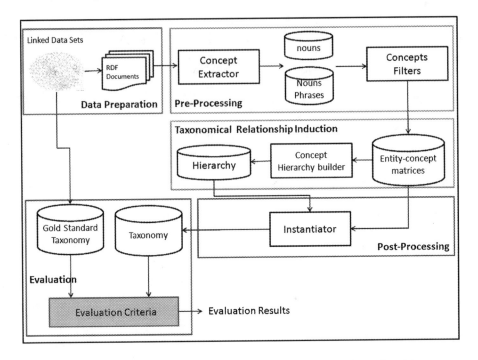

Fig. 1. The framework of building taxonomical relationship of entities

Pre-processing is designed to extract important nouns and noun phrases as candidate concepts of the RDF documents. First, the nouns and the noun phrases are extracted by a Part Of Speech (POS) tagger. Second, after lemmatizing and stemming, the nouns and the noun phrases are considered as candidate concepts in building the concept-entity matrix. Third, the concept-entity matrix is used to extract important concepts based on Vector Space Model (VSM) and Latent Semantic Analysis (LSA) [15]. The components of the matrix are calculated by Term Frequency and Inverted Document Frequency (TF-IDF) [29], and the matrix is decomposed by using Singular Value Decomposition (SVD) [2]. The important concepts are extracted by decreasing the context dimension by using LSA, and the important concepts are used to generate the new concept-entity matrix after pre-processing.

Taxonomical Relationship Induction is designed to construct the hierarchy based on the concept-entity matrix from the Pre-processing step. We implemented UPGMA, Subsumption and EXT introduced in Sect. 2 to build the taxonomy.

Post-processing is designed to instantiate the concepts of the hierarchy created in Taxonomical Relationship Induction. The concepts in the hierarchy are considered as classes in the ontology and the documents of an entity containing the concepts are considered as the instances of the classes.

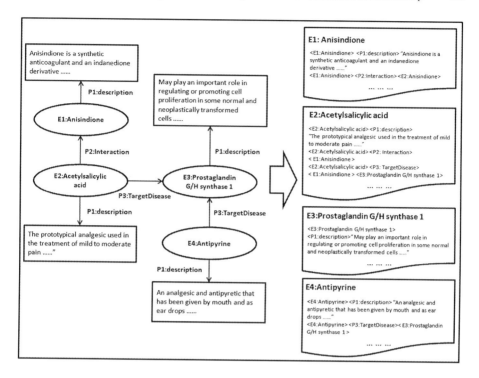

Fig. 2. An example of data preparation

4 Methodology

4.1 Data Preparation and Pre-processing

The Linked Data and other RDF resources are stored as either triples or RDF documents [6]. To adapt the existing hierarchical clustering algorithms, we need to process each entity as a document, which in effect partitions the whole data graph into small sub-RDF documents. For example, in Fig. 2, a data graph containing four entities is partitioned into four sub-RDF documents. We extracted nouns and noun phrases from each RDF document using the POS tagger after removing the punctuation marks and stop-words as mentioned in papers [27,39]. The nouns and the noun phrases tagged with NN (singular noun), NNP (proper noun), NNS (plural noun) and NNPS (plural proper nouns) are put into noun sequences. We generated candidate concepts from the noun sequences by three types of n-gram: unigrams, bigrams and trigrams [38]. The candidate concepts are lemmatized and stemmed for building the concept-entity matrix, which is used for finding the most important concepts that represent the entity.

We adapted VSM to represent documents. Each RDF document of the collection is defined as a single multidimensional vector $d = [w_1, w_2, \ldots, w_i]$, where each component w_i is a weight of the dimension i and each dimension reflects a candidate concept. We used TF-IDF to weight each component in the d. The TF-IDF is calculated as follows:

$$w_{i,d} = tf_{i,d} * log\frac{\#documents}{\#(documents\ contaning\ concept\ i)} \qquad (1)$$

where the $tf_{i,d}$ is the normalized frequency of concept i in document d.

A vector of the single multidimensional vectors d is used to build an entity-concept matrix that is transposed into a concept-entity matrix. We extracted important concepts based on LSA after decomposing the matrix using SVD. SVD [2] is considered as a data reduction method that approximates the original concept-entity matrix in a lower dimension. With the help of the dimension reduction, the concept vector can be represented by the value of a new context, that is, a lower dimension instead of the original dimension reflecting the number of the entities. The concept-entity matrix A can be broken down into three matrices: an orthogonal matrix known as U, a diagonal matrix known as S, and a transpose of the orthogonal matrix known as V. SVD can be presented as follows:

$$A = U * S * V^T \qquad (2)$$

where the columns of U are orthogonal eigenvectors of AA^T, the columns of V are orthogonal eigenvectors of A^TA, and S is a diagonal matrix containing the square roots of eigenvalues from U or V in descending order. The U can be considered as the representation of the concepts in the context S. LSA [15] finds a low-rank k approximation A_k to the concept-entity matrix A by selecting the k largest singular values of S, the first k columns of U and V^T. Hence U_k, formed by the first k columns of U, can be regarded as the representation of the important concepts in top k [39]. In practice, we extracted the important concepts that reflect the most weighted values of the U_k in each column. Then the entity can be represented with the most important concepts extracted. The important concepts in top k and the entities form a new concept-entity matrix M. A simple example of the procedure of pre-processing is shown in Fig. 3.

4.2 Taxonomical Relationship Induction

In this section, we present how to construct a concept hierarchy for the three methods using the new concept-entity matrix M.

UPGMA. UPGMA [20] constructs a concept hierarchy based on the similarity of two concepts. The computation of the similarity of two concepts is based on VSM as shown below:

$$Sim(c_1, c_2) = \frac{\sum_1^n (w_{(e_i,c_1)} * W_{(e_i,c_2)})}{\sqrt{\sum_1^n (w_{(e_i,c_1)})^2} * \sqrt{\sum_1^n (w_{(e_i,c_2)})^2}} \qquad (3)$$

where $w_{(e_i,c_j)}$ is the IF-IDF value of the entity e_i and the concept c_j computed by Eq. 1. The new node, the superclass of the two most similar concepts, is created

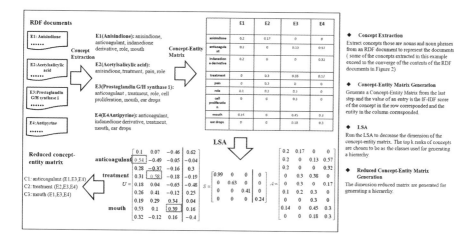

Fig. 3. An example of the procedure of the pre-processing

as a virtual class that has the common entities shared by the two concepts. We generate the concept hierarchy by using the Algorithm 1 as shown below:

Algorithm 1. UPGMA-based tree generation

Input: concept-entity matrix M and the index I of the RDF documents
Output: concept hierarchy

1: put all the concepts in Queue Q
2: put *root* into a hierarchy H
3: **while** the size of $Q > 0$ **do**
4:    **for all** each concept c_i in Q **do**
5:       **for all** each concept c_j in Q **do**
6:          compute the similarity of c_i and c_j using the Equation 3
7:       **end for**
8:    **end for**
9:    create new class node of the super class C_{i+j} of the most similar concepts c_i and c_j in H
10: **end while**

Subsumption. Subsumption [31] defines the is_A relationship based on the co-occurrences in different RDF documents of entities. Given $c1$ and $c2$, $c1$ is said to subsume $c2$ if the following conditions are satisfied:

$$P(c1|c2) > 0.8, \ P(c2|c1) < 1 \tag{4}$$

We generate the concept hierarchy using the Algorithm 2 shown as follows:

EXT. EXT [19] uses an extensible greedy algorithm to put the most important concept into the subclass of the most similar concept in the current level of the concept hierarchy. The sequence in which to choose the most important concept is decided by the centrality of the similarity graph created from the concept-entity matrix. The centrality of the concepts in the similarity graph is computed as follows:

Algorithm 2. Subsumption-based tree generation

Input: concept-entity matrix M and the index I of the RDF documents
Output: concept hierarchy

    for all each concept c_i in M **do**
2:    **for all** each concept c_j in M **do**
        if c_i and c_j satisfies the condition in Equation 4 **then**
4:        put c_j to the subclass of c_i
        end if
6:    **end for**
    end for
8: sort concepts by descending order of the number of subclass in Queue Q
    put $root$ into a hierarchy H
10: **for all** each concept c_i in Q **do**
    for all each subclass c_j **do**
12:    put the class node of c_j to the subclass of class node of c_i in H
    end for
14: **end for**

$$Cen(c_x) = \sum_{c_x \neq c_y \neq c_z \in C} \frac{\delta_{c_y c_z}(c_x)}{\delta_{c_y c_z}} \qquad (5)$$

where $\delta_{c_y c_z}$ is the total number of shortest paths from concept c_y to concept c_z, and $\delta_{c_y c_z}(c_x)$ is the number of concepts that pass through c_x. The similarity graph consists of vertices that are concepts, and edges that are added if the similarity of the two concepts is above the threshold α. The similarity of the two concepts is calculated by using Eq. 3, and whether a concept c_i should be put as the subclass of a concept c_j depends on the threshold β. Concept hierarchy is generated by using Algorithm 3 as follows:

Algorithm 3. EXT-based tree generation

Input: concept-entity matrix M and the index I of the RDF documents, threshold1 α, and threshold2 β
Output: concept hierarchy

    put all concepts into a similarity graph G
    for all each concept c_i in M **do**
3:    **for all** each concept c_j in M **do**
        if the similarity of c_i and c_j is above α **then**
        put an edge connecting c_i with c_j into G
6:    **end if**
    end for
    end for
9: sort concepts by descending order of centrality computed by Equation 5 in Queue Q
    put $root$ into a hierarchy H
    while the size of $Q > 0$ **do**
12:    **for all** each concept c_i in Q **do**
        for all each class c_j in H **do**
        compute the similarity of c_i and c_j
15:    **end for**
        if the most similarity of the most similar class c_j of the c_i is above β **then**
        put the c_i as the subclass of c_i in H
18:    **else**
        put the c_i as the subclass of $root$ in H
        end if
21:    **end for**
    end while

4.3   Post-processing

We put all the entities into the generated concept hierarchy based on the concept-entity matrix. For example, a class named "anticoagulant" in Fig. 3 will be assigned with three instances "E1:anisindione", "E3:porstaglandin G/H synthase 1", and "antipyrine". Notice that, since UPGMA and Subsumption support multiple inheritance [31,34], we also allow multiple inheritance for the assignment of instances during the post-processing step.

5   Preliminary Experiment

We implemented our framework based on JDK_1.6 using the Intel, I-5 CPU with 4 GB RAM and 1TB hard disk on a Ubuntu[4] system.

5.1   Data Set and Gold standard

We used Diseasome[5] from Linked Life Data[6] as our source data for our experiment. The Diseasome supplies entities about diseases and also describes schemas relationships. We separated the data set into two parts: data that contains the triples of entities, and a gold standard that contains the triples about the schema. For example, given two triples "$< Diseasome : diseases/272 >< Diseasome : class >< Diseasome : diseaseClass/Endocrine >$" and "$< Diseasome : diseases/2013 >< Diseasome : subtypeOf >< Diseasome : diseases/272 >$", we can get an is_A relationship in which a class named "Endocrine" has a subclass named "diseases:272" with an instance named "diseases:2013". The Diseasome contains 1308 instances of diseases, and we used them to create a disease ontology in our experiment.

5.2   Data Preparation and Pre-Processing

We partitioned the data into star-shaped RDF documents and parsed the documents using the POS tagger[7]. The nouns and noun phrases extracted were stemmed and lemmatized by the Lucene String Parser[8] to form a concept-entity matrix. We used JAMA[9] to run SVDs to choose important concepts for the entities. The SVDs are computed with the different values of k such as "100, 300 and 500", each of which means the number of concepts extracted to build an ontology. The three different concept-entity matrices, dimensionally reduced, cover different entities of the original data set, affecting the recall for the constructed ontology. Figure 4 shows the coverage of the entities by the different values of k.

---

[4] http://www.ubuntu.com/
[5] http://diseasome.eu/
[6] http://linkedlifedata.com/
[7] http://nlp.stanford.edu/software/tagger.shtml
[8] http://lucene.apache.org/
[9] http://math.nist.gov/javanumerics/jama/

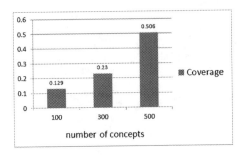

Fig. 4. The coverage of using different number of concepts for building an ontology

The coverage is computed by $coverage(k) = \frac{\#entities(k)}{\#entities}$, where $entities(k)$ is the entities remained by reducing the dimension into k. Using 100 concepts only covers 13 % of the entities and using 500 51 % of the entities. We noticed that coverage increases as the number of concepts increases.

5.3   Criteria for Evaluation

In order to evaluate the taxonomies generated by the three approaches, we compared generated ontologies with the gold standard. Since class names in the taxonomies vary in the three methods, lexical comparisons of the class names are inappropriate. We adapted the Taxonomic Precision (TP) and Taxonomic Recall (TR) [10,25] to measure the quality of the generated ontologies.

TP and TR are based on the Semantic Cotopy (SC), which defines super-sub concepts (is-A) relations in an ontology:

$$SC(c, o) = \{c_i | c_i \in C \wedge (c_i \leqslant c \vee c \geqslant c_i)\} \tag{6}$$

The Common Semantic Cotopy (CSC) that avoids the influence of lexical precision in the taxonomic measurement can be defined as:

$$CSC(c_i, O_1, O_2) = \{c_j | c_i \in C_1 \cap C_2 \wedge (c_j \leqslant C_1 c_i \vee c \geqslant C_1 c_j)\} \tag{7}$$

The TP_{CSC} and TR_{CSC} can be computed as:

$$TP_{CSC}(O_1, O_2) = \frac{1}{|CO_1 \cap CO_2|} \sum_{c \in CO_1 \cap CO_2} TP_{CSC}(c, c, O_1, O_2) \tag{8}$$

$$TR_{CSC}(O_1, O_2) = TP_{CSC}(O_2, O_1) \tag{9}$$

Taxonomic F-measure (TF) calculates the harmonic mean of TP_{CSC} and TR_{CSC}:

$$TF(O_1, O_2) = \frac{2 * TR_{CSC}(O_1, O_2) * TP_{CSC}(O_1, O_2)}{TR_{CSC}(O_1, O_2) + TP_{CSC}(O_1, O_2)} \tag{10}$$

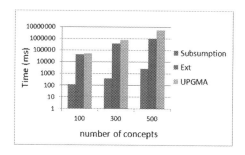

Fig. 5. The running time of the three algorithms

5.4   Results

We used UPGMA, Subsumption, and EXT to generate ontologies in our experiment. For EXT, the α and the β are both set to 0.5.

Figure 5 shows the running times of the three algorithms. Subsumption is the fastest algorithm: 116 ms for 100 concepts, 391 ms for 300, and 2683 ms for 500. The running times of the three methods increase as the number of concepts used increases.

Figure 6 shows ontology quality as the number of concepts involved changes. For TR, UPGMA gets the highest score among the three algorithms: 11 % for 500 concepts. The low TRs obtained by the algorithms are in part due to the decomposition of the concept-entity matrix, which makes the converge of entities for building the ontologies limited. For TP, Subsumption and EXT show almost the same precision that reaches 24 % when using 500 concepts. The F-measure demonstrates that quality increases as more concepts are used.

We used only those entities of the gold standard that were covered by the new decomposed concept-entity matrix to get the F-measure shown in Fig. 7. The figure shows that UPGMA performs poorer than the other two algorithms on F-measure but performs the best on Recall. EXT performs better than both Subsumption and UPGMA on F-measure. Subsumption received the best score

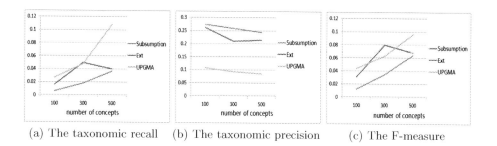

(a) The taxonomic recall   (b) The taxonomic precision   (c) The F-measure

Fig. 6. The quality of the ontologies generated by the three algorithms

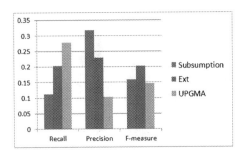

Fig. 7. The average TP, TR, and F-measure of the ontologies generated by the three algorithms, which are evaluated by the sub-structure in the gold standard. The sub-structure only contains the entities used after reducing dimensions of RDF documents

on Precision. Considering the running times of the three algorithms, the authors regarded EXT as the best algorithm for our purpose.

Each hierarchical document clustering method adapted in this paper builds an ontology that has more layers than the ontology generated from the gold standard has, affecting the quality of the former ontology. However, the number of layers in the ontology can be reduced by adjusting parameters in the algorithms - an improvement work reserved for the future.

6   Conclusion

In this paper, we introduced a framework for generating ontology from Linked Data or RDF resources based on hierarchical document clustering algorithms that are used to generate concept hierarchies from retrieved documents or social tags. We implemented three most classic and popular methods for hierarchical document clustering to generate ontology in our framework. We evaluated the three algorithms with a preliminary experiment and the experiment shows EXT is the best algorithm to build the ontology for RDF resources. We also learned that the quality of the ontology generated can be affected by the number of concepts that are used to represent the entities and to formalize the classes in the ontology, and suggested an ontology quality improvement that reduces the number of layers in the ontology.

Acknowledgments. This research was funded by the MSIP(Ministry of Science, ICT & Future Planning), Korea in the ICT R&D Program 2013. We appreciate Ashley Hess for the proof reading.

References

1. Alani, H., Kim, S., Millard, D.E., Weal, M.J., Hall, W., Lewis, P.H., Shadbolt, N.R.: Automatic ontology-based knowledge extraction from web documents. IEEE Intell. Syst. **18**(1), 14–21 (2003)

2. Baker, K.: Singular Value Decomposition Tutorial. The Ohio State University, Columbus (2005)
3. Berners-Lee, T., et al.: Linked data-the story so far. Int. J. Semant. Web Inf. Syst. **5**(3), 1–22 (2009)
4. Blanco, R., Mika, P., Vigna, S.: Effective and efficient entity search in RDF data. In: Aroyo, L., Welty, C., Alani, H., Taylor, J., Bernstein, A., Kagal, L., Noy, N., Blomqvist, E. (eds.) ISWC 2011, Part I. LNCS, vol. 7031, pp. 83–97. Springer, Heidelberg (2011)
5. Blum D., Cohen, S.: Generating RDF for application testing. In: ISWC Posters & Demos (2010)
6. Bron, M., Balog, K., de Rijke, M.: Example based entity search in the web of data. In: Serdyukov, P., Braslavski, P., Kuznetsov, S.O., Kamps, J., Rüger, S., Agichtein, E., Segalovich, I., Yilmaz, E. (eds.) ECIR 2013. LNCS, vol. 7814, pp. 392–403. Springer, Heidelberg (2013)
7. Cerbah, F.: Mining the content of relational databases to learn ontologies with deeper taxonomies. In: 2008 IEEE/WIC/ACM International Conference on Web Intelligence and Intelligent Agent Technology, pp. 553–557. IEEE (2008)
8. Cimiano, P., Hotho, A., Staab, S.: Learning concept hierarchies from text corpora using formal concept analysis. J. Artif. Intell. Res. (JAIR) **24**, 305–339 (2005)
9. De Knijff, J., Frasincar, F., Hogenboom, F.: Domain taxonomy learning from text: the subsumption method versus hierarchical clustering. Data Knowl. Eng. **83**, 54–69 (2013)
10. Dellschaft, K., Staab, S.: On how to perform a gold standard based evaluation of ontology learning. In: Cruz, I., Decker, S., Allemang, D., Preist, C., Schwabe, D., Mika, P., Uschold, M., Aroyo, L.M. (eds.) ISWC 2006. LNCS, vol. 4273, pp. 228–241. Springer, Heidelberg (2006)
11. Ding, L., DiFranzo, D., Graves, A., Michaelis, J.R., Li, X., McGuinness, D.L., Hendler, J.A.: TWC data-gov corpus: incrementally generating linked government data from data. gov. In: Proceedings of the 19th International Conference on World Wide Web, pp. 1383–1386. ACM (2010)
12. Drymonas, E., Zervanou, K., Petrakis, E.G.M.: Unsupervised ontology acquisition from plain texts: the *OntoGain* system. In: Hopfe, ChJ, Rezgui, Y., Métais, E., Preece, A., Li, H. (eds.) NLDB 2010. LNCS, vol. 6177, pp. 277–287. Springer, Heidelberg (2010)
13. Fisher, D.: Improving inference through conceptual clustering. In: Proceedings of AAAI Conference 1987, pp. 461–465 (1987)
14. Fung, B.C.M., Wang, K., Ester, M.: Hierarchical document clustering using frequent itemsets. In: Proceedings of SIAM International Conference on Data Mining, pp. 59–70 (2003)
15. Garcia, E.: Latent semantic indexing (lsi) a fast track tutorial (2006)
16. Han, J., Cai, Y., Cercone, N.: Knowledge discovery in databases: an attribute-oriented approach. In: Proceedings of the International Conference on Very Large Data Bases, pp. 547–547. Citeseer (1992)
17. Han, J., Fu, Y.: Dynamic generation and refinement of concept hierarchies for knowledge discovery in databases. In: Proceedings of AAAI, vol. 94, pp. 157–168 (1994)
18. Heath, T., Bizer, C.: Linked data: evolving the web into a global data space. Synth. Lect. Semant. Web: Theory Technol. **1**(1), 1–136 (2011)
19. Heymann, P., Garcia-Molina, H.: Collaborative Creation of Communal Hierarchical Taxonomies in Social Tagging Systems. Stanford Infolab Publications, Heidelberg (2006)

20. Jain, A.K., Dubes, R.C.: Algorithms for Clustering Data. Prentice-Hall Inc., New Jersey (1988)
21. Kummamuru, K., Lotlikar, R., Roy, S., Singal, K., Krishnapuram, R.: A hierarchical monothetic document clustering algorithm for summarization and browsing search results. In: Proceedings of the 13th International Conference on World Wide Web, pp. 658–665. ACM (2004)
22. Lawrie, J.D., Croft, W.B.: Generating hierarchical summaries for web searches. In: Proceedings of the 26th Annual International ACM SIGIR Conference on Research and Development in Informaion Retrieval, pp. 457–458. ACM (2003)
23. Li, C.: Knowledge discovery in database. J. Northwest.Univ.: Nat. Sci. Ed. **29(1)**, 114–119 (1999)
24. Parundekar, R., Knoblock, C.A., Ambite, J.L.: Linking and building ontologies of linked data. In: Patel-Schneider, P.F., et al. (eds.) ISWC 2010. LNCS, vol. 6496, pp. 598–614. Springer, Heidelberg (2010)
25. Paukkeri, M.-S., García-Plaza, A.P., Fresno, V., Unanue, R.M., Honkela, T.: Learning a taxonomy from a set of text documents. Appl. Soft Comput. **12**(3), 1138–1148 (2012)
26. Pivk, A.: Automatic ontology generation from web tabular structures. AI Commun. **19**(1), 83–85 (2006)
27. Paolo-Ponzetto, S., Strube, M.: Taxonomy induction based on a collaboratively built knowledge repository. Artif. Intell. **175**(9), 1737–1756 (2011)
28. Sahoo, S.S., Halb, W., Hellmann, S., Idehen, K., Thibodeau, Jr., T., Auer, S., Sequeda, J., Ezzat, A.: A survey of current approaches for mapping of relational databases to RDF. W3C RDB2RDF Incubator Group Report (2009)
29. Salton, G., Buckley, C.: Term-weighting approaches in automatic text retrieval. Inf. Process. Manag. **24**(5), 513–523 (1988)
30. San Martín, M., Gutierrez, C.: Representing, querying and transforming social networks with RDF/SPARQL. In: Aroyo, L., et al. (eds.) ESWC 2009. LNCS, vol. 5554, pp. 293–307. Springer, Heidelberg (2009)
31. Sanderson, M., Croft, B.: Deriving concept hierarchies from text. In Proceedings of the 22nd Annual International ACM SIGIR Conference on Research and Development in Information Retrieval, pp. 206–213. ACM (1999)
32. Santoso, H.A., Haw, S.C., Abdul-Mehdi, Z., et al.: Ontology extraction from relational database: concept hierarhy as background knowledge. Knowl.-Based Syst. **61(8)**, 729–741 (2010)
33. Schmitz, P.: Inducing ontology from flickr tags. Collaborative Web Tagging Workshop at WWW2006 **50**, 210–214 (2006)
34. Steinbach, M., Karypis, G., Kumar, V., et al.: A comparison of document clustering techniques. In: KDD Workshop on Text Mining, pp. 525–526. Bibliometric, Boston (2000)
35. Tho, Q.T., Hui, S.C., Fong, A.C.M., Cao, T.H.: Automatic fuzzy ontology generation for semantic web. IEEE Trans. Knowl. Data Eng. **18(6)**, 842–856 (2006)
36. Tijerino, Y.A., Embley, D.W., Lonsdale, D.W., Ding, Y., Nagy, G.: Towards ontology generation from tables. World Wide Web **8**(3), 261–285 (2005)
37. Tsui, E., Wang, W.M., Cheung, C.F., Lau, A.S.M.: A concept-relationship acquisition and inference approach for hierarchical taxonomy construction from tags. Inf. Process. Manag. **46**(1), 44–57 (2010)
38. Yang, H.: Personalized concept hierarchy construction. Ph.D. thesis, University of Southern California (2011)

39. Zheng, H.-T., Borchert, C., Kim., H.-G.: A concept-driven automatic ontology generation approach for conceptualization of document corpora. In: IEEE/WIC/ACM International Conference on Web Intelligence and Intelligent Agent Technology, WI-IAT'08, vol. 1, pp. 352–358. IEEE (2008)
40. Zong, N., Im, D.-H., Yang, S., Namgoon, H., Kim, H.-G.: Dynamic generation of concepts hierarchies for knowledge discovering in bio-medical linked data sets. In: Proceedings of the 6th International Conference on Ubiquitous Information Management and Communication, p 12. ACM (2012)

Author Index

Printed in the United States
By Bookmasters